INTRODUCTION TO NUMBER THEORY

INTRODUCTION TO NUMBER THEORY

William W. Adams
Larry Joel Goldstein

Department of Mathematics
University of Maryland, College Park

Prentice-Hall, Inc.
Englewood Cliffs, New Jersey

Library of Congress Cataloging in Publication Data

ADAMS, WILLIAM W
Introduction to number theory.

Includes index.
1. Numbers, Theory of. I. Goldstein, Larry Joel, joint author. II. Title.
QA241.A24 512'.7 75-12686
ISBN 0-13-491282-9

10 9 8 7 6 5 4 3 2 1

Printed in the United States of America

PRENTICE-HALL INTERNATIONAL, INC., *London*
PRENTICE-HALL OF AUSTRALIA, PTY. LTD., *Sydney*
PRENTICE-HALL OF CANADA, LTD., *Toronto*
PRENTICE-HALL OF INDIA PRIVATE LIMITED, *New Delhi*
PRENTICE-HALL OF JAPAN, INC., *Tokyo*
PRENTICE-HALL OF SOUTHEAST ASIA (PTE.) LTD., *Singapore*

To Elizabeth, Ruth and Sarah
and
To my parents, Martin and Rose Goldstein

Contents

3

Congruences 39

4

The Law of Quadratic Reciprocity 100

5

Arithmetic Functions 136

6

A Few Diophantine Equations 156

11

The Representation of Integers by Binary Quadratic Forms 307

Preface

The present book is an introduction to number theory which focuses on the theory of Diophantine equations. It is intended, first of all, as a text for the standard elementary number theory course taught to majors in mathematics and mathematics education. Therefore, we have tried to proceed at a very slow pace in the initial stages of the book, and have done considerably more "talking" about number theory, its structure and its goals than is customary in books at this level. We have tried to make that portion of the book corresponding to a one-semester first course in number theory as accessible as possible to the broadest spectrum of students without compromising the content. To do this we have, for example, indulged in a considerable amount of numerical calculation, and assumed almost no background beyond a course in high school algebra. The theory of Diophantine equations provides us with a common theme around which to organize our discussions. This (we hope) makes number theory seem less a disjointed collection of miscellaneous topics than an organized discipline with goals, one of which is the study of Diophantine equations.

Chapters 1-6 provide a great deal of flexibility in organizing a one semester course in number theory, as can be seen from the following table of logical dependence.

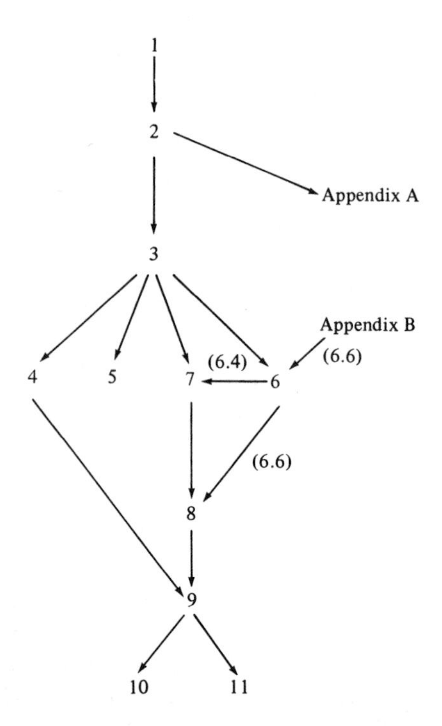

Chapters 7-11 give an introduction to algebraic number theory by studying quadratic Diophantine equations, which lead to the study of quadratic fields. Beginning at Chapter 8 the book is written at a higher level and also presupposes a first undergraduate course in abstract algebra. In Chapter 8, we develop an algorithm for solving any quadratic Diophantine equation of the form $ax^2 + bxy + cy^2 = n$. Chapters 7 and 8 provide a brief introduction to quadratic fields. In Chapter 9, we develop, in some depth, the arithmetic of modules in a quadratic number field and prove the unique factorization theorem for modules. From Chapter 9, the reader can proceed either to Chapter 10 or Chapter 11. Chapter 10 is devoted to a study of various Diophantine equations, namely the Bachet equation $y^2 = x^3 + k$, Fermat's Last Theorem for $n = 3$, and norm form equations. Chapter 11 is devoted to the study of the representation of integers by binary quadratic forms.

We have included well over a thousand problems. It is our conviction that number theory is most fun when approached with a spirit of adventure and discovery. Therefore, our problems include simple numerical calculations, experimentation designed to let the student discover his own conjectures, problems which amplify the theory in the text, and problems for the afficionado. We have designated the latter sort of problem with an asterisk.

We employ the following numbering scheme in the book. When referring to other sections than the current section of the book, we will do the following: We will use a single digit to refer to other sections in the current chapter. For example, Section 3 refers to Section 3 of the current chapter. On the other hand, Section 11.3 refers to Section 3 of Chapter 11. When referring to items inside the sections, single digits refer to items in the current section. Thus, for example, Lemma 2 would refer to Lemma 2 of the current section. But Lemma 3.2 would refer to Lemma 2 of Section 3 *of the current chapter*. We will use 3-digit numbers only when we need to refer outside the current chapter. In this case Lemma 8.3.2 would refer to Lemma 2 of Section 3 of Chapter 8. All logarithms in this book are assumed to be natural and will be denoted log.

The authors wish to thank the many colleagues whose conversations and suggestions have helped improve this book. To Drs. Adam Kleppner and James Schafer go our thanks for class-testing the manuscript. To Drs. Thomas Apostol, Bruce Berndt and Ralph Greendberg go our thanks for reading the manuscript, and making many helpful suggestions. Our special thanks and gratitude is extended to Drs. Ethan Bolker and Emil Grosswald, whose incisive reading and critique of the manuscript went far beyond the call of either duty or friendship. Our typists, Debbie Curran and Paula Verdun, did a magnificent job of preparing the manuscript. And we thank the Department of Mathematics of the University of Maryland for providing the typing services. We are very grateful to Elizabeth Adams for her patient and careful assistance in reading the proofs. Finally, we wish to express our appreciation to the staff of Prentice-Hall, especially Penny Linskey, for their support and for the professional manner in which they have produced this book.

William W. Adams
College Park, Maryland Larry Joel Goldstein

INTRODUCTION TO NUMBER THEORY

1

Introduction

1.1 What Is Number Theory?

Contemporary number theory is such a vast subject that it is not easy to describe the sort of mathematics which does or does not belong to it. However, at the risk of oversimplification, let us just say that number theory is the branch of mathematics concerned with the study of the properties of the *integers*:

$$\ldots -4, -3, -2, -1, 0, 1, 2, 3, 4, \ldots.$$

The integers (or at least the positive integers) and the rules of arithmetic associated with them are among the oldest and most fundamental products of the human thought process. One ancient civilization after another came to grips with the need to count—for barter, for telling time, for calculating calendars, for measuring lengths and areas, and for building. Out of this necessity to count grew the number concept and the basic rules of arithmetic. Over 5000 years ago, the Egyptians and Chinese made regular use of arithmetic in their daily lives.

When calculating with the integers, it is hard not to notice the many patterns and properties which they exhibit. And man's curiosity made him wonder whether or not the patterns which are observed to hold for the first thousand (or first million) integers persist as general patterns of all the integers. Such questions belong to the province of number theory. They were already asked by the ancient Egyptians and Chinese. However, among the ancient civilizations, it was primarily the Greeks, especially the Pythagoreans, who gave modern number theory its start.

To get a more specific idea of the sort of problems considered by number theorists, as well as some idea of this book's approach to the subject, let us consider some problems considered by the ancient Greeks.

Problem 1: Find all right triangles the lengths of whose sides are integers.

If the sides of such a triangle have lengths x, y, z, with z the length of the hypotenuse, then the Pythagorean theorem implies that

$$x^2 + y^2 = z^2. \tag{1}$$

Thus, we see that Problem 1 is equivalent to determining all triples of integers (x, y, z) satisfying Eq. (1). Such triples are called *Pythagorean triples.* Examples are (3, 4, 5) and (5, 12, 13). Moreover, if a, b, c are any integers and if

$$x = \pm(a^2 - b^2)c, \qquad y = \pm 2abc, \qquad z = \pm(a^2 + b^2)c,$$

then (x, y, z) is a Pythagorean triple. (Do the algebra.) Moreover, it was proved by the Greek mathematician Diophantus that every Pythagorean triple is of this form for suitable integers a, b, c.

Problem 2: Show that $\sqrt{2}$ is not the ratio of two integers (that is, $\sqrt{2}$ is irrational).

The solution of Problem 2 is one of the nicest accomplishments of Greek mathematics. It goes like this: If $\sqrt{2} = x/y$, x and y integers, then $2 = x^2/y^2$ or

$$x^2 - 2y^2 = 0. \tag{2}$$

We may assume that the fraction x/y is reduced, so that not both of x, y can be even. But since $x^2 = 2y^2$, x must be even, say $x = 2t$. Therefore, $4t^2 = 2y^2$ and $2t^2 = y^2$, so y is even also, contradicting the fact that not both x and y are even.

Problem 3: Suppose that a, b, c are given integers. Find all integers x, y such that $ax + by = c$.

We shall present a complete solution of Problem 3 in Chapter 2. To appreciate the subtlety of the problem, let us consider two special cases. First, consider the equation

$$2x + 3y = 5. \tag{3}$$

If a value for x is given, there is a value y for which the equation holds, namely $\frac{1}{3}(5 - 2x)$. However, if we choose x to be an integer, $y = \frac{1}{3}(5 - 2x)$

need not be. For example, if $x = 2$, $y = \frac{1}{3}$. Thus, solutions for which x and y are integers are somewhat subtle to describe. However, some elementary reasoning enables us to describe them all. For if integers x, y satisfy $2x + 3y = 5$, we must have that $2x = 5 - 3y$ is even. But then y must be odd. (Otherwise $3y$ would be even and $5 - 3y$ odd.) Let $y = 2t + 1$ for some integer t. Then

$$x = \tfrac{1}{2}(5 - 3y) = 1 - 3t,$$

so any solution of Eq. (3) for which x, y are integers is of the form

$$x = 1 - 3t, \qquad y = 2t + 1, \qquad t \text{ an integer.} \tag{4}$$

Conversely, if t is any integer, and x and y are given by Eq. (4), we have

$$2x + 3y = 2(1 - 3t) + 3(2t + 1) = 5,$$

so that (x, y) is a solution of Eq. (3). Thus, the solutions of (3) in integers are given by (4). For example, $t = 0, 1, -3$ yield the respective solutions $(x, y) = (1, 1), (-2, 3), (10, -5)$.

As a second special case of Problem 3, consider the equation $3x + 6y = 7$. This equation has *no* solutions in integers (x, y), since if x and y are integers, $3x + 6y = 3(x + 2y)$ is an integer divisible by 3. But 7 is not divisible by 3, so $3x + 6y = 7$ has no solutions in integers.

Note that although Problems 1–3, on the surface, have no connection with one another, all of them boil down to determining all the solutions (or lack of solutions) in integers of certain equations, namely $x^2 + y^2 = z^2$, $x^2 - 2y^2 = 0$, $2x + 3y = 5$, and $3x + 6y = 7$. The general problem of determining integer solutions to equations is one of the central themes in number theory. In fact, a large number of problems in number theory can be thought of as problems concerning the solution of equations in integers. What makes number theory interesting and gives it a special flavor is that such problems are much more delicate than determining all solutions to an equation. Indeed, as we showed above, $x^2 - 2y^2 = 0$ has no solutions in nonzero integers x, y. Nonetheless, it has many solutions (e.g., $x = \sqrt{2}$, $y = 1$). Also, given any value for x and y, a solution to $x^2 + y^2 = z^2$ can be determined by setting $z = \pm\sqrt{x^2 + y^2}$. However, the solutions in integers (x, y, z) have quite a complicated description, especially when compared with the simplicity of the original equation.

All the equations we have considered above are examples of what are called *Diophantine equations*. More precisely, a Diophantine equation is a polynomial* equation (in any number of unknowns) whose solutions in integers are to be determined. Such equations are named for the Greek mathematician Diophantus of Alexandria, who first made them the object

*Number theorists often consider more general Diophantine equations, e.g., $2^x + y^2 = 3$, but we will restrict ourselves to polynomial equations in this book.

of systematic study. Since ancient times, Diophantine equations have been one of the focal points for number-theoretic investigations. In fact, much of what we know about the integers has been obtained in efforts to solve Diophantine equations. Therefore, we have chosen to make Diophantine equations the central theme of this book.

In this book, we shall explore various properties of the integers. But time and again, we shall return to Diophantine equations to illustrate how the information we have obtained about the integers can be applied. Although it is certainly not possible to discuss all of number theory in the context of Diophantine equations, the techniques necessary to solve even a limited class of Diophantine equations illustrate a wide range of number-theoretic methods and thus provide an ideal context in which to study number theory.

This book can be divided into two parts. Chapters 1–7 require only the minimal prerequisites outlined in Section 1.2 and cover what is usually called *elementary number theory*. In Chapters 8–11, a greater level of sophistication is assumed in the form of a first course in abstract algebra. The content of the second half of the book includes an introduction to *algebraic number theory* in the special case of quadratic fields. Our goal is to use algebraic ideas to unravel the theory of quadratic Diophantine equations of the form

$$ax^2 + bxy + cy^2 = m.$$

Over the centuries, the theory of such equations has been the testing ground for many new number-theoretic ideas by some of the greatest number theorists, including Fermat, Legendre, Lagrange, Gauss, and Dirichlet. There still remain many unsolved problems concerning them, and therefore they are an exciting subject to cover in a course in number theory at the undergraduate level.

As further illustrations of interesting Diophantine equations, let us give two examples studied by the seventeenth-century French jurist and amateur mathematician Pierre Fermat.

Problem 4: Determine all those positive integers a which are a sum of two perfect squares. In other words, for which a is the Diophantine equation

$$x^2 + y^2 = a$$

solvable in integers x, y?

We shall solve Problem 4 completely in Chapter 6. However, for now, let us concern ourselves only with the case $a =$ a prime, that is, the case where a has no factors other then ± 1 and $\pm a$. In this special case, Fermat proved that $x^2 + y^2 = a$ can be solved in integers if and only if $a - 1$ is divisible by 4 or $a = 2$. Thus, for example, $x^2 + y^2 = 7$ has no solutions in integers ($7 - 1 = 6$ is not divisible by 4), whereas $x^2 + y^2 = 13$ has solutions in integers ($13 - 1 = 12$ is divisible by 4). It would be instructive for you to check the first 20 or so examples of Fermat's result.

Fermat left a legacy in the form of an (as yet) unsolved Diophantine equation, which has become known as *Fermat's Last Theorem.* Fermat learned much of what he knew about number theory from reading a Latin translation of Diophantus' works. In one section of the book, Diophantus discussed the Pythagorean equation (Problem 1). In a marginal note written in this section, Fermat wrote that in contrast to the Pythagorean equation, we have

Problem 5: The Diophantine equation $x^n + y^n = z^n$, $n \geq 3$, has no solutions in integers x, y, z except for the solutions obtained by setting one of x, y, z equal to 0.

Moreover, Fermat claimed to have a marvelous proof of his assertion but that the margin was too narrow to contain it. Mathematicians have searched exhaustively for Fermat's "marvelous proof" for three and a half centuries. The consensus of mathematicians today is that Fermat thought that he had a proof but that the proof was incorrect. In any case, the statement which Fermat made remains. Is it true or false? Mathematicians have proved Fermat to be correct for many special values of n. For example, it is now known that Fermat's assertion is, indeed, correct for all n up to 30,000†. Moreover, high-speed computers have been employed searching for examples where Fermat's statement is false, but no such example has ever been found. We shall discuss Fermat's Last Theorem further in Chapters 6 and 10, where we shall show it to be true for $n = 3, 4$.

The above examples of number-theoretic questions illustrate at least one important feature of number theory: Many number-theoretic problems begin with the observation of a phenomenon displayed by the integers. Very often, a suspected property of the integers is tested experimentally by verifying a large number of cases. Of course, this procedure cannot, in general, hope to yield any sort of proof. But it does serve many purposes. First, the experimental evidence can lend credence to a suspected truth being investigated. Second, the experimental evidence can show that the phenomenon under study does not always occur by exhibiting counterexamples to the phenomenon. Third, the experimental evidence can lead to a guess of a correct method of proof. The process of experimentally verifying suspected results was at one time a manual process. Now, however, the high-speed computer is an important tool for the contemporary number theorist.

---

†See the article by Wells Johnson in Math. Computation, Jan. 1975.

The reader should not get the impression that number theory is purely empirical. The observational part of number theory is only the beginning. After formulating a statement about the integers which stands up to the test of experiment, the number theorist then is faced with the problem of proving or disproving the statement. And supplying proofs for facts which are suspected to be true is often an extraordinarily difficult business. Very often, an easily stated conjecture can be extensively supported by experimental evidence but nevertheless cannot be proved or disproved. For example, Goldbach, in 1742, stated that every even integer greater than 2 is the sum of two primes. No correct proof has ever been given.

Another surprising feature of number theory is that when it is possible to prove a number-theoretic conjecture, the proof often will incorporate ideas which are seemingly far removed from the original conjecture. For example, many proofs of number-theoretic results rely on geometric, algebraic, or analytic (i.e., calculus) techniques which are not at all suggested in the statements of the theorems, which, after all, are concerned only with integers.

Before we begin any substantive discussion of number theory, it seems appropriate to ask the question: Why study number theory? There are many reasons, but let us be content to quote a few. The integers arise in the daily activities of human beings. And human curiosity poses questions which demand answers simply by virtue of the fact that they are asked. This puts number theory in the same category as all the other "pure" sciences which seek answers to questions about natural phenomena, and this is already sufficient reason to warrant its study. However, there are other equally cogent reasons. Number theory can be considered as an art form, since its results can be viewed from an aesthetic point of view. When viewed in this way, number theory merits study because it is enjoyable and pleasing to our collective aesthetic sense. Finally, another very important reason for studying number theory stems from the central position which number theory occupies in mathematics. By studying number theory, it is possible to obtain an overview of much of contemporary mathematics, for there are few fields of mathematics which have absolutely no connection with number theory. Also, the study of number theory often leads to the creation of whole fields of mathematics. It is for all these reasons that number theory was dubbed by Gauss the *Queen of Mathematics.*

1.2 Prerequisites

We shall denote the sets of integers, rational numbers, real numbers, and complex numbers by **Z, Q, R,** and **C**, respectively. To simplify the exposition, we shall make the following convention:

Throughout Chapters 1–6, all lowercase italic letters (e.g., a, b, c, m, n, x, y, z) *will stand for integers. In Chapters 7–11, this convention will be relaxed to include rational numbers.*

Let us now make a list of the prerequisites necessary for reading this book.

1. We assume that the reader is familiar with the notation of set theory.
2. We shall assume that the reader is familiar with the algebraic properties of the integers. That is, we shall assume that he is familiar with the properties of addition, subtraction, multiplication, and division. Also we shall assume that the reader can manipulate inequalities involving integers. Whenever these assumed facts are called for, they will be used without further comment. Similarly, the reader may solve exercises posed in this book by using any algebraic manipulations among integers which he is accustomed to from, say, high school algebra.
3. We shall assume two properties of the integers which will form the starting point for many of our proofs. The first of these properties is the *principle of mathematical induction*, with which the reader is probably already familiar. The second is the so-called *well-ordering principle.*

Principle of Mathematical Induction: Suppose that for each positive integer n there is given a proposition $P(n)$. Further, suppose that $P(1)$ is true and that whenever $P(n)$ is true so is $P(n + 1)$. Then $P(n)$ is true for all positive integers n.

Well-Ordering Principle: Let S be a nonempty collection of positive integers. Then S contains a smallest element. In other words, S contains a positive integer n such that $n \leq x$ for all x in S.

We shall often use the well-ordering principle in the following way. Suppose that we wish to prove that all positive integers have a certain property. We form the set S of all positive integers *not* having the property. By the well-ordering principle, there is a smallest element z of S. That is, z is the smallest positive integer not having the given property. This z usually has sufficiently strange properties to imply a contradiction. Thus, S must be empty; i.e., the given property holds for all positive integers n. The reader will see numerous examples of this idea of proof in Chapter 2.

4. From Chapter 8 on, we shall assume that the reader has had a course in abstract algebra covering the definitions and most elementary facts about groups, rings, and fields (e.g., groups, subgroups, quotient groups, order of an element, Lagrange's theorem, equivalence relations, rings, ideals, factorization theory, fields).

1.2 Exercises

1. Use induction to prove the formula

$$1 + 2 + \cdots + n = \frac{n(n+1)}{2}.$$

2. Use induction to prove the formula

$$1^2 + 2^2 + \cdots + n^2 = \frac{n(n+1)(2n+1)}{6}.$$

3. Prove

$$1^3 + 2^3 + \cdots + n^3 = (1 + 2 + \cdots + n)^2.$$

4. Prove

$$(x+y)^n = x^n + \binom{n}{1}x^{n-1}y + \binom{n}{2}x^{n-2}y^2 + \cdots + \binom{n}{n-1}xy^{n-1} + y^n,$$

where

$$\binom{n}{k} = \frac{n(n-1)\cdots(n-k+1)}{k(k-1)\cdots 2\cdot 1}.$$

5. Consider the following proof that all billiard balls are the same color. Use induction on n, the number of billiard balls. If $n = 1$, the result is clear. So suppose that $n > 1$. Line up the billiard balls. By induction the first $n - 1$ billiard balls all have the same color. Also by induction the last $n - 1$ billiard balls all have the same color. Thus, it must be that they all have the same color. What is the fallacy?

6. Prove that the following alternative version of the principle of induction is equivalent to the version given in the text. Suppose that for each positive integer n there is a given proposition $P(n)$. Suppose that $P(1)$ is true. Further, suppose that whenever $P(m)$ is true for all positive integers $m \leq n$, then $P(n+1)$ is true. Then $P(n)$ is true for all positive integers n.

1.3 How to Use this Book

Let us close this chapter with a few hints for the student. We hope that these hints will make the book easier to use and will make this first course in number theory more enjoyable.

First, the student should do lots of exercises. There are a large number provided. Some are routine calculations which illustrate theorems. Others develop theories which extend the scope of the text. Others are difficult and are meant as a challenge for the most ingenious readers. They will usually be labeled with one or more asterisks.

The student should make up numerical examples to illustrate the definitions and theorems. Often, a numerical example will provide considerably more insight into a discussion than that provided by a study of the logical arguments alone. Also, by following a proof using a numerical example, the mechanics of the proof are often clearly exposed to view.

Finally, the student should experiment with the integers, calculate, make tables, conjecture properties of the integers, and try to prove or disprove these conjectures. If you doubt that a conjecture is true, test the first thousand (or million) cases on any high-speed computer which is available to you. Remember that empirical reasoning is the mother of number-theoretic results and that by carrying out some experiments you can join in the long chain of distinguished scientists, amateur and professional alike, who have contributed to number theory.

2

Divisibility and Primes

2.1 Introduction

In this chapter, we shall discuss how the integers are structured from the point of view of multiplication. To get some idea of the sort of structure we have in mind, let us consider first the structure of the integers from the point of view of addition. The number 1 is a very special integer. For from 1, we get by successive additions $2 = 1 + 1$, $3 = 1 + 1 + 1$, $4 = 1 + 1 + 1 + 1$, etc. In other words, every positive integer is obtained by adding together an appropriate number of ones. Do the integers exhibit the same phenomenon with respect to multiplication which they exhibit with respect to addition? That is, does there exist a single integer n such that every positive integer greater than 1 is obtained by multiplying n by itself an appropriate number of times? It is easy to see that no such integer n exists. For example, if $n = 2$, then the sequence $2, 2^2, 2^3, \ldots$ misses $3, 5, \ldots$, and, in fact, misses "most" integers. Similar reasoning works for any n (Exercise). However, a phenomenon similar to the additive phenomenon does occur. Let us say that an integer p is a *prime* if $p > 1$ and the only factors of p are ± 1 and $\pm p$. We shall show in this chapter that every integer greater than 1 can be written as a product of primes, e.g., $4 = 2 \cdot 2$, $5 = 5$, $6 = 2 \cdot 3$, $12 = 2 \cdot 2 \cdot 3$, $100 = 2 \cdot 2 \cdot 5 \cdot 5$. Thus, instead of having one building block for the multiplicative structure of the integers, we have many building blocks, the primes. (In fact, we shall prove in this chapter that there are an infinite number of them.)

Example 1: The factorization of an integer as a product of primes is used in finding least common denominators of fractions. For example, let us compute $\frac{5}{162} + \frac{31}{60}$. Since $162 = 2 \cdot 3^4$ and $60 = 2^2 \cdot 3 \cdot 5$, the least common denominator of the two fractions is $2^2 \cdot 3^4 \cdot 5 = 1620$. Thus,

$$\frac{5}{162} + \frac{31}{60} = \frac{5}{162} \cdot \frac{2 \cdot 5}{2 \cdot 5} + \frac{31}{60} \cdot \frac{3^3}{3^3} = \frac{50}{1620} + \frac{837}{1620} = \frac{887}{1620}.$$

Not only shall we prove that every integer >1 can be written as a product of primes, but we shall also show that such a factorization is unique up to rearrangement of the factors (i.e., we do not count $2 \cdot 3$ and $3 \cdot 2$ as different factorizations of 6). The uniqueness of factorization into primes is very important. Indeed, as we shall see in this chapter, the uniqueness of factorization in the integers is one of the most subtle properties of the integers and is one of the primary facts which makes number theory "work."

Example 2: To appreciate the fact that the uniqueness of factorization has interesting consequences, let us use it to find all positive factors of the integer $b = 4667544 = 2^3 \cdot 3^5 \cdot 7^4$. If a is a factor of b, then $b = ac$ for some positive integer c. If $a = p_1 \cdots p_r$ and $c = q_1 \cdots q_s$ are factorizations of a and c, respectively, as products of primes, then $p_1 \cdots p_r q_1 \cdots q_s = ac = b$ is a factorization of b as a product of primes. By the uniqueness of factorization, $p_1, \ldots, p_r$ must be primes taken from the factorization $2^3 \cdot 3^5 \cdot 7^4$ with 2 appearing at most three times, 3 appearing at most five times, and 7 appearing at most four times. Thus, the positive factors of b are the numbers $2^k 3^\ell 7^m$ for $0 \leq k \leq 3$, $0 \leq \ell \leq 5$, $0 \leq m \leq 4$.

We shall establish our unique factorization theorem (known as the *fundamental theorem of arithmetic*) in Section 2.4. To do so, it will first be necessary to prove certain facts about divisibility of integers. We shall derive these facts using only the principle of mathematical induction and the well-ordering principle.

The theory of divisibility which we shall develop will allow us to take our first steps in solving Diophantine equations. The underlying reason for this is as follows: In solving ordinary equations, we are accustomed to use the algebraic processes of addition, subtraction, multiplication, and division. However, if we are attempting to find solutions of an equation in integers, usually we may not divide, for the quotient of two integers is not necessarily an integer. Thus, it should seem reasonable that in order to solve Diophantine equations, it is necessary to study divisibility properties of integers. As a consequence of the results of this chapter, we shall be able to solve completely any Diophantine equation of the form $ax + by = c$.

2.2 Divisibility

Definition 1:* We say that a *divides* b (written $a\,|\,b$) if and only if there is a c such that $b = ca$. In this case we also say that a is a *factor* of b or that b is *divisible* by a or that b is a *multiple* of a.

For example, $3\,|\,12$, $7\,|\,245$, $41\,|\,2009$, and $588\,|\,4667544$.

If a does not divide b, we write $a \nmid b$. For example, $3 \nmid 5$, $4 \nmid 5$, $7 \nmid 5$, since the only integers dividing 5 are ± 1 and ± 5.

We shall record some trivial properties of the relation of divisibility between two integers.

Proposition 2: (i) $a\,|\,b$ and $a\,|\,c$ imply that $a\,|\,bx + cy$.
(ii) $a\,|\,b$ implies that $a\,|\,bc$.
(iii) $a\,|\,b$ and $b\,|\,c$ imply that $a\,|\,c$.
(iv) Let $a > 0$ and $b > 0$. Then $a\,|\,b$ implies that $a \leq b$.
(v) $a\,|\,b$ and $b\,|\,a$ imply that $a = \pm b$.

Proof:

(i) If $a\,|\,b$ and $a\,|\,c$, then there exist s and t such that $b = as$, $c = at$. But then $bx + cy = asx + aty = a(sx + ty)$, so that $a\,|\,bx + cy$.

(ii) This follows from part (i) with $x = c$, $y = 0$.

(iii) If $a\,|\,b$ and $b\,|\,c$, then $b = as$, $c = bt$ for some s, t. Therefore, $c = ast$ and $a\,|\,c$.

(iv) To prove part (iv), write $b = ac$. Since a and b are positive, we see that c is positive, and hence $c \geq 1$. Thus, $b = ac \geq a$.

(v) First note that if $a\,|\,b$, we have $|a|\,\big|\,|b|$ since if $b = ac$, then $|b| = |a|\cdot|c|$. Thus, if $a\,|\,b$ and $b\,|\,a$, we have $|a|\,\big|\,|b|$ and $|b|\,\big|\,|a|$, so that by part (iv), $|a| \leq |b|$ and $|b| \leq |a|$. These imply that $|a| = |b|$ and that $a = \pm b$, as desired.† ■

Part (i) of Proposition 2 implies, for example, that since $2\,|\,4$ and $2\,|\,6$, we must have $2\,|\,4\cdot 3 + 6\cdot 7$ or $2\,|\,54$. We shall often use the statement of part (i) in the following form:

(i′) *Suppose that* $c = ax + by$ *and that* $d\,|\,b$ but that $d \nmid c$. *Then* $d \nmid a$.

Indeed, if d were to divide a, part (i) of Proposition 2 would imply that $d\,|\,c$. Thus (i′) is proved.

---

*Recall the convention we made in Chapter 1 that all lowercase italic letters denote integers.

†The inequalities used in this proof are among our assumptions concerning the integers.

Warning: The following assertion is *not* always valid: If $c = ax + by$, $d \mid b$, and $d \nmid a$, then $d \nmid c$. Why not?

One way to check whether a given integer divides another integer is by long division. For example, to determine whether $2437 \mid 51329$, let's divide it out:

$$
\begin{array}{r}
21 \\
2437\,\overline{)51329} \\
\underline{4874} \\
2589 \\
\underline{2437} \\
152
\end{array}
$$

Thus, $51329/2437 = 21 + (152/2437)$, or, equivalently, $51329 = 21(2437) + 152$. In particular, we see that $2437 \nmid 51329$, since there is a remainder of 152 when 51329 is divided by 2437. What we wish to show next is that we may always carry out the above process of division to obtain a quotient and remainder and that the quotient and remainder are unique. This result is known as the *division algorithm*.

Theorem 3: Let a and b be integers with $a > 0$. Then there exist unique integers q and r satisfying

$$b = qa + r, \qquad 0 \le r < a.$$

The division algorithm is equivalent to the assertion $b/a = q + (r/a)$, $0 \le r/a < 1$. In the above example, $b = 51329$, $a = 2437$, $q = 21$, and $r = 152$.

It is immediate that $a \mid b$ if and only if $r = 0$. In other words, $a \mid b$ if and only if the remainder on division of b by a is zero.

To understand our proof of Theorem 3, it helps to draw a picture of the real line. The integers $na (n = 0, \pm 1, \pm 2, \ldots)$ are represented by equally spaced points along the line, at distance a apart. For large enough n, the number na lies to the right of b. The integer qa just before b defines the quotient q, and the distance remaining from qa to b defines the remainder r. The case $a = 6$, $b = 27$ is illustrated in Fig. 2.1.

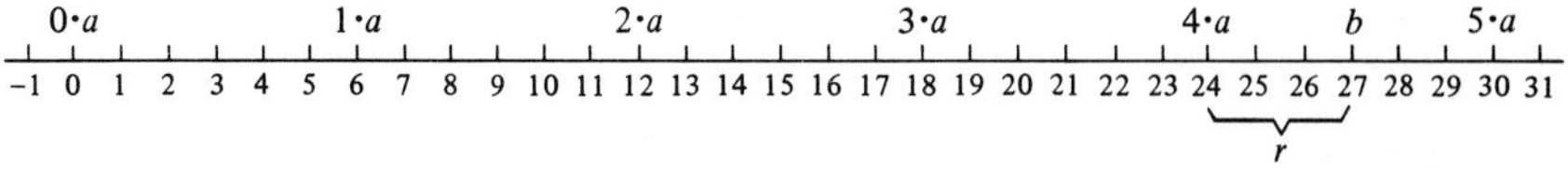

$27 = 4 \cdot 6 + 3$, $q = 4$, $r = 3$.

Figure 2.1

Proof of Theorem 3: We first assume that $b \ge 0$. It is clear that there is a natural number n such that $na > b$ (e.g., $n = b + 1$). Let $q + 1$ be the least such integer (well-ordering). Then

$$(q+1)a > b \geq qa.$$

Let $r = b - qa$. Then $b \geq qa$ implies that $r = b - qa \geq 0$. Finally $(q+1)a = qa + a > b$ implies that $r = b - qa < a$.

We leave the case where $b < 0$ as an exercise.

To show that q and r are unique, suppose that

$$b = qa + r = q_1a + r_1, 0 \leq r, r_1 < a.$$

Either $r \geq r_1$ or $r_1 \geq r$. Suppose, for the sake of argument, that $r \geq r_1$. Then

$$0 \leq r - r_1 < a, \tag{*}$$

and

$$(q_1 - q)a = r - r_1.$$

Thus, $a \mid r - r_1$. If $r - r_1 > 0$, then Proposition 2, (iv) would imply that $a \leq r - r_1$, contradicting (*). Thus, $r - r_1 = 0$ and $r = r_1$. Therefore, $(q_1 - q)a = 0$ and $q = q_1$. ∎

2.2 Exercises

1. Which of the following divisibility relations are true?

   $2\mid 2$; $2\mid 6$; $3\mid 17$; $-7\mid 14$; $8\mid 0$; $17\mid 135$;
   $10\mid(-120)$; $-17\mid(-68)$; $-23\mid(-117)$;
   $3481\mid 437289$; $3481\mid 435125$.

2. List all the divisors of 12, 13, 72, and 260.
3. In the following cases, divide b by a to obtain the quotient and remainder:
   (a) $a = 17, b = 23$.
   (b) $a = 17, b = -23$.
   (c) $a = 14, b = 364$.
   (d) $a = 376, b = 43581$.
   (e) $a = 43581, b = 376$.
4. Draw the figure corresponding to Fig. 2.1 for the following cases:
   (a) $a = 17, b = 40$.
   (b) $a = 3, b = 40$.
   (c) $a = 5, b = 40$.
   (d) $a = 7, b = -40$.
5. How many integers from 1 to 100 are divisible by 9? How many integers from 1 to 2000 are divisible by 9? If n, a are positive integers, how many integers from 1 to n are divisible by a?
6. How many integers between 25 and 250 are divisible by 11? How many

integers between 250 and 25000 are divisible by 11? If n, m, a are positive integers with $n > m$, how many integers between m and n are divisible by a?

7. If a, b, c are nonzero integers, prove that $ac \mid bc$ if and only if $a \mid b$.
8. Prove or disprove the following statement: If $d \neq 0$ and $c = ax + by$ and $d \mid c$ and $d \mid b$, then $d \mid a$.
9. Show that for an integer n, n^2 cannot be of the form $3k + 2$ but can be of the form $3k$ or $3k + 1$. (That is, if we divide a square integer by 3, 2 cannot be a remainder.) (*Hint:* Divide n by 3 and consider separately the three possible remainders.)
10. Show that for all integers n, $4 \nmid n^2 + 2$.
11. Show that for every integer n, $2 \mid n^2 - n$, $6 \mid n^3 - n$.
12. Prove or disprove the following statement: $a^2 \mid b^3$ implies that $a \mid b$.
13. (a) Show that for every integer n there are integers k and r such that $n = 3k + r$ and $r = -1, 0$, or 1.
    (b) Can you generalize this exercise?
14. (a) Show that given any integer k there is an integer n such that $5 \mid n^3 + k$. (*Hint:* Write $k = 5q + r$, where $0 \leq r < 5$.)
    (b) Is the same statement true if we replace 5 by 7?
15. Show that the last digit of a perfect square can only be 0, 1, 4, 5, 6, or 9.
16. Show that a positive integer is a perfect square if and only if it has an odd number of positive divisors. (*Hint:* Experiment with as many numerical examples as you need to see the point in general.)
17. Show that for any positive integer n, $2 \mid 3^n - 1$, $3 \mid 4^n - 1$, $4 \mid 5^n - 1$, etc. (*Hint:* Make use of the polynomial identity $x^n - 1 = (x - 1)(x^{n-1} + x^{n-2} + \cdots + x + 1)$.)
18. Use the polynomial identity $x^n + 1 = (x + 1)(x^{n-1} - x^{n-2} + x^{n-3} - \cdots + x^2 - x + 1)$ when n is an odd positive integer to prove that if $2^n + 1$ has no positive divisors except itself and 1 (i.e., is prime), then n must be a power of 2. (*Hint:* If n is odd, then $3 \mid 2^n + 1$; if $n = 2k$ and k is odd, then $5 \mid 2^n + 1$, etc.) Primes of the form $2^n + 1$ are called *Fermat primes.*
19. Looking at Problems 17 and 18, can you give a general statement about the use of polynomial identities in questions of divisibility?
20. Show that $1 + 2 + \cdots + n \mid 3(1^2 + 2^2 + \cdots + n^2)$ for all $n \geq 1$.
21. Suppose that m, n are odd.
    (a) Show that $8 \mid m^2 - n^2$.
    (b) Show that $8 \mid m^4 + n^4 - 2$.

22. Show that the binomial coefficients

$$\binom{n}{k} = \frac{n!}{k!(n-k)!} \qquad (0 \leq k \leq n)$$

are integers, where $x! = 1 \cdot 2 \cdots x$.

23. Let $n, n_1, n_2, \ldots, n_\ell$ be positive integers such that $n = n_1 + \cdots + n_\ell$. Show that

$$\frac{n!}{n_1! \cdots n_\ell!}$$

is an integer.

24. The Greeks studied so-called *figurate numbers*, defined as follows: The nth triangular number is defined to be the number of dots in the nth triangle in the following sequence:

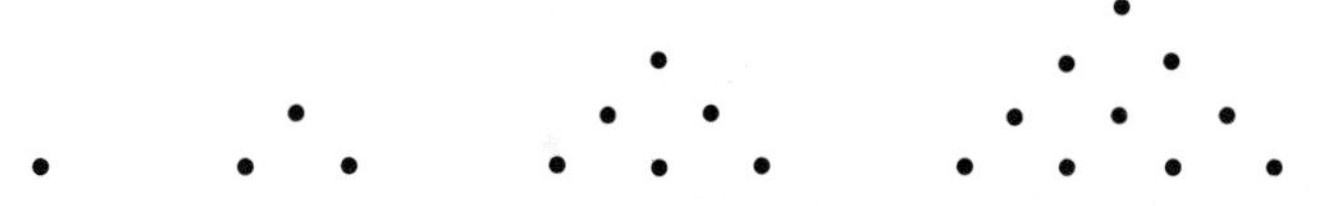

Then nth square number is the number of dots in the nth square in the following sequence:

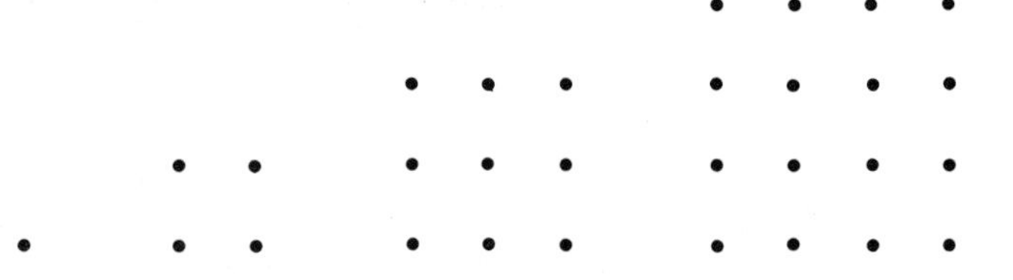

The nth pentagonal number is the number of dots in the nth pentagon in the following sequence:

and so forth for hexagonal, septagonal, etc., numbers.

(a) Show that the nth triangular number is $\frac{1}{2}(n^2 + n)$.
(b) Show that the nth square number is n^2.
(c) Show that the nth pentagonal number is $\frac{1}{2}(3n^2 - n)$.
(d) Find a general formula for the nth k-gonal number.
(e) Show that a pentagonal number cannot have 3, 4, 8, or 9 as its last digit.

25. (a) Let n be a positive integer >1. Show that every positive integer a can be written uniquely in the form $a = a_0 + a_1 n + a_2 n^2 + \cdots + a_k n^k$, where $0 \le a_i \le n - 1$. For $n = 10$, $a_0, \ldots, a_k$ are the digits of the decimal representation of a. The general case is what is called the *representation of a to base n*.
    (b) Find the representations of $a = 57$, $a = 139$, and $a = 199$ to base 7.

2.3 The Greatest Common Divisor

The concept of a greatest common divisor is probably the second most important one in the discussion of divisibility. (The concept of a prime is the most important.)

Definition 1: Let a and b be integers. We call d the *greatest common divisor* (gcd) of a and b if and only if

(i) $d > 0$;
(ii) $d \mid a$ and $d \mid b$ (that is, d is a *common divisor* of a and b);
(iii) Whenever $e \mid a$ and $e \mid b$, we have $e \mid d$.

Write $d = \gcd(a, b)$. If $\gcd(a, b) = 1$, then we say that a and b are *relatively prime*.

Example 2: Let us give a few examples of greatest common divisors:

(i) $a = 2$, $b = 3$, $d = 1$.
(ii) $a = 12$, $b = 15$, $d = 3$.
(iii) $a = 25$, $b = 85$, $d = 5$.
(iv) $a = 100$, $b = 475$, $d = 25$.

By way of illustration, let us prove part (ii). The divisors of 12 are ± 1, ± 2, ± 3, ± 4, ± 6, and ± 12. The divisors of 15 are ± 1, ± 3, ± 5, and ± 15. The common divisors of 12 and 15 are ± 1 and ± 3. And, indeed, 3 is the only positive common divisor which all the common divisors divide. It is very instructive to contemplate the precise role which unique factorization plays in the above computation of $\gcd(12, 15) = 3$.

Let's make some comments about the definition of greatest common divisor. First, requirement (i) asserts that $\gcd(a, b)$ must be a positive integer. Requirement (ii) is that $\gcd(a, b)$ be a divisor of both a and b; that is, $\gcd(a, b)$ is a common divisor of a and b. Finally, requirement (iii) says that $\gcd(a, b)$ has the property that all common divisors of a and b divide $\gcd(a, b)$.

It is not at all clear that a and b have a gcd. Proving that gcd's exist will be one of the major tasks of this section.

Not only does Definition 1 leave the existence of the greatest common divisor of a and b in doubt, but it also leaves open the possibility that a and b have two greatest common divisors. Actually, however, it is easy to show that there can be at most one, for if d_1 and d_2 satisfy the conditions of Definition 1, then requirement (ii) implies that $d_1 \mid a$ and $d_1 \mid b$, $d_2 \mid a$ and $d_2 \mid b$. Therefore, by requirement (iii) applied successively to $e = d_1$, $d = d_2$ and $e = d_2$, $d = d_1$, we see that $d_1 \mid d_2$ and $d_2 \mid d_1$. Therefore, from Proposition 2.2 (v), we deduce that $d_1 = \pm d_2$. However, since both d_1 and d_2 are positive, we see that $d_1 = d_2$. Thus, we have shown that a and b have at most one greatest common divisor. This fact will be recorded in the statement of Theorem 3 below.

Since the concept of a greatest common divisor is so important, we shall give two proofs of its existence. The first proof is less "messy," while the second is constructive in the sense that it gives an explicit method for finding greatest common divisors. Both proofs are somewhat difficult conceptually. Indeed, the existence of a gcd is the most subtle point of this chapter. Any proof of its existence must contain a crafty idea or two. However, someone else has already thought up these clever ideas, and we can sit back and appreciate their beauty.

There is one more point to make before proceeding with the proofs. As an unexpected, even miraculous, by-product of the proofs we obtain a complete solution of the Diophantine equation $ax + by = c$.

Theorem 3: Let a and b be integers, at least one of which is nonzero. Then $d = \gcd(a, b)$ exists and is unique.

Proof 1: Consider the set S of all integers of the form $ax + by$, where x and y range through all integers. That is,

$$S = \{ax + by \mid x, y \text{ integers}\}.$$

S contains a positive integer since S contains the integers a, $-a$, b, and $-b$ ($a = a \cdot 1 + b \cdot 0$, $-a = a \cdot (-1) + b \cdot 0$, etc.). Let d be the *least* positive integer in S. From the way in which S was defined, we may write $d = ax_0 + by_0$ for some integers x_0 and y_0. We claim that this d is the gcd of a and b. (This is, insofar as we are concerned, very crafty.) Well, $d > 0$ by definition, and so requirement (i) of Definition 1 holds. If $e \mid a$ and $e \mid b$, then $e \mid d$ by Proposition 2.2 (i). To show $d \mid a$ we use the division algorithm (Theorem 2.3) to write

$$a = qd + r, \qquad 0 \leq r < d,$$

and show that $r = 0$. This follows since

$$\begin{aligned} r &= a - qd = a - q(ax_0 + by_0) \\ &= a - aqx_0 - bqy_0 = a(1 - qx_0) + b(-qy_0). \end{aligned}$$

Thus, r belongs to S ($r = ax + by$ with $x = 1 - qx_0$, $y = -qy_0$). Since

$0 \leq r < d$ and d is the least positive integer in S, it follows that $r = 0$, as desired. Thus, $d | a$. We show that $d | b$ by a similar argument. Thus requirement (ii) of Definition 1 holds and $d = \gcd(a, b)$. ■

Note: The proof that $d | a$ is one which occurs over and over, and you should become very familiar with the principle: When a given integer is defined to be the least positive integer with a given property and you want to show that it divides another integer, divide and take the remainder. Show that this remainder has the same property and thus must be zero.

Please note the following amazing corollary of the proof.

Corollary 4: If $d = \gcd(a, b)$, then there are integers x and y such that

$$d = ax + by.$$

Example 5: (See Example 2 and Corollary 4.)

(i)	$a = 2, b = 3;$	$x = -1, y = 1;$	$1 = 2(-1) + 3(1).$
(ii)	$a = 12, b = 15;$	$x = -1, y = 1;$	$3 = 12(-1) + 15(1).$
(iii)	$a = 25, b = 85;$	$x = 7, y = -2;$	$5 = 25(7) + 85(-2).$
(iv)	$a = 100, b = 475;$	$x = 5, y = -1;$	$25 = 100(5) + 475(-1).$

We shall now give the second proof of Theorem 3. It is called the *Euclidean algorithm* (Euclid, *Elements*, Book VII, Proposition 2). It gives an explicit method for computing the gcd of two integers.

Proof 2 of Theorem 3: We shall assume that $a > 0$. (The case where $a \leq 0$ will be left to the reader.) By Theorem 2.3, we may write

$$\begin{aligned}
b &= q_1 a + r_1 & 0 < r_1 < a \\
a &= q_2 r_1 + r_2 & 0 < r_2 < r_1 \\
r_1 &= q_3 r_2 + r_3 & 0 < r_3 < r_2 \\
&\ \ \vdots & \vdots \\
r_{n-2} &= q_n r_{n-1} + r_n & 0 < r_n < r_{n-1} \\
r_{n-1} &= q_{n+1} r_n & \text{(no remainder)}.
\end{aligned}$$

Since the remainders decrease and are greater than or equal to zero, eventually we must have zero as a remainder. Of course, we stop the process the first time we obtain a zero remainder. Here n may be $0, 1, 2, 3, \ldots$, and we define $r_0 = a$.

In this proof, we shall show that $r_n = \gcd(a, b)$. By definition, $r_n > 0$ so requirement (i) in Definition 1 is satisfied. By definition, $r_n | r_{n-1}$. Therefore, $r_n | r_n$ and $r_n | r_{n-1}$ imply that $r_n | r_{n-2}$ (Proposition 2.2 (i) again) using the second to last equation above. Then $r_n | r_{n-1}$ and $r_n | r_{n-2}$ imply that $r_n | r_{n-3}$. We continue this way and work our way up the above set of equations. Eventually, we get $r_n | r_3$, and $r_n | r_2$ which imply that $r_n | r_1$; and so $r_n | a$; and so $r_n | b$. Thus, requirement (ii) in Definition 1 is also satisfied. Finally, if $e | a$ and $e | b$, then, as before, $e | r_1$ and so $e | r_2$, etc. Continuing down the chain of equations eventually leads to $e | r_n$, and so requirement (iii) of Definition 1 is also satisfied. ■

You should compute d by this method for the examples in Example 2. We shall work out gcd(2437, 51329). We have already performed the first division in Section 2.

$$\begin{aligned} 51329 &= 21 \cdot 2437 + 152 \\ 2437 &= 16 \cdot 152 + 5 \\ 152 &= 30 \cdot 5 + 2 \\ 5 &= 2 \cdot 2 + 1 \\ 2 &= 2 \cdot 1 + 0. \end{aligned}$$

Therefore, gcd(2437, 51329) = 1. Thus, 2437 and 51329 are relatively prime.

We note that this proof gives a constructive proof of Corollary 4 also. Namely $d = r_n$ and so

$$\begin{aligned} d &= r_{n-2} - q_n r_{n-1} \\ &= r_{n-2} - q_n(r_{n-3} - q_{n-1} r_{n-2}) \\ &= r_{n-2}(1 + q_n q_{n-1}) - r_{n-3} q_n. \end{aligned}$$

Now substitute $r_{n-2} = r_{n-4} - q_{n-2} r_{n-3}$, etc. We work our way up the entire string of equations, and we finally obtain d in the form $ax + by$.

You should carry out this procedure for the examples in Example 5 (*Note:* You may not get the same solution given in Example 5). We shall illustrate the procedure by finding x and y such that

$$2437x + 51329y = 1. \tag{1}$$

$$\begin{aligned} 1 &= 5 - 2 \cdot 2 \\ &= 5 - 2(152 - 30 \cdot 5) = -2 \cdot 152 + 61 \cdot 5 \\ &= -2 \cdot 152 + 61(2437 - 16 \cdot 152) \\ &= 61 \cdot 2437 - 978 \cdot 152 \\ &= 61 \cdot 2437 - 978(51329 - 21 \cdot 2437) \\ &= -978 \cdot 51329 + 20599 \cdot 2437. \end{aligned}$$

Therefore, $x = 20599$, $y = -978$.

This may seem like a fair amount of computation, but if you look at the size of the solution and if you think about solving (1) in integers x and y, you will see that the method is indeed quite efficient.

We shall now record a few more facts about gcd's.

Theorem 6: Let $d = \gcd(a, b)$.

(i) $d = 1$ if and only if there exist integers x, y such that $ax + by = 1$.
(ii) $\gcd(a/d, b/d) = 1$.
(iii) $a \mid bc$ and $d = 1$ imply that $a \mid c$.
(iv) $a \mid bc$ implies that $(a/d) \mid c$.
(v) $\gcd(ma, mb) = md$, provided $m > 0$.

These are all very important facts, with part (iii) the most fundamental. The above order is for convenience in proving them.

Proof:

(i) Corollary 4 is precisely the statement that $d = 1$ implies the existence of x and y. Conversely, if $ax + by = 1$ and $d \mid a$ and $d \mid b$, we see that $d \mid 1$, so that $d = \pm 1$. Thus, since $d > 0$, we have $d = 1$.

(ii) Again from Corollary 4, there exist integers x and y such that

$$ax + by = d,$$

and so

$$\frac{a}{d}x + \frac{b}{d}y = 1.$$

Recall that a/d and b/d are integers, and thus from part (i), $\gcd(a/d, b/d) = 1$, as desired.

(iii) Since $d = 1$, there exist integers x and y such that $ax + by = 1$. Multiplying through by c yields

$$c = acx + bcy.$$

Therefore, $a \mid a$ and $a \mid bc$ imply that $a \mid c$, as desired.

(iv) It is clear that $a \mid bc$ implies that $a/d \mid (b/d)c$. Now simply apply parts (ii) and (iii).

(v) It is clear that $md \mid ma$ and $md \mid mb$. Suppose that $e \mid ma$ and $e \mid mb$. We must show that $e \mid md$. Write

$$d = ax + by.$$

Then

$$md = max + mby,$$

and so clearly $e \mid md$. Thus, from Definition 1, $md = \gcd(ma, mb)$. ∎

Note: In the proof of part (iii) above, you have seen and hopefully will absorb a second fundamental principle for proofs in number theory. (A first was given in a similar note on p. 19.) If you want to prove something about a and b and if $\gcd(a, b) = 1$, write $1 = ax + by$ and see if it helps. It very often does.

Now we are able to tell the complete story concerning the solutions to $ax + by = c$, as we promised.

Theorem 7: If $a, b\ c$ are integers and at least one of a, b is nonzero, set $d = \gcd(a, b)$. Then we can solve the Diophantine equation

$$ax + by = c \tag{2}$$

in integers x and y if and only if $d \mid c$. In this case, let $x = x_0$, $y = y_0$ be one solution. Then the most general solution to (2) is

$$x = x_0 + \frac{b}{d}k, \qquad y = y_0 - \frac{a}{d}k, \qquad k = 0, \pm 1, \pm 2, \ldots.$$

Proof: If integers x and y exist so that $ax + by = c$, then, as usual, $d \mid a$ and $d \mid b$ imply that $d \mid c$. Conversely, suppose that $d \mid c$. From Corollary 4 there are integers x' and y' such that

$$ax' + by' = d.$$

Multiplying through by c/d, we see that $x = (c/d)x'$, $y = (c/d)y'$ is a solution to our original equation (2). (The point of the assumption $d \mid c$ is that $x = (c/d)x'$ and $y = (c/d)y'$ are then integers.)

For the second part of the theorem, we start with a solution x_0, y_0 of (2). By substituting $x = x_0 + (b/d)k$, $y = y_0 - (a/d)k$ directly into (2), we see that x, y is a solution of (2). Conversely, suppose that x, y is an arbitrary solution of (2). We assume that $a \neq 0$. (If $a = 0$, then $b \neq 0$ by assumption, and we proceed similarly.) Therefore,

$$ax + by = c = ax_0 + by_0$$

$$a(x - x_0) = b(y_0 - y).$$

Thus,

$$b \mid a(x - x_0) \quad \text{and hence} \quad \frac{b}{d} \,\Big|\, x - x_0$$

by Theorem 6(iv). Thus, $x - x_0 = (b/d)k$ for some integer k; that is,

$$x = x_0 + \frac{b}{d}k.$$

Substituting this value of x into (2) and solving for y we obtain

$$a\left(x_0 + \frac{b}{d}k\right) + by = c,$$

so that

$$by = c - ax_0 - a\frac{b}{d}k$$

$$= by_0 - b\frac{a}{d}k,$$

and thus

$$y = y_0 - \frac{a}{d}k,$$

as desired. ■

As an example, we derived on p. 20 that a solution to

$$2437x + 51329y = 1$$

is $x_0 = 20599$ and $y_0 = -978$. Since here $d = 1$, the most general solution to this equation is

$$x = 20599 + 51329k$$
$$y = -978 - 2437k,$$

for $k = 0, \pm 1, \pm 2, \ldots$. So, for example, $k = -1$ yields

$$x = -30730 \quad \text{and} \quad y = 1459.$$

2.3 Exercises

1. Find the gcd's of the following pairs of numbers and prove that your answers are correct by using the definition of gcd:
   (a) $a = 15$, $b = 20$.
   (b) $a = 21$, $b = 315$.
   (c) $a = 54$, $b = 8$.
   (d) $a = 24$, $b = 49$.
2. Find the gcd's of the following numbers by using the Euclidean algorithm:
   (a) $a = 10587$, $b = 534$.
   (b) $a = 9800$, $b = 180$.
   (c) $a = 1587645$, $b = 6755$.
3. For each of the parts of Exercises 1 and 2, write $\gcd(a, b)$ in the form $ax + by$.
4. Solve the following linear Diophantine equations:
   (a) $8x + 3y = 27$
   (b) $2x + 11y = 34$.
   (c) $3x + 83y = -4$.
5. Prove that $\gcd(n, n + 1) = 1$ for all n.
6. Suppose that k has the property that $\gcd(n, n + k) = 1$ for all $n > 0$. Prove that $k = 1$ or -1.
7. For what k is it true that $\gcd(n, n + k) = 2$ for all $n > 0$?
8. Prove that $\gcd(a, b) = \gcd(a + kb, b)$ for all k.
9. Prove that if $a \mid c$ and $b \mid c$ with $\gcd(a, b) = 1$, then $ab \mid c$. (*Hint:* Follow the note in the text concerning relatively prime integers.)
10. Find two fractions whose denominators are 11 and 13, respectively, and such that their sum is 67/143.

11. The Smiths run a restaurant which charges a flat fee of \$11 per adult and \$7 per child (inflation!). At the end of an evening the total in the cash register is \$657. What is the smallest number of people who could have dined that day?

12. Find all integers x which have the following property: x leaves remainder 6 when divided by 11, and x leaves remainder 3 when divided by 7.

13. (a) Show that the Diophantine equation $ax + by = c$, where $a > 0$, $b > 0$, $c > 0$, has only a finite number of positive solutions.
    (b) Find the number of positive solutions of $16x + 27y = 390$.

14. Let a, b be positive integers. By a *least common multiple* of a and b, denoted $\operatorname{lcm}(a, b)$, we mean an integer m such that (i) $m > 0$, (ii) $a \mid m$ and $b \mid m$, and (iii) if n is such that $a \mid n$ and $b \mid n$, then $m \mid n$.
    (a) Let $a = 5$, $b = 3$. Show that 15 is a least common multiple of a and b.
    (b) Let $a = 16$, $b = 24$. Find $\operatorname{lcm}(a, b)$. Answer the same question for $a = 12$, $b = 15$.
    (c) Prove that a and b have at most one least common multiple.
    (d) Prove that a and b have a least common multiple. In fact, prove that $ab/\gcd(a,b)$ is a least common multiple of a and b, so that $\operatorname{lcm}(a, b)\gcd(a, b) = ab$. (*Hint*: Apply Exercise 9.)

15. Suppose that $\gcd(a, 4) = 2$ and $\gcd(b, 4) = 2$. Prove that $\gcd(a + b, 4) = 4$.

16. Let a, b, c be integers.
    (a) Formulate a definition of the greatest common divisor of a, b, c.
    (b) Prove that if a, b, c are not all zero, then they have a greatest common divisor (denoted $\gcd(a, b, c)$) and only one.
    (c) Show that $\gcd(a, b, c) = \gcd(\gcd(a, b), c)$.
    (d) Show that $\gcd(a, b, c)$ can be written in the form $ax + by + cz$.

17. Solve the same problem as in Exercise 16, except generalize all results to n integers $a_1, a_2, \ldots, a_n$.

18. Find all solutions of the Diophantine equation $3x + 5y + 4z = 6$.

19. Solve the system of simultaneous Diophantine equations

$$\begin{aligned} x + 2y + 3z &= 4 \\ 2x - z &= -1. \end{aligned}$$

20. Prove that the system of simultaneous Diophantine equations

$$\begin{aligned} 3x + 6y + z &= 2 \\ 4x + 10y + 2z &= 3 \end{aligned}$$

has no solutions.

21. (a) Let a, b, c, d be any integers. Show that the Diophantine equation
$$ax + by + cz = d$$
is solvable if and only if $\gcd(a, b, c) \mid d$.
(b) Assume that the equation in part (a) is solvable and let (x_0, y_0, z_0) be one solution. Determine all solutions.

22. In this exercise, we shall determine an upper bound for the number of steps in the Euclidean algorithm. Suppose that $b \geq a > 0$ and that
$$\begin{array}{ll} b = q_1 a + r_1, & 0 < r_1 < a \\ a = q_2 r_1 + r_2, & 0 < r_2 < r_1 \\ \vdots & \vdots \\ r_k = q_{k+2} r_{k+1} + r_{k+2}, & 0 < r_{k+2} < r_{k+1} \\ \vdots & \vdots \\ r_{n-1} = q_{n+1} r_n. & \end{array}$$
Then $r_1 > r_2 > \cdots > r_n$.
(a) Show that $b \geq 2r_1$, $a \geq 2r_2$.
(b) Show that for $k \geq 1$, we have $r_k \geq 2r_{k+2}$.
(c) Show that $b \geq 2^{n/2}$ so that $n \leq 2(\log b/\log 2)$.
(d) Show that for $b \leq 10000$, the Euclidean algorithm must terminate in at most 28 steps.

23. Write a computer program for calculating the gcd of a and b. You should refer to Exercise 22 to find out how many times you may have to repeat the division algortihm in your program.

24. Let $a_1, a_2, \ldots, a_n$ be nonzero numbers. We say that m is a *least common multiple* of $a_1, \ldots, a_n$ provided (i) $m > 0$, (ii) $a_i \mid m$ for $i = 1, \ldots, n$, and (iii) whenever $b_i \mid m_1$ for $i = 1, \ldots, n$, we have $m \mid m_1$.
(a) Prove that $a_1, \ldots, a_n$ have one and only one least common multiple, denoted $\operatorname{lcm}(a_1, \ldots, a_n)$.
(b) Experiment with various values of $a_1, \ldots, a_n$, and in each case compute $\operatorname{lcm}(a_1, \ldots, a_n)$.
(c) From your experimentation, show that it is not always true that
$$\gcd(a_1, \ldots, a_n) \cdot \operatorname{lcm}(a_1, \ldots, a_n) = a_1 \cdots a_n$$
if all a_i are positive. Can you come up with a generalization of Exercise 14(d) which is valid for gcd's and lcm's of more than two numbers?

*25. Let us define the Farey sequence of fractions of order n to be the set of all reduced fractions a/b, $0 \le a/b \le 1$, $1 \le b \le n$. Thus, the first three Farey sequences are $\frac{0}{1}, \frac{1}{1}$; $\frac{0}{1}, \frac{1}{2}, \frac{1}{1}$; and $\frac{0}{1}, \frac{1}{3}, \frac{1}{2}, \frac{2}{3}, \frac{1}{1}$.
Prove:

(a) If a_1/b_1, a_2/b_2 are fractions which are adjacent to one another in a Farey sequence, then $a_2b_1 - a_1b_2 = 1$.

(b) If a_1/b_1, a_2/b_2, a_3/b_3 are adjacent in a Farey sequence, then $a_2 = a_1 + a_3$, $b_2 = b_1 + b_3$.

26. Let m, n be integers such that $\gcd(m, n) = 1$.

(a) When does $\gcd(m, n) = \gcd(m - n, m + n)$ hold?

(b) Let a, b, c, d, m, n be integers such that $ad - bc = 1$, $mn \ne 0$. Show that $\gcd(am + bn, cm + dn) = \gcd(m, n)$.

*27. Let $a \ge 1$ and let m, n be distinct positive integers. Show that

$$\gcd(a^{2^m} + 1, a^{2^n} + 1) = \begin{cases} 1 & \text{if } a \text{ is even,} \\ 2 & \text{if } a \text{ is odd.} \end{cases}$$

28. Let us define the Fibonacci sequence of integers as follows: $F_0 = 1$, $F_1 = 1$, $F_2 = 1 + 1 = 2$, $F_3 = 1 + 2 = 3$, $F_4 = 2 + 3 = 5$, $F_5 = 3 + 5 = 8, \ldots, F_{n+2} = F_{n+1} + F_n$. This sequence was introduced in the Middle Ages by the Italian mathematician Fibonacci.

(a) Show that two consecutive terms in the Fibonacci sequence are relatively prime.

(b) Show that $F_n = \binom{n-1}{0} + \binom{n-2}{1} + \binom{n-3}{2} + \cdots$.

(Eventually, the terms are zero.)

29. What is the greatest common divisor of the binomial coefficients $\binom{n}{1}, \binom{n}{2}, \ldots, \binom{n}{n-1}$? (See Exercise 17.)

30. Let a, b be distinct integers. Show that there exists an infinite number of integers x such that $\gcd(a + x, b + x) = 1$.

31. Assume that a, b, c, d are integers, $b \ne d$, and that $\gcd(a, b) = \gcd(c, d) = 1$. Show that $(a/b) + (c/d)$ is not an integer.

32. Let a, b be integers, $b \ne 0$. Suppose that we apply the Euclidean algorithm

$$a = bq_1 + r_1$$
$$b = r_1q_2 + r_2$$
$$r_1 = r_2q_3 + r_3$$
$$\vdots$$

Show that we have the following continued fraction expansion for a/b:

$$\frac{a}{b} = q_1 + \cfrac{1}{q_2 + \cfrac{1}{q_3 + \cdots}}$$

33. Show that every even integer n can be written in the form $a - b$ with $\gcd(a, n) = \gcd(b, n) = 1$.

*34. Let m, n be positive integers, $d = \gcd(m, n)$. Show that $\gcd(2^m - 1, 2^n - 1) = 2^d - 1$.

2.4 Unique Factorization

Now, at last, we are prepared to prove the main result of this chapter, the uniqueness of factorization of integers into prime factors. First we define what our "building blocks" are.

Definition 1: A *prime* is an integer p such that

(i) $p > 1$;
(ii) if $a \mid p$, then $a = \pm 1$ or $\pm p$.

For example, $p = 2, 3, 5, 7, 11$, etc., are primes. We shall state some preliminary results as lemmas.

Lemma 2: If $n > 1$, then n is a product of primes.

It should be emphasized that a prime is considered to be a product of primes, namely the product of one prime.

Proof of Lemma 2: If there are integers $n > 1$ which are not products of primes, then there is a least such integer (well-ordering). Call it m. Then m cannot be a prime. And thus from the definition of a prime, there is a divisor a of m ($a \mid m$) such that $a \neq \pm 1$ and $a \neq \pm m$. We may assume that $a > 0$. Write $m = ab$. Then $1 < a, b < m$ (Proposition 2.2 (iv)). Since m is the least integer >1 which is not a product of primes, a and b must be products of primes, say $a = p_1 p_2 \cdots p_r$ and $b = q_1 q_2 \cdots q_s$, where $p_1, \ldots, p_r, q_1, \ldots, q_s$ denote primes. Then

$$m = ab = p_1 \cdots p_r q_1 \cdots q_s$$

is a product of primes. This contradicts our choice of m and concludes the proof of Lemma 2. ■

Lemma 3: (Euclid's lemma). Let p be a prime. Then $p \mid ab$ implies that $p \mid a$ or $p \mid b$.

Proof: This is an immediate consequence of Theorem 3.6 (iii), since $p \nmid a$ implies that $\gcd(p, a) = 1$, and so $p \mid b$. ■

Corollary 4: Let p be a prime. Then $p \mid a_1 a_2 \cdots a_r$ implies that $p \mid a_i$ for some i, $1 \leq i \leq r$.

Proof: Exercise. Use induction. ■

Theorem 5: (Unique Factorization). Let $n > 1$ be an integer. Then

$$n = p_1 p_2 \cdots p_r$$

is a product of primes and this factorization is unique. The uniqueness is understood to mean that if

$$n = p_1 p_2 \cdots p_r = q_1 q_2 \cdots q_s,$$

where $p_1, \ldots, p_r, q_1, \ldots, q_s$ are primes, then $r = s$ and, after rearranging $q_1, \ldots, q_s$, we have $p_1 = q_1, \ldots, p_r = q_r$.

That we must allow for rearrangement of the q's to make the corresponding factors equal is clear, since, for example,

$$12 = 2 \cdot 2 \cdot 3 = 2 \cdot 3 \cdot 2 = 3 \cdot 2 \cdot 2.$$

Proof of Theorem 5: We know from Lemma 2 that each $n > 1$ can be factored into a product of primes. If there were integers $n > 1$ for which such factorization was not unique, then there would be a least such integer (well-ordering); call it m. Suppose that

$$m = p_1 p_2 \cdots p_r = q_1 q_2 \cdots q_s.$$

Then $p_1 \mid q_1 q_2 \cdots q_s$, and so by Corollary 4, $p_1 \mid q_i$ for some i. By rearranging the q's, we may assume that $i = 1$. Therefore, $p_1 \mid q_1$. Since q_1 is a prime, we have $p_1 = \pm 1$ or $p_1 = \pm q_1$. Since $p_1 > 1$, we have $p_1 = q_1$. We may thus cancel $p_1 = q_1$ in the expression for m to obtain

$$m_1 = p_2 \cdots p_r = q_2 \cdots q_s.$$

But $m_1 < m$ (Proposition 2.2 (iv)). Thus, since m is the least integer >1 without unique factorization, we have $r - 1 = s - 1$, and, after rearranging the q's, $p_2 = q_2, \ldots, p_r = q_s$. Thus, $r = s$ and $p_1 = q_1, \ldots, p_r = q_r$. Thus, the factorization of m must have been unique after all. This contradiction establishes Theorem 5. ■

Corollary 6: If $n < -1$, then there are primes $p_1, \ldots, p_r$ such that

$$n = -p_1 \cdots p_r$$

and the factorization is unique.

Proof: Indeed, $-n$ satisfies the hypothesis of Theorem 5. ■

At this point we might mention something which tends to confuse students. Why, in Definition 1, did we exclude 1 from being a prime? Of course, in mathematics there is nothing sacred about the label we put on things. However, these labels are carefully chosen to make the ideas as transparent as possible. In our present case, if we had allowed 1 to be a prime, then Theorem 5 would be false. For example, $6 = 2 \cdot 3 = 2 \cdot 3 \cdot 1 = 1 \cdot 2 \cdot 3 \cdot 1 = 1 \cdot 2 \cdot 3 \cdot 1 \cdot 1$, etc. Therefore, the uniqueness of the factorization is destroyed if we call 1 a prime.

A comment concerning the proof of Theorem 5 is also in order. If you look at the proof that factorization is always possible (Lemma 2) and the proof that factorization is unique, you might be misled into believing that the proofs are about equal in difficulty. This is not so, and, as we have remarked before, the uniqueness is by far the deeper result. If you examine the two proofs closely, you will see that the proof of existence used nothing but the well-ordering principle and trivial facts about divisibility. On the other hand, the crucial point in the proof of uniqueness relied on the statement that $p_1 | q_1 \cdots q_s$ implies that $p_1 | q_i$ for some i. This was essentially Lemma 3, that is, Theorem 3.6 (iii), and the latter result relied on the rather nontrivial material concerning greatest common divisors in Section 3.

We shall close this section by collecting a few more facts about primes and factorization. The first ones concern themselves with the question of how many primes there are and how they are distributed. These questions lead to some of the deepest and most beautiful results in number theory and are still the subject of much research. We shall content ourselves with Corollaries 7 and 8 below. The student who is interested should look at Appendix A, which contains an alternative proof of Corollary 7 and an improvement of it. The proofs rely on calculus and serve to give the reader a small glimpse into the deep and marvelous subject of applications of calculus to number theory (called *analytic number theory*).

Corollary 7: (*Euclid*). There are an infinite number of primes.

Proof: If this were false, then there would be a finite number, k, of primes, which we denote by $p_1, \ldots, p_k$. Set

$$n = p_1 p_2 \cdots p_k + 1.$$

We know (Lemma 2 or Theorem 5) that there is a prime p such that $p | n$. Since p is a prime and since $p_1, \ldots, p_k$ are all the primes, there must be an i such that $p = p_i$. Thus, $p | p_1 p_2 \cdots p_k$. This fact, together with $p | n$, implies that $p | 1$. This contradiction establishes the result. ■

Corollary 8: There are arbitrarily large gaps between successive primes.*

*On the other hand, it is an unsolved problem as to whether there are an infinite number of prime pairs $(p, p+2)$ with p and $p+2$ primes (here the gap is 2). Examples are (3, 5), (5, 7), (11, 13), (17, 19), and (29, 31).

Proof: The assertion is that for any positive integer n there is a sequence of n successive integers none of which is prime. Indeed, if we denote by $n!$ the product $1 \cdot 2 \cdot 3 \cdots n$, then

$$(n+1)! + 2, (n+1)! + 3, \ldots, (n+1)! + (n+1)$$

is a sequence of n integers. That none of them can be prime is clear since for $2 \leq k \leq n+1$

$$k \mid (n+1)! + k$$

and $2 \leq k < (n+1)! + k$. Thus, $(n+1)! + k$ has a divisor other than ± 1 and itself and so is not prime. ■

Finally we shall give a method for factoring integers and generating primes. The latter is called the *sieve of Eratosthenes*. It is called a sieve because the procedure is like taking a sieve filled with integers and dropping out most of the integers through holes in the sieve. When we are done only primes are left in the sieve. We start with

Proposition 9: If $n > 1$ is an integer and is not a prime, then there is a prime p such that

$$p \mid n \quad \text{and} \quad p \leq \sqrt{n}.$$

Proof: Since n is not a prime, there are integers a, b such that

$$n = ab \quad \text{and} \quad 2 \leq a \leq b < n.$$

Thus, $n = ab \geq a^2$ or $a \leq \sqrt{n}$. Let p be a prime such that $p \mid a$. Then $p \mid n$ and $p \leq a \leq \sqrt{n}$. ■

For example, 18 is not a prime, and $2 \mid 18$ and $2 \leq \sqrt{18}$. Also 25 is not a prime, and $5 \mid 25$ and $5 \leq \sqrt{25}$.

Thus, to check whether a given integer $n > 1$ is a prime it suffices to check whether n is divisible by any of the primes $p \leq \sqrt{n}$. If not, n is a prime. This in general saves a lot of work.

For example, is $n = 271$ a prime? Well $16 < \sqrt{271} < 17$, and so 271 is a prime or there is a prime $p \leq 16$ such that $p \mid 271$. The primes ≤ 16 are 2, 3, 5, 7, 11, and 13. We now must do some work to see if any of these primes divide into 271 evenly. They do not (do the division). Thus, 271 is prime.

The above idea can be put to work to generate primes. The resulting procedure is the sieve of Eratosthenes. Suppose that we wish to find all the primes $\leq n$. Well, any nonprime is divisible by a prime $\leq \sqrt{n}$. Thus, list the integers from 2 through n. Strike out all multiples of 2 except 2 itself. The next integer in the list is the prime 3. Strike out all multiples of 3 except 3 itself. The next integer in the list that has not been struck out is the prime 5 (4 was struck out as a multiple of 2). Strike out all multiples of 5 except 5 itself. Continue doing this. Suppose that we have just struck out all multiples of p except p itself. The next integer left in the list will be the next prime after p; call it q. We continue doing this until $q > \sqrt{n}$, and then we quit. All that is left are the primes between 2 and n, as all other integers have been struck out as multiples of primes $\leq \sqrt{n}$.

Example 10: $n = 30$. Since $5 < \sqrt{30} < 6$, we must strike out all multiples of 2, 3, and 5:

2 3 ~~4~~ 5 ~~6~~ 7 ~~8~~ ~~9~~ ~~10~~
11 ~~12~~ 13 ~~14~~ ~~15~~ ~~16~~ 17 ~~18~~ 19 ~~20~~
~~21~~ ~~22~~ 23 ~~24~~ ~~25~~ ~~26~~ ~~27~~ ~~28~~ 29 ~~30~~

So the list of primes ≤ 30 is 2, 3, 5, 7, 11, 13, 17, 19, 23, and 29.

Notice that to find all the primes up to 100 we would only have to strike out the multiples of one more prime, namely 7, so the method is remarkably efficient.

2.4 Exercises

1. Factor the following integers into products of primes: 13, 16, 28, 144, 169, 44, 100.
2. Determine all the primes p with $p \leq 100$ from the sieve of Erathosthenes. (It should not take you more than about 1 hour to use the sieve to find all primes $p \leq 500$ and so check Table 1 at the end of this book. This depends a great deal on how immune you are to writer's cramp.)
3. Write a computer program to determine all the primes $p \leq n$ using the sieve of Eratosthenes. If you have a computer available, compile a list of all primes ≤ 1000.
4. Given a nonzero integer n, set A_n equal to the set of primes $p \mid n$. So $A_1 = \varnothing$, $A_2 = \{2\}$, $A_6 = \{2, 3\}$, and $A_{12} = \{2, 3\}$. Prove the following useful statements:
   (a) $\gcd(n, m) = 1$ if and only if $A_n \cap A_m = \varnothing$.
   (b) $n \mid m$ implies that $A_n \subseteq A_m$ (and so $A_n \nsubseteq A_m$ implies that $n \nmid m$).
   (c) If $d = nx + my$, then $A_d \subseteq A_m \cap A_n$.
   (d) If $d = nm$, then $A_d = A_n \cup A_m$.
   (e) If $d = \gcd(n, m)$, then $A_d = A_n \cap A_m$.
5. Given any pair of integers m, n both larger than 1, show that it is possible to write

$$m = p_1^{a_1} p_2^{a_2} \cdots p_k^{a_k}$$
$$n = p_1^{b_1} p_2^{b_2} \cdots p_k^{b_k},$$

where $p_1, \ldots, p_k$ are distinct primes and $a_1, \ldots, a_k, b_1, \ldots, b_k$ are integers ≥ 0 (e.g., $12 = 2^2 \cdot 3 \cdot 5^0$ and $15 = 2^0 \cdot 3 \cdot 5$).

6. Using the expressions of Exercise 5, show that
   (a) $m \mid n$ if and only if $a_1 \leq b_1, \ldots, a_k \leq b_k$.
   (b) $\gcd(n, m) = p_1^{c_1} p_2^{c_2} \cdots p_k^{c_k}$, where $c_i = \text{minimum}(a_i, b_i)$ $(1 \leq i \leq k)$.
   (c) $\text{lcm}(n, m) = p_1^{d_1} p_2^{d_2} \cdots p_k^{d_k}$, where $d_i = \text{maximum}(a_i, b_i)$ $(1 \leq i \leq k)$.
   (The lcm is defined in Exercise 14 of Section 2.3.)

7. Use Exercise 6 to show that

$$(\text{lcm}(n, m))(\gcd(n, m)) = nm,$$

   and thus give another proof of Exercise 14(d) of Section 2.3.

8. Prove that for positive integers m, n the proposition

$$a^m \mid b^n \quad \text{implies that} \quad a \mid b$$

   is true if and only if $m \geq n$.

9. Let $a \geq 2$, $n \geq 2$. Show that if $a^n - 1$ is a prime, then $a = 2$ and n is a prime. (*Hint:* Look at Exercise 17 of Section 2.2.) Primes of this form are called *Mersenne primes.*

10. (This exercise is for those with access to a computer.) Let $\pi(x) =$ the number of primes $\leq x$. Use the program you wrote for Exercise 3 to compute (a) $\pi(x) - (x/\log x)$ and (b) $\pi(x)/(x/\log x)$ for $x = 100, 200, 300, 400, \ldots, 10000$. On the basis of the numerical data derived, can you make any conjectures? (These same calculations were performed by Gauss.) *Note:* $\log x$ means the natural log of x.

11. Let $f(x) = a_0 x^n + \cdots + a_n$ be a nonzero polynomial with integer coefficients. Show that $f(k)$ is composite for infinitely many integers k. (*Hint:* Reduce the problem to the case where $|a_n| > 1$.)

*12. Show that for $n > 1$, the number $1 + \frac{1}{2} + \cdots + 1/n$ is not an integer.

*13. Show that for $n \geq 1$ the number $1 + \frac{1}{3} + \cdots + 1/(2n + 1)$ is not an integer.

14. Let a/b be a reduced fraction (that is, $\gcd(a, b) = 1$). Assume that a/b is a zero of the polynomial $a_0 x^n + \cdots + a_n$, $a_0, \ldots, a_n$ integers, $a_0 \neq 0$. Show that $a \mid a_n$, $b \mid a_0$.

15. Let p be a prime and a an integer such that $1 < a < p$. Show that $\binom{p}{a}$ is divisible by p.

16. Let p be a prime, and a, b integers. Show that $a^p - b^p$ is either relatively prime to p or $p^2 \mid (a^p - b^p)$. (*Hint:* Apply Exercise 15.)

17. Prove that $n^{34} - 9$ is never prime for any n.

18. (Hilbert) Let R denote the collection of all positive integers of the form $4k + 1$, k an integer.
    (a) Show that the product of two elements of R is again an element of R.
    (b) Say that an element m of R is a *prime element* if the only elements of R which divide m are 1 and m. Find all prime elements ≤ 100.
    (c) Show that every integer of R can be written as a product of prime elements.
    (d) Show that the factorization in part (c) is not necessarily unique.

19. Let p_n denote the nth prime.
    (a) Show that $p_{n+1}^2 < p_1 p_2 \cdots p_n$ for $n \geq 4$.
    (b) Show that $p_n < 2^{2^n}$.

20. A famous theorem known as Bertrand's postulate asserts that for $n \geq 1$ there exists a prime p such that $n \leq p \leq 2n$.
    (a) Verify Bertrand's postulate for $n \leq 100$.
    (b) Assume Bertrand's postulate and verify that $p_n \leq 2^n$, where p_n denotes the nth prime.

*21. Show that the only possible positive integers which are not sums of consecutive integers are powers of 2.

22. Let $d_1(n)$ denote the number of divisors of n of the form $4k + 1$ and let $d_3(n)$ denote the number of divisors of n of the form $4k + 3$. Show that $d_1(n) \geq d_3(n)$.

23. (a) Show that $\sqrt{5}$, $\sqrt[3]{3}$, and $\sqrt[7]{2}$ are not rational numbers.
    (b) Let a be a positive integer and let $a = p_1^{e_1} \cdots p_t^{e_t}$ be the factorization of a into a product of powers of distinct primes $p_1, \ldots, p_t$. Show that $\sqrt[m]{a}$ is a rational number if and only if $m \mid e_1$, $m \mid e_2$, $\ldots, m \mid e_t$.

Appendix A

Euler's Proof of the Infinitude of Primes

In this appendix, we shall give another proof that there are infinitely many primes. The proof given here is due to the eighteenth-century Swiss mathematician Leonhard Euler. Its novelty is that it connects the ideas of calculus with the arithmetic of the integers. Let us consider the function of s defined by

$$\zeta(s) = \sum_{n=1}^{\infty} \frac{1}{n^s}.$$

This function is called the *Riemann zeta function*, after the German mathematical giant Bernhard Riemann, who, in 1859, systematically studied the deeper properties of the function. Actually, the infinite series $\zeta(s)$ was first introduced by Euler, nearly 100 years before Riemann's work.

By the integral test, the series for $\zeta(s)$ converges for $s > 1$ and therefore defines a function of s for $s > 1$. Moreover, since

$$\frac{1}{n^s} \geq \int_n^{n+1} \frac{dx}{x^s},$$

we see that

$$\sum_{n=1}^{\infty} \frac{1}{n^s} \geq \int_1^{\infty} \frac{dx}{x^s} = \frac{1}{s-1}.$$

Thus, we see that

$$\lim_{s \to 1+} \zeta(s) = \infty.$$

The graph of $\zeta(s)$, as you may readily check, is as given in Fig. A.1.

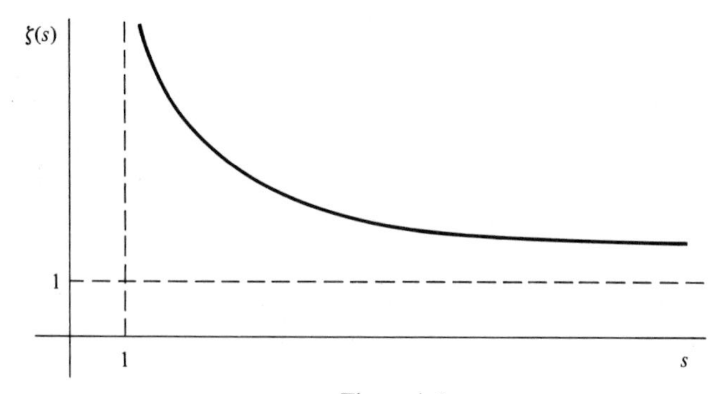

Figure A.1

We will need the concept of an infinite product. This is simple to define and is very similar to an infinite sum. Namely if $a_1, a_2, a_3, \ldots$ is a sequence of real numbers, define*

$$\prod_{n=1}^{\infty} a_n = \lim_{N \to \infty} \prod_{n=1}^{N} a_n = \lim_{N \to \infty} a_1 a_2 \cdots a_N.$$

The connection of $\zeta(s)$ with primes is given in the following theorem.

Theorem 1: (Product Formula)

$$\zeta(s) = \prod_p \left(1 - \frac{1}{p^s}\right)^{-1} \quad \text{for } s > 1.$$

*This definition of an infinite product is at variance with the usual definition given in analysis books. For technical reasons, one usually requires that the limit be nonzero. However, our definition will suffice for our purposes.

We should make a comment concerning the notation in Theorem 1. The symbol $\prod_p$ means take the product over all primes p. Thus, if $p_1 < p_2 < p_3 < \cdots$ denotes all the primes listed in order, then Theorem 1 asserts that

$$\zeta(s) = \prod_{m=1}^{\infty} \left(1 - \frac{1}{p_m^s}\right)^{-1} = \lim_{N\to\infty} \prod_{m=1}^{N} \left(1 - \frac{1}{p_m^s}\right)^{-1}$$

if there are an infinite number of primes, or, if there are just M primes,

$$\zeta(s) = \prod_{m=1}^{M} \left(1 - \frac{1}{p_m^s}\right)^{-1}.$$

Proof of Theorem 1: Since $0 < 1/p_m^s < 1$, we have the geometric series formula

$$\frac{1}{1 - \dfrac{1}{p_m^s}} = 1 + \frac{1}{p_m^s} + \frac{1}{p_m^{2s}} + \frac{1}{p_m^{3s}} + \cdots. \tag{*}$$

This sum contains just some of the terms in the series

$$\zeta(s) = 1 + \frac{1}{2^s} + \frac{1}{3^s} + \frac{1}{4^s} + \cdots = \sum_{n=1}^{\infty} \frac{1}{n^s},$$

namely those for $n = p_m^k$, powers of p_m. Let's see what happens when we multiply the first two series (*) together* ($p_1 = 2$, $p_2 = 3$):

$$\left(1 - \frac{1}{2^s}\right)^{-1}\left(1 - \frac{1}{3^s}\right)^{-1} = \left(1 + \frac{1}{2^s} + \frac{1}{2^{2s}} + \cdots\right)\left(1 + \frac{1}{3^s} + \frac{1}{3^{2s}} + \cdots\right)$$
$$= \sum \frac{1}{n^s},$$

where the sum is over all n with only 2's and 3's as factors.

Similarly,

$$\left(1 - \frac{1}{2^s}\right)^{-1}\left(1 - \frac{1}{3^s}\right)^{-1}\left(1 - \frac{1}{5^s}\right)^{-1} = \sum \frac{1}{n^s}, \tag{1}$$

where the sum is over all n with just 2's, 3's, and 5's as factors. And, in general,

$$\prod_{m=1}^{N} \left(1 - \frac{1}{p_m^s}\right)^{-1} = \sum \frac{1}{n^s}, \tag{2}$$

where the sum is over all n with just $p_1, p_2, \ldots, p_N$ as factors. We have glossed over an important point. Let's look again at Eq. (1), for example. How do we know that every integer n with only 2's, 3's, and 5's as factors occurs once and only once. That $n = 2^a 3^b 5^c$ occurs is evident; the only way such an n could occur two or more times would be for n to have another expression of the form $2^{a'} 3^{b'} 5^{c'}$, which we know cannot happen because of

---

*We assume that the reader is familiar with the following result about infinite series: If $\sum_n a_n$ and $\sum_m b_m$ are convergent series of positive terms, then $\sum_{m,n} a_n b_m$ converges and equals $(\sum_n a_n)(\sum_m b_m)$.

unique factorization. The same considerations hold for Eq. (2). Indeed, Theorem 1 may be viewed as an analytic statement of the unique factorization theorem.

If there are only finitely many primes, say M of them, then Eq. (2) for $N = M$ would complete the proof of the theorem. Otherwise there are an infinite number of primes and $p_N \longrightarrow \infty$ as $N \longrightarrow \infty$. To complete the proof we must show that

$$\lim_{N\to\infty} \left| \zeta(s) - \prod_{m=1}^{N} \left(1 - \frac{1}{p_m^s}\right)^{-1} \right| = 0. \tag{3}$$

From (2),

$$\left| \zeta(s) - \prod_{m=1}^{N} \left(1 - \frac{1}{p_m^s}\right)^{-1} \right| = \Sigma \frac{1}{n^s}$$

$$\leq \sum_{n=p_N+1}^{\infty} \frac{1}{n^s},$$

where the first sum is over all n with at least one prime factor $p > p_N$. It is a standard fact from calculus that the "tail end" of a convergent infinite series must tend toward zero, and so

$$\lim_{N\to\infty} \sum_{n=p_N+1}^{\infty} \frac{1}{n^s} = 0.$$

This implies (3), and so Theorem 1 is proved. ■

Theorem 1 suffices to show there are an infinite number of primes.

Corollary 2: (Euclid). There are an infinite number of primes.

Proof (Euler): If not, let $p_1, \ldots, p_M$ be all the primes. Then by Theorem 1,

$$\zeta(s) = \prod_{m=1}^{M} \left(1 - \frac{1}{p_m^s}\right)^{-1},$$

a finite product. Therefore, clearly the expression on the right makes sense for every real $s > 0$ and defines a continuous function. However, we observed before that

$$\lim_{s\to 1_+} \zeta(s) = \infty,$$

which contradicts the statement

$$\lim_{s\to 1_+} \prod_{m=1}^{M} \left(1 - \frac{1}{p_m^s}\right)^{-1} = \prod_{m=1}^{M} \left(1 - \frac{1}{p_m}\right)^{-1} < \infty.$$ ■

Our goal is to prove a more precise version of Corollary 2, namely

$$\sum_p \frac{1}{p} = \infty.$$

Note that this equation asserts more than just the infinitude of primes.

For example, there are an infinite number of squares n^2 also, but

$$\sum_{n=1}^{\infty} \frac{1}{n^2} < \infty.$$

Thus, in some sense "there are more primes than squares." More precisely, the primes are denser than the squares in the set of all integers.

Theorem 3: (Euler). $\sum_p \frac{1}{p} = \infty.$

Proof: We look at $\log \zeta(s)$ for $s > 1$. Since $\zeta(s) > 0$ for $s > 1$, $\log \zeta(s)$ is defined; and since $\lim_{s \to 1_+} \zeta(s) = \infty$, we must have

$$\lim_{s \to 1_+} \log \zeta(s) = \infty. \tag{4}$$

By Theorem 1,

$$\begin{aligned} \log \zeta(s) &= \log \left(\lim_{N \to \infty} \prod_{k=1}^{N} \left(1 - \frac{1}{p_k^s}\right)^{-1} \right) \\ &= \lim_{N \to \infty} \left(\log \prod_{k=1}^{N} \left(1 - \frac{1}{p_k^s}\right)^{-1} \right). \end{aligned}$$

(Since $\log x$ is a continuous function, the fact that $a_N \longrightarrow a$ as $N \longrightarrow \infty$ implies that $\log a_N \longrightarrow \log a$ as $N \longrightarrow \infty$.) Thus,

$$\log \zeta(s) = \lim_{N \to \infty} \sum_{k=1}^{N} \left(-\log\left(1 - \frac{1}{p_k^s}\right) \right).$$

Recall the infinite series for $\log(1 - x)$: For $|x| < 1$,

$$-\log(1 - x) = x + \frac{x^2}{2} + \frac{x^3}{3} + \cdots = \sum_{m=1}^{\infty} \frac{x^m}{m}.$$

Since, for $s > 1$, $0 < 1/p_k^s < 1$,

$$\begin{aligned} -\log\left(1 - \frac{1}{p_k^s}\right) &= \sum_{m=1}^{\infty} \frac{1}{m p_k^{ms}} = \frac{1}{p_k^s} + \sum_{m=2}^{\infty} \frac{1}{m p_k^{ms}} \\ &< \frac{1}{p_k^s} + \sum_{m=2}^{\infty} \frac{1}{p_k^m} \\ &= \frac{1}{p_k^s} + \frac{1}{p_k^2} \sum_{m=0}^{\infty} \frac{1}{p_k^m} \\ &= \frac{1}{p_k^s} + \frac{1}{p_k^2} \frac{1}{1 - \frac{1}{p_k}} \\ &= \frac{1}{p_k^s} + \frac{1}{p_k(p_k - 1)}. \end{aligned}$$

(by the formula for a geometric series, $\sum_{m=0}^{\infty} r^m = 1/(1 - r)$ for $|r| < 1$)

Thus,

$$\log \zeta(s) \leq \lim_{N \to \infty} \sum_{k=1}^{N} \left(\frac{1}{p_k^s} + \frac{1}{p_k(p_k - 1)} \right).$$

We note that

$$\sum_{k=1}^{\infty} \frac{1}{p_k(p_k - 1)} \leq \sum_{n=2}^{\infty} \frac{1}{n(n-1)} < \infty,$$

and for $s > 1$,

$$\sum_{k=1}^{\infty} \frac{1}{p_k^s} \leq \sum_{n=1}^{\infty} \frac{1}{n^s} < \infty.$$

Thus,

$$\sum_{k=1}^{\infty} \frac{1}{p_k(p_k - 1)} \quad \text{and} \quad \sum_{k=1}^{\infty} \frac{1}{p_k^s}$$

converge, by the comparison test. Set

$$\alpha = \sum_{k=1}^{\infty} \frac{1}{p_k(p_k - 1)},$$

a constant independent of s. Then

$$\log \zeta(s) \leq \sum_{k=1}^{\infty} \frac{1}{p_k^s} + \alpha.$$

We immediately deduce from Eq. (4) that

$$\lim_{s \to 1_+} \sum_{k=1}^{\infty} \frac{1}{p_k^s} = \infty. \tag{5}$$

Furthermore, since $1/p_k > 1/p_k^s$ for all $s > 1$, we see that

$$\sum_{k=1}^{N} \frac{1}{p_k} > \sum_{k=1}^{N} \frac{1}{p_k^s},$$

and so for all $s > 1$,

$$\sum_{k=1}^{\infty} \frac{1}{p_k} > \sum_{k=1}^{\infty} \frac{1}{p_k^s}.$$

Finally then, from Eq. (5), we conclude that

$$\sum_{k=1}^{\infty} \frac{1}{p_k} = \infty,$$

as desired. ∎

3

Congruences

3.1 Introduction

In this chapter, we shall introduce yet another fundamental notion in number theory, the idea of a congruence. To motivate the concept of congruence, let us consider the process of measuring time using a clock. For the sake of simplicity, let us assume that we are interested in telling the correct hour (and not the number of minutes past the hour). Then the process of measuring time which we use in our daily lives amounts to the following: To each hour, we assign a number which indicates the number of hours which have elapsed since the last noon or midnight. After 12 hours pass, we begin counting all over again. Thus, the only times considered are those given by numbers from 1 to 12, and two times are assigned the same number if they differ by a multiple of 12 hours. Thus, we measure time by "disregarding multiples of 12 hours." We observe a similar phenomenon on the odometer of a car. Most odometers will record elapsed mileage up to 99,999 miles. At 100,000 miles, the odometer resets to 0 and begins to run through the numbers 0 through 99,999 again. Thus, the odometer reads the same at two points if the mileage between those points is a multiple of 100,000.

We can generalize the examples of a clock and an odometer as follows: Suppose that we have a dial on which are inscribed the numbers $0, 1, \ldots, n-1$, and imagine that the dial is attached to a device designed to count something (hours, miles, people, etc.). For every instance of the observed phenomenon the counter advances by 1. Then it is easy to see that the counter counts by disregarding multiples of n. Thus, when the actual count is 1, $n + 1, 2n + 1, 3n + 1, \ldots$, the dial will read 1; when the actual count is 2, $n + 2, 2n + 2, 3n + 2, \ldots$, the dial will read 2, and so forth. Or, in general, the dial will read the same for counts x and y, provided $x - y$ is divisible by n. This idealized "odometer" illustrates perfectly the abstract notion of a congruence.

Definition 1: Let n be a positive integer. We say that x and y are *congruent modulo n*, denoted $x \equiv y (\mathrm{mod}\ n)$, provided that $x - y$ is divisible by n. If x is not congruent to y modulo n, we write $x \not\equiv y (\mathrm{mod}\ n)$.

Thus, for example, $8 \equiv 3 (\mathrm{mod}\ 5)$, $4 \equiv 2 (\mathrm{mod}\ 2)$, $57 \equiv 43 (\mathrm{mod}\ 7)$. In our clock example above, the clock reads the same at two times precisely when the times are congruent modulo 12. Our odometer reads the same at mileages which are congruent to each other modulo 100,000. And our idealized odometer reads the same for numbers which are congruent modulo n. If we think in terms of our idealized odometer and group together all integers for which the odometer has a given reading, we see that the integers fall into classes as follows*:

	Odometer reading
$\ldots, -2n, \quad -n, \quad 0, n, \quad 2n, \quad 3n, \quad 4n, \ldots$	0
$\ldots, -2n + 1, -n + 1, 1, n + 1, 2n + 1, 3n + 1, 4n + 1, \ldots$	1
$\ldots, -2n + 2, -n + 2, 2, n + 2, 2n + 2, 3n + 2, 4n + 2, \ldots$	2
$\vdots$	$\vdots$
$\ldots, -n - 1, -1, n - 1, 2n - 1, 3n - 1, 4n - 1, 5n - 1, \ldots$	$n - 1$

For example, if $n = 2$, then there are two groups,

$\ldots, -4, -2, 0, 2, 4, \ldots$ are congruent to $0 (\mathrm{mod}\ 2)$, and
$\ldots, -3, -1, 1, 3, 5, \ldots$ are congruent to $1 (\mathrm{mod}\ 2)$,

so that congruence modulo 2 allows us to distinguish between the even and odd integers. If $n = 3$, then there are three groups, namely

$\ldots, -6, -3, 0, 3, 6, \ldots$ are congruent to $0 (\mathrm{mod}\ 3)$,
$\ldots, -5, -2, 1, 4, 7, \ldots$ are congruent to $1 (\mathrm{mod}\ 3)$, and
$\ldots, -4, -1, 2, 5, 8, \ldots$ are congruent to $2 (\mathrm{mod}\ 3)$.

Now that we have explained the notion of congruence, let us account for the importance of this notion in number theory. Let us first observe that $n \mid a$

*Negative integers can be read by our odometer dial turning backwards.

if and only if $a \equiv 0 \pmod{n}$. (Indeed, $n \mid a$ if and only if $a - 0 = kn$ for some k.) Thus, the notion of divisibility can be phrased in the language of congruences. Actually, congruences modulo n do much more than determine divisibility by n. Congruences modulo n keep track of the remainders which occur on divisibility by n. Thus, the theory of congruences can be looked at as a refinement of the theory of divisibility. But why the necessity of this refinement? One reason is that congruences can be treated in much the same manner as equations. Thus, for example, it makes sense to ask for all solutions x of the congruences

$$3x \equiv 2 \pmod{5}$$

$$5x^2 + 3x + 8 \equiv 0 \pmod{17}.$$

We shall see in the next section that one can perform almost all the algebraic manipulations used in high school algebra on congruences. Therefore, we may view congruences as giving us an algebraic machine with which to study divisibility.

The theory of congruences developed in this chapter has applications to Diophantine equations. To understand the general principle behind such applications, let us consider, for simplicity, the case of a Diophantine equation,

$$f(x, y) = 0 \tag{1}$$

in unknowns x and y, where $f(x, y)$ is a polynomial in two variables with integer coefficients. If (x, y) is a solution of the Diophantine equation (1) then since $n \mid 0$ for every integer n, we see that $n \mid f(x, y)$ for every n, so that

$$f(x, y) \equiv 0 \pmod{n}.$$

Thus, if (x, y) is a solution of Eq. (1), then (x, y) is also a solution of the congruence

$$f(x, y) \equiv 0 \pmod{n} \tag{2}$$

for all n. In particular, we have the following result:

Theorem 2: If the Diophantine equation

$$f(x, y) = 0$$

has a solution, then the congruences

$$f(x, y) \equiv 0 \pmod{n}$$

have solutions for all n.

We may use Theorem 2 to prove that certain Diophantine equations have no solutions by restating it in the following form:

Theorem 2′: Suppose that for some integer n the congruence

$$f(x, y) \equiv 0 \pmod{n}$$

has no solutions. Then the Diophantine equation

$$f(x, y) = 0$$

has no solutions.

This seemingly innocent principle is very powerful. As an example, let us consider the Diophantine equation

$$x^2 - 4y^2 = 2. \tag{3}$$

In this case, we may set $f(x, y) = x^2 - 4y^2 - 2$. Let us show that this Diophantine equation has no solutions by taking $n = 4$ and showing that the congruence

$$x^2 - 4y^2 - 2 \equiv 0 (\text{mod } 4) \tag{4}$$

has no solutions. Indeed, if we have integers x and y satisfying Eq. (4), then we have the equivalent statement that $4 \mid x^2 - 4y^2 - 2$. However, since $4 \mid 4y^2$, we have $4 \mid x^2 - 4y^2 - 2 + 4y^2$ (Proposition 2.2.2) and thus $4 \mid x^2 - 2$. But could $x^2 - 2$ be divisible by 4? We shall now show that it cannot, and we shall do this by considering the cases where x is even or odd separately. If x is even, say $x = 2t$, then $x^2 - 2 = 4t^2 - 2$, so that $4 \mid x^2 - 2$ implies that $4 \mid (4t^2 - 2) - 4t^2$, and thus $4 \mid -2$, which is ridiculous. If x is odd, say $x = 2t + 1$, then $x^2 - 2 = 4t^2 + 4t - 1$, so that $4 \mid x^2 - 2$ implies that $4 \mid (4t^2 + 4t - 1) - (4t^2 + 4t)$, and thus $4 \mid -1$, which is again ridiculous. Thus, the congruence (4) has no solutions and consequently neither does the corresponding Diophantine equation (3).

Thus, we see that if we could prove that congruences have no solutions, we could make corresponding negative statements about Diophantine equations. Even if a congruence has solutions, it can often give important information about the nature of the solutions (if any) of the corresponding Diophantine equations. We shall take up this topic later in the chapter. We shall also learn how to work with congruences, so that the argument concerning the congruence (4) can be given in one line.

3.1 Exercises

1. Decide whether the following statements are true or false.
   (a) $2 \equiv 4 (\text{mod } 2)$.
   (b) $11 \equiv 6 (\text{mod } 5)$.
   (c) $33 \equiv 18 (\text{mod } 11)$.
   (d) $57 \equiv 21 (\text{mod } 6)$.
   (e) $k^2 \equiv k (\text{mod } k)$, $k > 0$.
   (f) $11 \equiv -14 (\text{mod } 17)$.

2. (a) Show that every integer can be written in precisely one of the following forms: $4k$, $4k + 1$, $4k + 2$, $4k + 3$, where k is an integer.
   (b) Show that every integer is congruent to exactly one of the integers 0, 1, 2, or $3 (\text{mod } 4)$.
   (c) Show that the congruence $y^2 - x^2 - 2 \equiv 0 (\text{mod } 4)$ has no solutions and hence that the Diophantine equation $y^2 = x^2 + 2$ has no solutions. (*Hint:* Use part (a).)

3. Show that if the Diophantine equation $y^2 = x^3 + 2$ has a solution, then x and y must both be odd. (*Hint:* Consider congruences modulo 4 and use Exercise 2, part (a).)

4. Let x, y, z be integers. Prove the following:
   (a) $x \equiv x (\text{mod } n)$.
   (b) If $x \equiv y (\text{mod } n)$, then $y \equiv x (\text{mod } n)$.
   (c) If $x \equiv y (\text{mod } n)$ and $y \equiv z (\text{mod } n)$, then $x \equiv z (\text{mod } n)$.

3.2 Basic Properties of Congruences

In this section, we shall develop some of the fundamental facts concerning congruences. Our fundamental theme will be that a congruence is a form of 'equality' and that the congruence $a \equiv b (\text{mod } n)$ can, for most purposes, be viewed as an analogue of the equation $a = b$. Our main purpose in this section is to determine how far this analogy can be pushed. More specifically, we wish to determine which manipulations—for example, addition, subtraction, multiplication—admissible in dealing with equations are still admissible in dealing with congruences. *Throughout this chapter, n will denote a positive integer*. The only fact which we shall assume from the preceding section is the definition of congruence modulo n.

Before we go any further, let us prove two trivial results concerning congruences modulo n.

Proposition 1: Let a, b, c be integers. Then

(i) $a \equiv a (\text{mod } n)$.
(ii) If $a \equiv b (\text{mod } n)$, then $b \equiv a (\text{mod } n)$.
(iii) If $a \equiv b (\text{mod } n)$ and $b \equiv c (\text{mod } n)$, then $a \equiv c (\text{mod } n)$.

Proof:

(i) $n \mid a - a$.
(ii) If $n \mid a - b$, then $n \mid b - a$.
(iii) If $n \mid a - b$ and $n \mid b - c$, then $n \mid (a - b) + (b - c)$, so that $n \mid a - c$. ■

The practical impact of Proposition 1 is to allow us to reverse the order of congruences and to string together congruences modulo n. From now on, we shall carry out these operations without comment.*

Proposition 2: Assume that $a \equiv b(\text{mod } n)$ and $c \equiv d(\text{mod } n)$. Then

(i) $a + c \equiv b + d(\text{mod } n)$.
(ii) $a - c \equiv b - d(\text{mod } n)$.
(iii) $ac \equiv bd(\text{mod } n)$.

Proof: By assumption, $n \mid a - b$ and $n \mid c - d$. Therefore, $n \mid (a - b) + (c - d)$ implies that $n \mid (a + c) - (b + d)$, which implies that $a + c \equiv b + d(\text{mod } n)$, which proves part (i). Part (ii) is left as an exercise. To prove part (iii), observe that since $n \mid a - b$, we have $n \mid c(a - b)$, so that $n \mid ac - bc$, and thus $ac \equiv bc(\text{mod } n)$. Similarly, since $n \mid c - d$, we have $bc \equiv bd(\text{mod } n)$. Thus, $ac \equiv bc \equiv bd(\text{mod } n)$, and part (iii) is proved. ∎

An elementary induction argument allows us to extend Proposition 2 as follows:

Corollary 3: Assume that $a_1 \equiv b_1(\text{mod } n)$, $a_2 \equiv b_2(\text{mod } n), \ldots, a_m \equiv b_m(\text{mod } n)$. Then

$$a_1 + \cdots + a_m \equiv b_1 + \cdots + b_m(\text{mod } n)$$

and

$$a_1 \cdots a_m \equiv b_1 \cdots b_m(\text{mod } n).$$

From Proposition 2 and Corollary 3, we see that we may add, subtract, and multiply corresponding terms in congruences modulo n, and thus we can manipulate congruences very much as if they were equations.

Proposition 2 (or rather Corollary 3) has the following practical impact. Assume that $f(x)$ is a polynomial with integer coefficients and that $a \equiv b(\text{mod } n)$. Then

$$f(a) \equiv f(b)(\text{mod } n). \tag{1}$$

Indeed, if $f(x) = a_0 + a_1 x + \cdots + a_m x^m$, then Eq. (1) merely asserts that

$$a_0 + a_1 a + \cdots + a_m a^m \equiv a_0 + a_1 b + \cdots + a_m b^m(\text{mod } n),$$

which easily follows by repeated application of Corollary 3.

To see the practical value of Eq. (1), let $f(x) = x^3 + 4x^2 + 5x - 1$, $n = 3$. Then, since $13 \equiv 1(\text{mod } 3)$, the congruence (1) asserts that

*We shall abbreviate $a \equiv b(\text{mod } n)$, $b \equiv c(\text{mod } n)$ by writing $a \equiv b \equiv c(\text{mod } n)$ and shall similarly abbreviate longer strings of congruences.

$$\begin{aligned} 13^3 + 4\cdot 13^2 + 5\cdot 13 - 1 &\equiv 1^3 + 4\cdot 1^2 + 5\cdot 1 - 1 \pmod 3 \\ &\equiv 1 + 4 + 5 - 1 \pmod 3 \\ &\equiv 9 \pmod 3 \\ &\equiv 0 \pmod 3. \end{aligned} \tag{2}$$

Note how much easier the right-hand side of Eq. (2) was to compute than the left-hand side.

We have already mentioned in Section 1 that one of the main goals of the theory of congruences is the determination of the solutions (if any) of the congruence

$$f(x) \equiv 0 \pmod n, \tag{3}$$

where $f(x)$ is a polynomial with integer coefficients. By Eq. (1), we see that in order to determine if x is a solution it suffices to check whether some x' which is congruent to x modulo n is a solution. Is it always possible to choose an x' which does not involve too much computation to check? Indeed, it is always possible, as the following proposition shows:

Proposition 4: An integer x is congruent to one and only one of the integers $0, 1, 2, \ldots, n - 1 \pmod n$.

Proof: The assertion is simply a restatement of the division algorithm (Theorem 2.2.3). For by the division algorithm, we may write x in the form $x = qn + r$ with $0 \leq r \leq n - 1$. And by the definition of a congruence, we have $x \equiv r \pmod n$. If we also have $x \equiv r' \pmod n$ with $0 \leq r' \leq n - 1$, then $x - r' = q'n$ for some q', so that $x = q'n + r' = qn + r$. Then by the uniqueness assertion of the division algorithm, we have $r = r'$. ■

If $n = 2$, then Proposition 4 asserts that every integer is congruent to exactly one of 0 or $1 \pmod 2$. This corresponds to the trivial fact that every integer is either even or odd, but not both. If $n = 3$, then Proposition 4 asserts that every integer is congruent to exactly one of 0, 1, or $2 \pmod 3$.

We shall have much to say about how we go about solving the congruence (3), but for now let us drop our discussion of (3) and proceed with some further developments suggested by Proposition 4.

Proposition 4 asserts that the set of integers $0, 1, 2, \ldots, n - 1$ has the property that every integer is congruent modulo n to one and only one element of the set. This property is the one which makes the set useful in connection with congruence (3). However, the set $0, 1, 2, \ldots, n - 1$ is not the only set with this property, and this fact suggests the following definition:

Definition 5: A *complete residue system modulo n* is a set of n integers $r_1, \ldots, r_n$ such that every integer x is congruent to one and only one of $r_1, \ldots, r_n$.

Example 6:

(i) Proposition 4 asserts that $0, 1, 2, \ldots, n - 1$ is a complete residue system modulo n.

(ii) The following are complete residue systems modulo 5:

$$0, \ 1, \ 2, \ 3, \ 4;$$
$$-2, \ -1, \ 0, \ 1, \ 2;$$
$$25, \ 26, \ 27, \ 28, \ 29;$$
$$-30, \ -29, \ -28, \ -27, \ -26;$$
$$4355, \ 2311, \ 117, \ 13, \ -196.$$

Let us give a proof for the second set. Note that

$$-2 \equiv 3(\text{mod } 5), \qquad -1 \equiv 4(\text{mod } 5), \qquad 0 \equiv 0(\text{mod } 5),$$
$$1 \equiv 1(\text{mod } 5), \qquad 2 \equiv 2(\text{mod } 5).$$

If x is any integer, then x is congruent to one and only one of the integers 0, 1, 2, 3, 4 by Proposition 4. And by our above computation, if $x \equiv r(\text{mod } 5)$ with $0 \leq r \leq 4$, then r is congruent to one and only one of the integers -2, -1, 0, 1, 2, so that x is congruent to one and only one of the integers -2, -1, 0, 1, 2. The proof that all the other sets are complete residue systems modulo 5 is similar. The only point is that in each set there is one and only one integer congruent to each of 0, 1, 2, 3, 4.

Complete residue systems modulo n will allow us to carry out explicit calculations with congruences modulo n. We shall need the general concept in order to prove Fermat's, Euler's, and Wilson's congruences in Section 3. However, for general calculation, we shall stick to the complete residue system $0, 1, \ldots, n - 1$, since this is the easiest to remember. Or we shall use one such as the complete residue system $-2, -1, 0, 1, 2$ modulo 5, which has the advantage that the numbers are smaller than the ones in the usual complete residue system 0, 1, 2, 3, 4 and thus make calculations easier. The reader may not appreciate the saving afforded by such a slight change. But consider the following example:

Example 6′: Let us compute the remainder when we divide 6^{48} by 13. Of course, we might consider calculating 6^{48} explicitly, dividing by 13 and seeing what the remainder is. However, the amount of calculation required should restrain all but the most zealous readers. Fortunately, the computation of the remainder is easy using congruences modulo 13. Let us use the following complete residue system modulo 13: $-6, -5, -4, -3, -2, -1, 0, 1, 2, 3, 4, 5, 6$. We may organize our calculation as follows:

$$6^2 = 36 \equiv -3(\text{mod } 13)$$
$$6^4 = (6^2)^2 \equiv (-3)^2 = 9 \equiv -4(\text{mod } 13)$$
$$6^8 = (6^4)^2 \equiv (-4)^2 = 16 \equiv 3(\text{mod } 13)$$

$$6^{16} = (6^8)^2 \equiv 3^2 = 9 \equiv -4 \pmod{13}$$
$$6^{32} = (6^{16})^2 \equiv (-4)^2 = 16 \equiv 3 \pmod{13}$$
$$6^{48} = 6^{32} \cdot 6^{16} \equiv 3 \cdot (-4) = -12 \equiv 1 \pmod{13}.$$

That is, $6^{48} \equiv 1 \pmod{13}$, or 6^{48} leaves the remainder 1 when divided by 13. The reader should try to carry out this example by using the complete residue system 0, 1, 2, . . . , 12 instead. The calculations are more tedious.

In Proposition 2, we saw that we could add, subtract, and multiply corresponding sides of congruences in the same way that we can add, subtract, and multiply corresponding sides of equations in algebra. Let us now turn to the problem of dividing both sides of a congruence by the same number.

The problem with dividing both sides of a congruence by a number is that we require that all numbers in a congruence be integers. To get some idea of the difficulties involved, consider the congruence

$$2x \equiv 3 \pmod 4.$$

If this were an equation, we would multiply both sides by $\frac{1}{2}$ to get $x = \frac{3}{2}$. Is there an integer which we could use to play the role of $\frac{1}{2}$ in order to solve the congruence? Unfortunately not, for the congruence has no solutions. Indeed, let us use the complete residue system 0, 1, 2, 3 modulo 4. If $x \equiv 0 \pmod 4$, then $2x \equiv 0 \pmod 4$; if $x \equiv 1 \pmod 4$, then $2x \equiv 2 \pmod 4$; if $x \equiv 2 \pmod 4$, then $2x \equiv 2 \cdot 2 \equiv 0 \pmod 4$; if $x \equiv 3 \pmod 4$, then $2x \equiv 2 \cdot 3 \equiv 2 \pmod 4$. In any case, $2x \equiv 0$ or $2 \pmod 4$, so $2x \not\equiv 3 \pmod 4$, and our congruence cannot be solved.

Thus, in general, we cannot divide both sides of a congruence by an integer. Or, to put it differently, we cannot always find an integer that "plays the role" of the reciprocal of an integer. Let us define what we mean by this somewhat more closely.

Definition 7: Let a be an integer. By an *arithmetic inverse of a modulo n*, we mean an integer a^* such that $aa^* \equiv 1 \pmod n$.

The arithmetic inverse is precisely the number-theoretic analogue of $1/a$. Indeed, the congruence $aa^* \equiv 1 \pmod n$ is the congruence analogue of the equation $a \cdot (1/a) = 1$. Not every integer has an arithmetic inverse modulo n. For example, 2 does not have an arithmetic inverse modulo 4 (exercise). On the other hand, we have the following result:

Proposition 8: Suppose that $\gcd(a, n) = 1$. Then a has an arithmetic inverse a^* modulo n.

Proof: Since $\gcd(a, n) = 1$, there exist x and y such that $ax + ny = 1$. Then $ax \equiv 1 \pmod n$, and we may take $a^* = x$. ■

Note that the proof of Proposition 8 shows that we may use the Euclidean algorithm for computing an arithmetic inverse of a.

We shall show in the exercises that all arithmetic inverses of a modulo n are congruent to one another modulo n; that is, they are all "the same" modulo n.

Example 9:

(i) Let $n = 3$, $a = 2$. Then we may take $a^* = 2$ since $2 \cdot 2 \equiv 1 \pmod 3$.

(ii) Let $n = 5$, $a = 3$. Then we may take $a^* = 2$ since $3 \cdot 2 \equiv 1 \pmod 5$.

As we have said, the arithmetic inverse a^* of a, when it exists, plays the role of the reciprocal of a. For example, by using the notion of an arithmetic inverse, we can prove the following cancellation law for congruences:

Proposition 10: Suppose that $\gcd(a, n) = 1$ and that

$$ax \equiv ay \pmod n.$$

Then

$$x \equiv y \pmod n.$$

Proof: Let a^* be an arithmetic inverse of a modulo n, which exists by Proposition 8 since we have assumed that $\gcd(a, n) = 1$. Then we have that

$$x \equiv 1 \cdot x \equiv (a^*a)x \equiv a^*(ax) \equiv a^*(ay) \equiv (a^*a)y \equiv 1 \cdot y \equiv y \pmod n,$$

since $a^*a \equiv 1 \pmod n$. ∎

In the situation of Proposition 10, if $\gcd(a, n)$ is greater than 1, it is not always possible to cancel the a. For example, $2 \cdot 1 \equiv 2 \cdot 3 \pmod 4$, but $1 \not\equiv 3 \pmod 4$. However, we can improve Proposition 10 as follows:

Proposition 11: Assume that $\gcd(a, n) = d$ and that

$$ax \equiv ay \pmod n.$$

Then

$$x \equiv y \left(\bmod \frac{n}{d}\right).$$

Proof: Note that since $ax \equiv ay \pmod n$, we have $n \mid ax - ay$, so that $(n/d) \mid (a/d)(x - y)$, and thus

$$\frac{a}{d}x \equiv \frac{a}{d}y \left(\bmod \frac{n}{d}\right).$$

Moreover, $\gcd(a/d, n/d) = 1$ (Theorem 2.3.6, (ii)), so that by Proposition 10 we have

$$x \equiv y \left(\bmod \frac{n}{d}\right).$$

∎

Note that Proposition 10 is just the special case $d = 1$ of Proposition 11. Also note that Proposition 11 is simply a restatement of part (iv) of Theorem 2.3.6 in the language of congruences.

Example 12:

(i) Since $3 \cdot 2 \equiv 3 \cdot 16 \pmod{14}$ and $\gcd(3, 14) = 1$, we have $2 \equiv 16 \pmod{14}$.

(ii) Since $6 \cdot 5 \equiv 6 \cdot 18 \pmod{26}$ and $\gcd(6, 26) = 2$, we have $5 \equiv 18 \pmod{13}$. Note that it is *not* true that $5 \equiv 18 \pmod{26}$.

We now have enough machinery to begin consideration of one of the fundamental problems treated in this book, namely the problem of solving polynomial congruences. Let $f(x)$ be a polynomial with integer coefficients. The question we seek to answer is, for what x does the congruence

$$f(x) \equiv 0 \pmod{n} \tag{4}$$

hold? In the preceding section, we have seen the connection between the answer to this question and the problem of determining the solutions to the Diophantine equation $f(x) = 0$. In general, congruence (4) may have no solutions, as is seen in the following example:

Example 13: The congruence $x^2 + 1 \equiv 0 \pmod{8}$ has no solutions. Indeed, if x is a solution, then $x^2 \equiv -1 \equiv 7 \pmod{8}$. However, by using the complete residue system $0, \pm 1, \pm 2, \pm 3, \pm 4$ modulo 8, we see that in the five cases, x^2 is congruent modulo 8 to 0, 1, 4, 1, 0, respectively. (Check the calculations.) In particular, in every case, $x^2 \not\equiv 7 \pmod{8}$, so the congruence $x^2 + 1 \equiv 0 \pmod{8}$ has no solutions. As a by-product of our calculation, we make the following interesting observation:

A perfect square is congruent to 0, 1, *or* 4 *modulo* 8. *In particular, an odd perfect square is congruent to* $1 \pmod{8}$.

It is possible to make a few general comments about congruence (4). If x is one solution and $x \equiv x' \pmod{n}$, then we have already observed in Eq. (1) that x' is a solution. Therefore, we may use the notion of a complete residue system modulo n to give the following simple scheme for finding all solutions of (4). Choose some complete residue system modulo n, say $r_1, \ldots, r_n$. Test each of $r_1, \ldots, r_n$ to determine which are solutions of (4). Suppose that the solutions are $a_1, \ldots, a_t$. If x is any solution of (4), then $f(x) \equiv 0 \pmod{n}$. But by the definition of a complete residue system, $x \equiv r_i$ for some i. But then $f(x) \equiv f(r_i) \pmod{n}$ by (1), so that $f(r_i) \equiv 0 \pmod{n}$ and r_i is a solution of (4). In other words, r_i is one of $a_1, \ldots, a_t$. Thus, we see that if x is any solution of (4), then $x \equiv a_j \pmod{n}$ for some j, $1 \leq j \leq t$. Thus, it suffices to check (4) for solutions in any fixed complete residue system modulo n. This was essentially the procedure we used in Example 13. This procedure is really just trial and error. We seek, however, a better method than this.

Now, from our present discussion, we know that the x satisfying Eq. (4) are those satisfying one of a set of congruences of the form

$$x \equiv a_1 (\text{mod } n), \quad \text{or} \quad x \equiv a_2 (\text{mod } n), \quad \ldots, \quad \text{or} \quad x \equiv a_t (\text{mod } n).$$

Since the a_j are taken from our complete residue system modulo n, we see that no two a_j are congruent to one another modulo n, and the x described by each of the above congruences are all different from one another. In this situation, let us say that the congruence (4) has *t different solutions.* (Strictly speaking, of course, (4) would have an infinite number of solutions, but we count as different solutions only those which are incongruent to one another modulo n.)

Although (4) is in general quite complicated, in case $f(x)$ is a linear polynomial, the notion of an arithmetic inverse will allow us to get things under control. Thus, let us now study the *linear congruence*

$$ax \equiv b (\text{mod } n). \tag{5}$$

Even such a simple congruence may have no solutions, as we saw when we considered the example $2x \equiv 3 (\text{mod } 4)$ above.

Let Eq. (5) have a solution x, and let s be any common factor of a and n. Since $n \mid (ax - b)$, we would have $b = ax - kn$ for some k, and thus s is a factor of b. If, in particular, we take $s = \gcd(a, n)$, we see that in order for (5) to have a solution, we must have $\gcd(a, n) \mid b$. Let us record this in a lemma.

Lemma 14: If the congruence $ax \equiv b (\text{mod } n)$ has a solution, then $\gcd(a, n) \mid b$.

Proposition 15: Suppose that $\gcd(a, n) = 1$. Let a^* be an arithmetic inverse of a modulo n. Then x satisfies $ax \equiv b (\text{mod } n)$ if and only if $x \equiv a^*b (\text{mod } n)$.

Proof: If $ax \equiv b (\text{mod } n)$, then $aa^*x \equiv a^*b (\text{mod } n)$, so that $x \equiv a^*b (\text{mod } n)$, since $aa^* \equiv 1 (\text{mod } n)$. If $x \equiv a^*b (\text{mod } n)$, then $ax \equiv aa^*b \equiv 1 \cdot b \equiv b (\text{mod } n)$. ■

Example 16:

(i) Let us solve $2x \equiv 5 (\text{mod } 3)$. Since $2^* = 2$, the solutions are $x \equiv 2 \cdot 5 \equiv 1 (\text{mod } 3)$. Thus, the solutions are all integers of the form $3t + 1$.

(ii) Let us solve $3x \equiv 7 (\text{mod } 5)$. In this case, $3^* = 2$, and so $x \equiv 2 \cdot 7 \equiv 4 (\text{mod } 5)$. Thus, the solutions are all integers of the form $5t + 4$.

In the above examples, we determined the arithmetic inverse a^* by inspection. But this method is practical only for small a. In general, a^* can be determined using the methods of Chapter 2, for a^* is an arithmetic inverse of a modulo n if and only if $aa^* - 1 = kn$ for some k; that is, a^* can be obtained as the first component of a solution (x, y) of the Diophantine equation $ax + ny = 1$, an equation which we treated exhaustively in Chapter 2. Thus, a^* can be determined with comparative ease.

Let us now return to the study of the general linear congruence $ax \equiv b(\text{mod } n)$. Let $d = \gcd(a, n)$. By Lemma 14, we may as well assume that $d \mid b$, for otherwise there are no solutions. Then $b = db_1$. Moreover, since $d \mid a$ and $d \mid n$, we may write $a = da_1$, $n = dn_1$, and our congruence may be written

$$a_1 dx \equiv b_1 d(\text{mod } n_1 d).$$

Since $d = \gcd(a, n)$, we know from Proposition 11 that x is a solution of the last congruence if and only if

$$a_1 x \equiv b_1(\text{mod } n_1).$$

Moreover, since $d = \gcd(a, n)$, we know by Theorem 2.3.6 (ii) that $\gcd(a_1, n_1) = 1$. Therefore, the solutions of $ax \equiv b(\text{mod } n)$ are the same as the solutions of $a_1 x \equiv b_1(\text{mod } n_1)$. And by Proposition 15, the solutions of the latter are of the form

$$x \equiv a_1^* b_1(\text{mod } n_1),$$

where a_1^* is an arithmetic inverse of a_1 modulo n_1. Thus, finally, we may state

Theorem 17: Let $d = \gcd(a, n)$. The congruence $ax \equiv b(\text{mod } n)$ is solvable if and only if $d \mid b$. If $d \mid b$, then the solutions consist of all those x for which

$$x \equiv a_1^* \frac{b}{d}\left(\text{mod} \frac{n}{d}\right),$$

where a_1^* is an arithmetic inverse of a/d modulo n/d. Thus, the distinct solutions modulo n are given by $x = a_1^*(b/d) + k(n/d)$, $0 \leq k \leq d - 1$.

Example 18: Consider the congruence $27x \equiv 3(\text{mod } 15)$. Here $d = \gcd(27, 15) = 3$ and $3 \mid 3$, so there are solutions. Here $a/d = 9$, $b/d = 1$, and $n/d = 5$, and an arithmetic inverse of a/d modulo n/d is 4. Thus, the solutions of the congruence are $x \equiv 4 \cdot 1 \equiv 4(\text{mod } 5)$. In terms of congruences modulo 15, the solutions can be written $x \equiv 4, 9$ or $14(\text{mod } 15)$. Thus, we see that in terms of our general discussion, the present congruence has three different solutions. Note, however, that if $\gcd(a, n) = 1$, then by Proposition 15 there is only one solution.

It is useful to note that the above discussion is relevant in solving the linear Diophantine equation in two variables

$$ax + by = c, \tag{6}$$

where a and b are nonzero integers. If $b < 0$, we could solve the equation $-ax - by = -c$ and obtain the same solutions as those of Eq. (6), so we assume $b > 0$. Then (6) is equivalent to

$$ax \equiv c(\text{mod } b). \tag{7}$$

This problem was completely dealt with above.

For example, let us solve

$$7x + 5y = 3. \tag{8}$$

This is equivalent to $7x \equiv 3 \pmod 5$ or $2x \equiv 3 \pmod 5$. Here $2^* \equiv 3 \pmod 5$, and so $x \equiv 3 \cdot 3 \equiv 9 \equiv -1 \pmod 5$. Thus, $x = -1 + 5k$. Plugging this result into Eq. (8) immediately gives $y = 2 - 7k$, and the equation is solved.

In all fairness, it must be pointed out that the above discussion is circular. Namely, solving Eq. (7) involves determining a^*, and the only general method we have for doing this is to solve $ax + by = 1$. Nevertheless, this congruence method is often the most convenient method for solving (6), especially when a^* can be determined by inspection. Also, a formula for a^* will be given in Section 3.

We saw in Proposition 8 that a has an arithmetic inverse modulo n provided that $\gcd(a, n) = 1$. Let us now ask the converse question. Which integers a have an arithmetic inverse a^*? This is equivalent to asking the question, for which a is the congruence

$$aa^* \equiv 1 \pmod n$$

solvable for a^*? Let $d = \gcd(a, n)$. By Theorem 17, a^* exists if and only if $d \mid 1$, that is, if and only if $d = 1$. Thus, we see that the a which have an arithmetic inverse modulo n are precisely those for which $\gcd(a, n) = 1$. Let us try to get a better description of these integers. To do this, we need a lemma.

Lemma 19: Suppose that $\gcd(a, n) = 1$ and that $a \equiv a' \pmod n$. Then $\gcd(a', n) = 1$.

Proof: Since $a \equiv a' \pmod n$, there exists a k such that $a = a' + kn$. Suppose that $\gcd(a', n) > 1$. Then there is a prime p such that $p \mid \gcd(a', n)$. Then $p \mid a'$ and $p \mid n$ by the definition of gcd. But then $p \mid a' + kn$, and so $p \mid a$. This implies that $p \mid \gcd(a, n)$, so that $\gcd(a, n) > 1$, a contradiction. ■

From the preceding lemma, we can easily describe the integers with arithmetic inverses. Let us consider the complete residue system modulo n given by $0, 1, \ldots, n-1$. Every integer x is congruent to precisely one of $0, 1, \ldots, n-1$. Suppose that x has an arithmetic inverse modulo n and that $x \equiv a \pmod n$, $0 \leq a \leq n-1$. Then by our discussion above, $\gcd(x, n) = 1$, and thus from Lemma 19 we see that $\gcd(a, n) = 1$. Conversely, if x is any integer such that $x \equiv a \pmod n$ $(0 \leq a \leq n-1)$ with $\gcd(a, n) = 1$, then x has an arithmetic inverse modulo n. Indeed, since $\gcd(a, n) = 1$, a has an arithmetic inverse a^*, and a^* is an arithmetic inverse for x since

$$xa^* \equiv aa^* \equiv 1 \pmod n.$$

Thus, we have completely proved the following result:

Proposition 20: Suppose that the integers a in the complete residue system $0, 1, \ldots, n-1$ such that $\gcd(a, n) = 1$ are given by $a_1, a_2, \ldots, a_t$. Then an integer x has an arithmetic inverse modulo n if and only if

$$x \equiv a_i (\text{mod } n)$$

for some i.

Example 21:

(i) $n = 3$. Then the integers a among 0, 1, 2 such that $\gcd(a, 3) = 1$ are 1 and 2, so that $t = 2$ and x has an arithmetic inverse modulo 3 if and only if $x \equiv 1$ or $2 (\text{mod } 3)$.

(ii) $n = 12$. The integers a among $0, 1, 2, \ldots, 11$ such that $\gcd(a, 12) = 1$ are 1, 5, 7, 11, so that $t = 4$ and x has an arithmetic inverse modulo 12 if and only if $x \equiv 1, 5, 7$, or $11 (\text{mod } 12)$.

(iii) $n = p$, a prime. The integers a among $0, 1, \ldots, p - 1$ such that $\gcd(a, p) = 1$ are $1, 2, \ldots, p - 1$, since p is a prime. Thus, x has an arithmetic inverse modulo the prime p if and only if $x \equiv 1, 2, 3, \ldots$, or $p - 1 (\text{mod } p)$.

Proposition 20 utilized the complete residue system modulo n given by $0, 1, \ldots, n - 1$. However, we leave it to the reader to verify that any complete residue system modulo n could be used instead and the proposition would still hold. Modulo this bit of checking, the reader should notice that the members a of a complete residue system modulo n which are relatively prime to n are very special. This motivates the following definition:

Definition 22: A *reduced residue system modulo n* is a set of integers $a_1, \ldots, a_t$ with the following property: Every integer x such that $\gcd(x, n) = 1$ is congruent to one and only one of $a_1, \ldots, a_t$.

The simplest way of obtaining a reduced residue system modulo n is to determine which integers a of a given complete residue system modulo n are relatively prime to n. The set of such integers is a reduced residue system modulo n. For suppose that $r_1, \ldots, r_n$ is a complete residue system modulo n and that the integers relatively prime to n from among $r_1, \ldots, r_n$ are $a_1, \ldots, a_t$. Then if x is an integer such that $\gcd(x, n) = 1$, we certainly have $x \equiv r_i (\text{mod } n)$ for some i (since $r_1, \ldots, r_n$ forms a complete residue system modulo n). Moreover, since $\gcd(x, n) = 1$, we have $\gcd(r_i, n) = 1$ by Lemma 19. Thus, r_i is one of $a_1, \ldots, a_t$, and so $x \equiv a_j (\text{mod } n)$ for some j. Finally, x is congruent to only one of $a_1, \ldots, a_t$, since the a_j are part of a complete residue system modulo n. Thus, $a_1, \ldots, a_t$ is a reduced residue system modulo n.

Example 23:

(i) $n = 3$. A reduced residue system modulo 3 is 1, 2.

(ii) $n = 12$. A reduced residue system modulo 12 is 1, 5, 7, 11. Another is $\pm 1, \pm 5$.

(iii) A reduced residue system modulo $n = p$, where p is a prime, is $1, 2, \ldots, p - 1$.

(iv) The reduced residue system derived from the complete residue system 25, 26, 27, 28, 29 modulo 5 is just 26, 27, 28, 29.

If $a_1, \ldots, a_t$ and $b_1, \ldots, b_s$ are two reduced residue systems modulo n, then each a_i is congruent to precisely one b_j modulo n and no two a_i are congruent to the same b_j. Moveover, each b_j is congruent to precisely one a_i and no two b_j are congruent to the same a_i. Therefore, $s = t$, and any two reduced residue systems modulo n contain the same number of elements. By using the reduced residue system derived from the complete residue system $0, 1, \ldots, n - 1$, we see that the number of integers in a reduced residue system modulo n equals the number of integers between 0 and $n - 1$ which are relatively prime to n.

Definition 24: Let $\phi(n)$ denote the number of integers between 0 and $n - 1$ which are relatively prime to n. Then $\phi(n)$ is called *Euler's phi function*, after the Swiss mathematician Leonhard Euler, who first studied its properties.*

You will find a table (Table 1) of the values of $\phi(n)$ for $n \leq 200$ at the end of the book.

From our above discussion, we have the following result:

Proposition 25: Every reduced residue system modulo n contains the same number of integers, and this number is precisely $\phi(n)$.

Example 26: From Example 23, we see that $\phi(3) = 2$, $\phi(12) = 4$, $\phi(p) = p - 1$ for p a prime, $\phi(5) = 4$. Note that it is not generally true that $\phi(n) = n - 1$; for example, $\phi(12) = 4 \neq 12 - 1$.

We shall meet both the notion of a reduced residue system modulo n and Euler's phi function later in this chapter as well as later in the book, so let us refrain from further discussion at this point.

To close this section, let us apply our congruence machinery to study a particular Diophantine equation. Let us consider the equation

$$x^3 + y^3 = z^3,$$

which is a special case of the famous Fermat equation. We shall show that

If (x, y, z) *is a solution of* $x^3 + y^3 = z^3$, *then at least one of* x, y, z *is divisible by* 3.

Proof: We shall in fact show that the congruence

$$x^3 + y^3 \equiv z^3 \pmod 9 \tag{9}$$

allows us to draw our conclusion.

---

*Note that by our definition we have $\phi(1) = 1$.

We first observe that if w is any integer, then

$$w^3 \equiv w(\text{mod } 3)$$

since $-1, 0, 1$ is a complete residue system mod 3, and these three cases are easily checked one by one. Since any congruence mod 9 is also a congruence mod 3, congruence (9) tells us that

$$x + y \equiv z(\text{mod } 3), \tag{10}$$

so

$$z = x + y + 3a \qquad \text{for some } a.$$

Plugging this back into Eq. (9) yields

$$\begin{aligned} x^3 + y^3 \equiv z^3 &= (x + y + 3a)^3 \\ &= x^3 + y^3 + 3x^2y + 3y^2x + 9x^2a + 9y^2a \\ &\quad + 27a^2x + 27a^2y + 27a^3 + 18xya \\ &\equiv x^3 + y^3 + 3x^2y + 3y^2x(\text{mod } 9). \end{aligned}$$

Subtracting $x^3 + y^3$ from both sides yields

$$3xy(x + y) \equiv 0(\text{mod } 9),$$

or from Proposition 11

$$xy(x + y) \equiv 0(\text{mod } 3).$$

Applying (10) again shows that

$$xyz \equiv 0(\text{mod } 3).$$

That is, $3 \mid xyz$, so x, y, or z is divisible by 3. ∎

3.2 Exercises

1. Let $f(x) = 11x^3 + 15x^2 + 9x - 2$.
   (a) Find the remainder when $f(2)$ is divided by 7.
   (b) Find the remainder when $f(6)$ is divided by 7.
   (c) Find the remainder when $f(97)$ is divided by 11.
2. Show that the following are complete residue systems modulo 11:
   (a) $-5, -4, -3, -2, -1, 0, 1, 2, 3, 4, 5$.
   (b) $0, 12, 24, 36, \ldots, 120$.
   (c) $20, 21, 22, 23, \ldots, 30$.
   (d) Any 11 consecutive numbers.
3. If n is a positive integer, define $n!$ (n factorial) by $n! = n(n - 1) \cdots 1$.
   (a) Which integer among the complete residue system $0, 1, \ldots, n - 1$ is $(n - 1)!$ congruent to modulo n for $n \leq 24$?
   (b) Can you make a general conjecture on the basis of part (a)?

4. Find the reduced residue system associated with the complete residue system $0, 1, \ldots, n-1$ for
   (a) $n = 9$.
   (b) $n = 15$.
   (c) $n = 24$.

5. On the basis of Exercise 4, compute $\phi(9)$, $\phi(15)$, $\phi(24)$.

6. Prove that $\phi(p^r) = p^r - p^{r-1}$, where p is a prime.

7. Find all solutions of the following congruences:
   (a) $3x \equiv 1 (\text{mod } 5)$.
   (b) $x^2 + 1 \equiv 0 (\text{mod } 4)$.
   (c) $x^3 + 2x + 1 \equiv 0 (\text{mod } 7)$.
   (d) $x^5 + x^4 + x^3 + x + 1 \equiv 0 (\text{mod } 5)$.

8. Find the remainder when 11^{p-1} is divided by p, where
   (a) $p = 2$.
   (b) $p = 3$.
   (c) $p = 5$.
   (d) $p = 7$.
   (e) $p = 11$.
   (f) $p = 13$.
   (g) Solve parts (a)–(f) with 11 replaced by 5.

9. Compute the remainder when $3^{\phi(n)}$ is divided by n, where
   (a) $n = 1$.
   (b) $n = 2$.
   (c) $n = 3$.
   (d) $n = 4$.
   (e) $n = 5$.
   (f) $n = 6$.
   (g) $n = 7$.
   (h) Solve parts (a)–(g) with 3 replaced by 5.
   (i) Can you make a conjecture based on parts (a)–(h)?

10. Solve the following congruences:
    (a) $3x \equiv 9 (\text{mod } 5)$.
    (b) $3x \equiv 9 (\text{mod } 24)$.
    (c) $5x \equiv 15 (\text{mod } 12)$.

11. Find the arithmetic inverses modulo 11 of 1, 2, 3, 7, 9, 10.

12. (a) Which integers have arithmetic inverses modulo 18?
    (b) Find arithmetic inverses for all integers which have them modulo 18.

13. Suppose that a^* and b^* are two arithmetic inverses of a modulo n. Prove that $a^* \equiv b^* (\text{mod } n)$.

14. Solve the following linear congruences:
    (a) $2x \equiv 5 \pmod 7$.
    (b) $19x \equiv 15 \pmod 8$.
    (c) $6x \equiv 12 \pmod{14}$.
    (d) $13x \equiv 27 \pmod{52}$.

15. (a) Show that 1143 has an arithmetic inverse modulo 1957.
    (b) Find the arithmetic inverse in part (a).

16. Show that a perfect cube is congruent to 0, 1, or $-1 \pmod 9$.

17. Show that a fourth power is congruent to 0 or $1 \pmod 5$.

18. Let $r_1, \ldots, r_n$ be a complete residue system modulo n.
    (a) Show that if a is any integer, then $r_1 + a, \ldots, r_n + a$ is a complete residue system modulo n.
    (b) Show that if $\gcd(a, n) = 1$, then $ar_1, \ldots, ar_n$ is a complete residue system modulo n.

19. Let $s_1, \ldots, s_{\phi(n)}$ be a reduced residue system modulo n.
    (a) Show that if $\gcd(a, n) = 1$, then $as_1, \ldots, as_{\phi(n)}$ is a reduced residue system modulo n.
    (b) Is it true that $s_1 + a, \ldots, s_{\phi(n)} + a$ is a reduced residue system modulo n for all integers a?

20. Let p be a prime and r be a positive integer. Suppose that a does not have an arithmetic inverse modulo p^r. Then show that some power of a is congruent to 0 mod p^r.

21. Let a be a given integer. Write a computer program to determine which integer among $0, 1, \ldots, n - 1$ is congruent to a modulo n.

22. Let $f(x)$ be a polynomial with integer coefficients. Write a computer program to compute which integer among $0, 1, \ldots, n - 1$ is congruent to $f(x)$ modulo n. Try to carry out your calculation as efficiently as you can, remembering that what is an easy calculation, in principle, may yield numbers too large for a computer to handle.

23. Find all solutions of the congruence $2x + 3y \equiv 5 \pmod 7$.

24. Find all solutions of the congruence $x^2 + y^2 - 5y - 2 \equiv 0 \pmod 9$.

*25. When can the congruence $a_1x_1 + \cdots + a_mx_m \equiv b \pmod n$ be solved for $x_1, \ldots, x_m$? If it can be solved, what will the solutions be?

26. (Casting out nines) Suppose that we are given an integer

$$x = a_n 10^n + a_{n-1}10^{n-1} + \cdots + a_0,$$

where the a_i are chosen from among $0, 1, \ldots, 9$. Then $a_n \cdots a_0$ is the decimal representation of the integer x.

(a) Show that

$$x \equiv a_n + a_{n-1} + \cdots + a_0 \pmod 9.$$

(b) Is 5,785,684 divisible by 9?

27. Prove Proposition 2, part (ii).

28. (a) Show that 2 does not have an arithmetic inverse modulo 4.
(b) Modulo what integers does 2 have an arithmetic inverse? What is the inverse?

29. Give a direct proof of Proposition 10 without using Proposition 8.

30. Let $r_1, \ldots, r_n$ be any complete residue system modulo n and suppose that the integers a among $r_1, \ldots, r_n$ such that $\gcd(a, n) = 1$ are $a_1, \ldots, a_t$. Then x has an arithmetic inverse modulo n if and only if $x \equiv a_i \pmod n$ for some i.

31. Prove that the linear congruence $ax \equiv b \pmod n$, where $\gcd(a, n) = 1$, has a solution using the fact that there exist x and y such that $ax + ny = 1$.

32. Prove that if $a \equiv a' \pmod n$, then $\gcd(a', n) = \gcd(a, n)$. (This is a generalization of Lemma 19.)

33. Let n be a given integer.
(a) Show that $x^2 - y^2 = n$ is solvable in integers if and only if n is odd or $4 \mid n$.
(b) If $x^2 - y^2 = n$ is solvable in integers, determine all the solutions.

34. Show that the number of reduced fractions $a/b, 0 < a/b \leq 1, 1 \leq b \leq n$, is just $\phi(1) + \phi(2) + \cdots + \phi(n)$.

35. Let m be odd and let $r_1, \ldots, r_k$ be any reduced residue system modulo m. Show that $r_1 + r_2 + \cdots + r_k \equiv 0 \pmod m$.

*36. (Euler) Show that $2^{2^5} + 1$ is divisible by 641. (*Hint:* $2^{32} = 2^4 \cdot 2^{28} = (641 - 5^4) \cdot 2^{28}$.)

37. Prove that $11^{2k} + 19^{2k}$ is divisible by 241 when k is odd.

38. (a) For $n = 2, 3, 4, 6, 8, 12, 18, 24, 30$, show that the integers between 1 and n and relatively prime to n (i.e., the standard reduced residue system mod n) consist of 1 and primes.
*(b) Show that for $n > 30$, the standard reduced residue system mod n always contains a composite.

39. Solve the system of congruences

$$2x + 7y \equiv 2 \pmod 5$$
$$3x - y \equiv 11 \pmod 5.$$

3.3 Some Special Congruences

In this section, we shall prove two special congruences which are of both historical and technical importance in the theory of numbers. The first of these is due to Euler and was a generalization of a theorem of Fermat. The latter is called *Fermat's Little Theorem* (to distinguish it from Fermat's Last Theorem). The second result is called *Wilson's theorem*. We shall prove both these results and present some applications of them.

In principle, it is always possible to test whether two integers x and y are congruent modulo n. We need only calculate $x - y$ and divide by n. However, if $x - y$ is large, it is extremely tedious to carry out the division. Both the theorems of Euler and Wilson allow one to deal with certain congruences where x and y are huge. Moreover, they have the advantage that they are "general" congruences.

We first state Fermat's theorem.

Theorem 1 (Fermat's Little Theorem): Let p be a prime and assume that $p \nmid a$. Then

$$a^{p-1} \equiv 1 (\text{mod } p).$$

Before proving Fermat's theorem we shall give some examples. We have $2^{10} \equiv 1(\text{mod } 11)$ and $99^{820} \equiv 1(\text{mod } 821)$. (821 is a prime.) Moreover, the calculation in Example 2.6′ is trivial from Fermat's theorem. We shall give another application which is typical. Let us determine the remainder when we divide 2^{1137} by 17. Since 17 is a prime and $17 \nmid 2$, Fermat's theorem implies that

$$2^{16} \equiv 1(\text{mod } 17).$$

We have

$$1137 = 16 \cdot 71 + 1.$$

Therefore,

$$2^{1137} = (2^{16})^{71} \cdot 2^1 \equiv 1^{71} \cdot 2 = 2(\text{mod } 17).$$

Thus, the remainder on dividing 2^{1137} by 17 is 2. Note that 2^{1137} is a number so large that actually carrying out the division would be impractical.

From Example 2.26, if $n = p$ is a prime, then $\phi(n) = p - 1$, and we see that Fermat's theorem is a special case of the following theorem due to Euler:

Theorem 2 (Euler): If $\gcd(a, n) = 1$, then

$$a^{\phi(n)} \equiv 1(\text{mod } n),$$

where $\phi(n)$ denotes Euler's phi function.

Proof: Let $r_1, \ldots, r_{\phi(n)}$ be a reduced residue system modulo n. (See Definition 2.22 and the subsequent discussion.) Since $\gcd(a, n) = 1$ and $\gcd(r_i, n)$

$= 1$, we see that $\gcd(ar_i, n) = 1$ for all i. Therefore, by the definition of a reduced residue system, each ar_i is congruent to one of the integers $r_1, \ldots, r_{\phi(n)}$. Moreover, no two ar_i are congruent to the same r_j, for if $ar_i \equiv r_j (\text{mod } n)$ and $ar_{i'} \equiv r_j (\text{mod } n)$, then

$$ar_i \equiv ar_{i'} (\text{mod } n),$$

so that $r_i \equiv r_{i'} (\text{mod } n)$ (Proposition 2.10), which implies that $r_i = r_{i'}$ since no two of $r_1, \ldots, r_{\phi(n)}$ are congruent modulo n. Thus, we see that $ar_1, \ldots, ar_{\phi(n)}$ are congruent to $\phi(n)$ different integers from among $r_1, \ldots, r_{\phi(n)}$. That is, $ar_1, \ldots, ar_{\phi(n)}$ are congruent to $r_1, \ldots, r_{\phi(n)}$ modulo n, in some order.

Thus,

$$(ar_1)(ar_2) \cdots (ar_{\phi(n)}) \equiv r_1 r_2 \cdots r_{\phi(n)} (\text{mod } n),$$

so that

$$r_1 r_2 \cdots r_{\phi(n)} a^{\phi(n)} \equiv r_1 r_2 \cdots r_{\phi(n)} (\text{mod } n).$$

Since $\gcd(r_1 \cdots r_{\phi(n)}, n) = 1$, Proposition 2.10 again implies that

$$a^{\phi(n)} \equiv 1 (\text{mod } n).$$

This completes the proof of Euler's theorem. ■

One interesting application of Euler's theorem is that it allows one to exhibit explicitly arithmetic inverses of integers modulo n. Suppose that $\gcd(a, n) = 1$. Then we see that $a^* = a^{\phi(n)-1}$ is an arithmetic inverse of a modulo n, for by Euler's theorem we see that

$$aa^* = a \cdot a^{\phi(n)-1} = a^{\phi(n)} \equiv 1 (\text{mod } n).$$

Thus, $a^{\phi(n)-1}$ is an arithmetic inverse of a modulo n.

Let us now turn to Wilson's theorem.

Theorem 3 (Wilson): Let p be a prime. Then

$$(p-1)! \equiv -1 (\text{mod } p).$$

Proof: If $p = 2$ or 3, the theorem may be checked directly since

$$1 \equiv -1 (\text{mod } 2)$$
$$1 \cdot 2 \equiv -1 (\text{mod } 3).$$

Thus, we may assume that $p > 3$.

The idea of the proof is to break up the integers $2, 3, \ldots, p-2$ into pairs of integers, each pair having product congruent to 1. Then only 1 and $p-1$ remain in the product $(p-1)!$ from which the result is clear.

For each integer r with $1 \leq r \leq p-1$, let r^* be its arithmetic inverse mod p lying in the interval $1 \leq r^* \leq p-1$. (An arithmetic inverse of r exists since $p \nmid r$; we surely can choose r^* so that $0 \leq r^* \leq p-1$ since $0, 1, \ldots, p-1$ is a complete residue system mod p; finally $r^* = 0$ implies that $1 \equiv rr^* = 0 (\text{mod } p)$, which is absurd.) We assign to each integer r its arithmetic inverse r^*. In particular r^* is assigned its arithmetic inverse, which we denote by r^{**}. But

$$r^*r^{**} \equiv 1 \pmod p$$

and

$$r^*r \equiv 1 \pmod p$$

imply that $r^*r \equiv r^*r^{**} \pmod p$ and thus by Proposition 2.10 we have $r^{**} \equiv r \pmod p$. Since both integers r and r^{**} lie between 1 and $p - 1$, we see that $r^{**} = r$. So we have assigned to the integer r^* the integer r. We have broken up the integers $1, 2, \ldots, p - 1$ into pairs of integers (r, r^*) except that there is the possibility that $r = r^*$. Now $r = r^*$ implies that $r \equiv r^* \pmod p$, and so $r^2 \equiv rr^* \equiv 1 \pmod p$; we have then that $p \mid r^2 - 1 = (r - 1)(r + 1)$ and so $p \mid r - 1$ or $p \mid r + 1$. Thus, $r \equiv 1 \pmod p$ or $r \equiv -1 \equiv p - 1 \pmod p$. Thus, if we exclude 1 and $p - 1$ from the list $1, 2, \ldots, p - 1$ we have divided up the $p - 3$ integers $2, 3, \ldots, p - 2$ into $s = (p - 3)/2$ pairs which have no elements in common. Let us list these pairs:

$$(r_1, r_1^*), \ldots, (r_s, r_s^*).$$

Then

$$\begin{aligned} 2 \cdot 3 \cdots (p - 2) &= r_1 r_1^* r_2 r_2^* \cdots r_s r_s^* \\ &\equiv 1 \cdot 1 \cdots 1 \pmod p \qquad (\text{since } r_i r_i^* \equiv 1 \pmod p)) \\ &\equiv 1 \pmod p. \end{aligned}$$

Thus,

$$\begin{aligned} 1 \cdot 2 \cdot 3 \cdots (p - 1) &\equiv (p - 1) \pmod p \\ &\equiv -1 \pmod p, \end{aligned}$$

whence Wilson's theorem. ■

Example 4: If $p = 7$, $(p - 1)! = 720 = 7 \cdot 102 + 6 \equiv -1 \pmod 7$. Note that the pairing described in the proof of Wilson's theorem in case $p = 7$ is $(2, 4), (3, 5)$.

Let us give a few quite surprising applications of Wilson's theorem and Fermat's theorem.

Theorem 5: Let p be an odd prime. The congruence

$$x^2 \equiv -1 \pmod p$$

has a solution if and only if $p \equiv 1 \pmod 4$. When $p \equiv 1 \pmod 4$, then $x = ((p - 1)/2)!$ is a solution.

Thus, we have a complete characterization of all primes p for which there is a square root of -1 mod p. For example, $x^2 \equiv -1 \pmod 7$ has no solutions, while $x^2 \equiv -1 \pmod 5$ does. (Check complete residue systems modulo 7 and 5, respectively.)

Proof of Theorem 5: First assume that $x^2 \equiv -1 \pmod p$ for some x. Then

$$x^{p-1} = (x^2)^{(p-1)/2} \equiv (-1)^{(p-1)/2} \pmod p.$$

One the other hand, by Fermat's theorem,

$$x^{p-1} \equiv 1 \pmod p.$$

Therefore, $1 \equiv (-1)^{(p-1)/2} \pmod p$, so that $p \mid 1 - (-1)^{(p-1)/2}$. If $1 - (-1)^{(p-1)/2} \neq 0$, then $1 - (-1)^{(p-1)/2} = 2$, which contradicts the fact that p is odd. Thus, $1 = (-1)^{(p-1)/2}$ so that $(p-1)/2$ is even; i.e., $p \equiv 1 \pmod 4$.

Conversely, assume that $p \equiv 1 \pmod 4$. Then

$$\begin{aligned}(p-1)! &= 1 \cdot 2 \cdots \frac{p-1}{2}(p-1)(p-2)\cdots\left(p - \frac{p-1}{2}\right) \\ &\equiv 1 \cdot 2 \cdots \frac{p-1}{2}(-1)(-2)\cdots\left(-\frac{p-1}{2}\right) \pmod p \\ &= (-1)^{(p-1)/2} \cdot 1^2 \cdot 2^2 \cdots \left(\frac{p-1}{2}\right)^2 \equiv \left(1 \cdot 2 \cdots \frac{p-1}{2}\right)^2 \pmod p\end{aligned}$$

(since $p \equiv 1 \pmod 4$).

On the other hand, by Wilson's theorem,

$$(p-1)! \equiv -1 \pmod p,$$

so that if we set $x = 1 \cdot 2 \cdots [(p-1)/2]$, we have

$$x^2 \equiv -1 \pmod p. \qquad \blacksquare$$

The preceding theorem is concerned with the solution of a special quadratic congruence. We shall study quadratic congruences in detail in the next chapter, and thus we shall refrain from putting Theorem 5 in its proper context now. However, let us draw a few elementary but startling conclusions from Theorem 5.

Corollary 6: Let x be an integer and let p be an odd prime dividing $x^2 + 1$. Then $p \equiv 1 \pmod 4$.

Proof: If $p \mid x^2 + 1$, then $x^2 \equiv -1 \pmod p$. Thus, we are done by Theorem 5. $\blacksquare$

Example 7: If $x = 6$, then $x^2 + 1 = 37 \equiv 1 \pmod 4$. If $x = 8$, then $x^2 + 1 = 65 = 5 \cdot 13$, and both 5 and 13 are $\equiv 1 \pmod 4$.

Example 8: Let us apply Theorem 5 to study the special Diophantine equation

$$y^2 = x^3 + 23. \qquad (1)$$

We shall show that there are no solutions. This equation is a special case of the *Bachet equation* $y^2 = x^3 + k$, which was discussed by Bachet* in 1621. The reader should note that our study of Eq. (1) uses only Theorem 5 and elementary facts concerning congruences. However, one cannot do this for general k. The argument for (1) makes use of special properties of the number 23.

We shall now show that (1) has no solutions. We shall do this by eliminating the four possible cases modulo 4; namely x must be congruent to one of 0, 1, 2, 3(mod 4). First we see that if $x \equiv 0$ or 2(mod 4), then $x^3 \equiv 0 \pmod 4$, and so

$$y^2 = x^3 + 23 \equiv 23 \equiv 3 \pmod 4.$$

But the squares are congruent to 0 or 1(mod 4), so that x cannot be congruent to 0 or 2(mod 4). Next, suppose that $x \equiv 3 \equiv -1 \pmod 4$. Then

$$y^2 = x^3 + 23 \equiv (-1)^3 + 23 = 22 \equiv 2 \pmod 4.$$

Again the squares are congruent to 0 or 1(mod 4), and so x cannot be congruent to 3(mod 4).

We now have only the possibility $x \equiv 1 \pmod 4$. This case is more difficult and requires an ingenious idea. We write (1) in the form

$$\begin{aligned} y^2 + 4 &= x^3 + 3^3 \\ &= (x + 3)(x^2 - 3x + 9). \end{aligned} \tag{2}$$

Then $x \equiv 1 \pmod 4$ implies that

$$x^2 - 3x + 9 \equiv 1 - 3 + 9 = 7 \equiv -1 \pmod 4. \tag{3}$$

Then there is a prime p such that $p \equiv -1 \pmod 4$ and $p \mid x^2 - 3x + 9$. This is true since it is easily seen that for all integers x, $x^2 - 3x + 9 \geq 2$, and so we may write

$$x^2 - 3x + 9 = p_1 p_2 \cdots p_t$$

as a product of primes. If each $p_i \equiv 1 \pmod 4$ $(1 \leq i \leq t)$, then

$$x^2 - 3x + 9 = p_1 p_2 \cdots p_t \equiv 1 \cdot 1 \cdots 1 \equiv 1 \pmod 4,$$

violating Eq. (3). So we have some p_i, call it p, dividing $x^2 - 3x + 9$ such that

*There is an incredibly rich literature concerning the Bachet equation. Special cases were discussed by many mathematicians during the eighteenth and nineteenth centuries. It was proved by L. J. Mordell in the early part of this century that the Diophantine equation $y^2 = x^3 + k$ (k given) has only a finite number of solutions. In 1966, A. Baker of Cambridge provided the first finite procedure for determining the solutions, although as of this writing the theory is not quite at the stage where it can be put on a computer. The Bachet equation is a historically important Diophantine equation, and we shall return to special cases of it throughout this book to illustrate the power of various theorems we shall prove.

$p \equiv -1 \pmod 4$. Then, in particular, Eq. (2) implies that

$$y^2 \equiv -4 = (-1)2^2 \pmod p. \tag{4}$$

Since $p \equiv -1 \pmod 4$, p is an odd prime, and thus 2 has an arithmetic inverse 2^* mod p. Thus, (4) implies that

$$(2^*y)^2 \equiv -1 \pmod p.$$

Thus, 2^*y is a solution of the congruence $z^2 \equiv -1 \pmod p$, which contradicts Theorem 5 since $p \equiv -1 \pmod 4$.

This completes the proof that Eq. (1) has no solutions.

3.3 Exercises

1. Find $a^k \pmod 7$ for $k = 0, 1, 2, 3, 4, 5, 6$ and
   (a) $a = 1$. (d) $a = 4$.
   (b) $a = 2$. (e) $a = 5$.
   (c) $a = 3$. (f) $a = 6$.
   In each case you should use the complete residue system $0, 1, 2, \ldots, 6$.

2. Continue Exercise 1.
   (a) Compute $a^k \pmod 7$ for *any* k and for $a = 1, 2, 3, 4, 5, 6$.
   (b) Can you make any conjectures about the powers of $a^k \pmod 7$?
   (c) Can you make any conjecture about the smallest value of $k > 0$ for which $a^k \equiv 1 \pmod 7$?
   (d) Test any conjectures you make in parts (b) and (c) by carrying out Exercises 1 and 2 for powers of integers modulo 11 and 13.

3. Show that if p is a prime, then $a^p \equiv a \pmod p$.

4. Carry out the proof of Wilson's theorem numerically in the cases $p = 5, 7$, and 11.

5. (a) Does the congruence $x^2 \equiv -1 \pmod{17}$ have a solution? If it has a solution, find it.
   (b) Repeat part (a) for the congruence $x^2 \equiv -1 \pmod{43}$.

6. Does the congruence $x^2 \equiv -1 \pmod{91}$ have a solution?

7. Does $x^2 \equiv -1 \pmod{65}$ have a solution?

8. Prove that if $a^{m-1} \equiv 1 \pmod m$ and $a^r \not\equiv 1 \pmod m$ for any $r \mid m - 1$, $r < m - 1$, then m is a prime.

9. Prove the converse of Wilson's theorem: Show that if m is an integer such that $(m - 1)! \equiv -1 \pmod m$, then m is prime. In fact, show that if m is not prime, then $(m - 1)! \equiv 0 \pmod m$.

10. Show that the Diophantine equation $x^2 + 1 = 23y$ has no solutions.

11. Let a be an odd integer. Show that if $k \geq 2$,
$$a^{2^{k-2}} \equiv 1(\text{mod } 2^k).$$
(*Hint:* Use induction on k. For $k = 3$, we considered this in Section 3.2.)

12. Show that $y^2 = x^3 - 73$ has no solutions in integers.

13. Let m be an integer having no prime factor $\equiv 3(\text{mod } 4)$, n an arbitrary integer. Show that the Diophantine equation
$$y^2 + 4m^2 = x^3 + (4n - 1)^3$$
has no solutions.

14. Write an efficient computer program to compute the arithmetic inverse a^* of a modulo n. (Your program should include a test to determine if a has such an inverse.)

15. Let p be a prime > 2, $0 < a < p - 1$. Show that
$$\binom{p-1}{a} \equiv (-1)^a(\text{mod } p).$$

*16. We showed that for $p \equiv 1(\text{mod } 4)$, we have
$$\left(\frac{p-1}{2}\right)!^2 \equiv -1(\text{mod } p).$$
Show that for $p \equiv 1(\text{mod } 4)$, we have
$$\left(\frac{p-1}{2}\right)!^2 \not\equiv -1(\text{mod } p^2).$$

17. Let p be a prime such that $n < p < 2n$. Then prove
 (a) $\binom{2n}{n} \equiv 0(\text{mod } p)$.
 (b) $\binom{2n}{n} \not\equiv 0(\text{mod } p^2)$.

18. Let p be a prime, a, b integers such that $p \nmid ab$. Show that $(a + b)^p \equiv a^p + b^p(\text{mod } p)$.

**19. (Wohlstenholme's theorem) Let $p > 3$ be a prime. Show the numerator of $1 + \frac{1}{2} + \frac{1}{3} + \cdots + 1/(p - 1)$ is divisible by p^2. (*Hint:* Note that the polynomial $(x - 1) \cdots (x - (p - 1))$ has $(p - 1)!$ as its constant term. Put $x = p$ and study the divisibility of the coefficients by p, and look at the coefficient of x.)

20. By Wilson's theorem $(p - 1)! + 1 = kp$ for some k.
 (a) When can $k = 1$?
 (b) When can $k = p$?

21. Let p be an odd prime.
   (a) Show that
$$2^2 \cdot 4^2 \cdot 6^2 \cdots (p-1)^2 \equiv (-1)^{(p+1)/2} (\text{mod } p).$$
   (b) Show that $1^2 \cdot 3^2 \cdots (p-2)^2 \equiv 2^2 \cdot 4^2 \cdot 6^2 \cdots (p-1)^2 (\text{mod } p)$.

3.4 Solving Polynomial Congruences, I

Let $f(x)$ be a nonzero polynomial with integer coefficients and let n be a positive integer. In this section, we shall take up the problem of determining the solutions to the congruence

$$f(x) \equiv 0(\text{mod } n). \tag{1}$$

We have previously considered a number of special cases of this problem. For example, we considered the linear congruence $ax \equiv b(\text{mod } n)$, which corresponds to $f(x) = ax - b$. Furthermore, we considered the congruence $x^2 \equiv -1(\text{mod } n)$, which corresponds to the polynomial $f(x) = x^2 + 1$. From these two special cases, we see that finding the solutions of polynomial congruences can be a tough business. In fact, mathematicians by no means have the complete story on solving polynomial congruences, and what we have to say on the subject will barely scratch the surface.

We have already seen that if x is a solution to congruence (1), then any y such that $x \equiv y(\text{mod } n)$ is also a solution. Thus, in order to keep track of the solutions to (1), we regard two solutions x and y to be "different" if $x \not\equiv y(\text{mod } n)$. Thus, (1) can have at most n different solutions because there are precisely n possible incongruent integers mod n. To find the solutions to (1) it would suffice to test the n integers in a fixed complete residue system (e.g., test $0, 1, 2, \ldots, n-1$). Suppose that the elements of the complete residue system which provide solutions to (1) are $a_1, \ldots, a_s$. Then x is a solution of (1) if and only if

$$x \equiv a_1(\text{mod } n), \quad \text{or} \quad x \equiv a_2(\text{mod } n), \quad \ldots, \quad \text{or} \quad x \equiv a_s(\text{mod } n).$$

Let us emphasize again that $0, 1, \ldots, n-1$ is often not the most convenient complete residue system for computations. If n is odd, then the complete residue system

$$-\frac{n-1}{2}, -\frac{n-1}{2} + 1, \ldots, -1, 0, 1, 2, \ldots, \frac{n-1}{2}$$

consists of numbers smaller in absolute value and so is easier to compute with. For example, if we were interested in the congruence

$$x^5 \equiv 2(\text{mod } 24)$$

and if we wished to determine whether $x \equiv 23 \pmod{24}$ was a solution, it would be senseless to use the number 23, for then we would have to raise 23 to the fifth power and decide whether 24 divides $23^5 - 2$. Since $23 \equiv -1 \pmod{24}$, it is much simpler to compute $(-1)^5 = -1$ from which it is immediately seen that $23^5 \not\equiv 2 \pmod{24}$.

Let us now take up the problem of solving congruence (1). Suppose that

$$n = p_1^{a_1} \cdots p_t^{a_t},$$

where $p_1, \ldots, p_t$ are distinct primes. Then if x is a solution of $f(x) \equiv 0 \pmod{n}$, x is also a solution of all the congruences

$$\begin{aligned} f(x) &\equiv 0 \pmod{p_1^{a_1}} \\ &\vdots \\ f(x) &\equiv 0 \pmod{p_t^{a_t}}. \end{aligned} \tag{2}$$

Conversely, suppose that x is a simultaneous solution of all congruences (2). Then $f(x)$ is divisible by $p_1^{a_1}$, by $p_2^{a_2}, \ldots$, by $p_t^{a_t}$. Since all the $p_i^{a_i}$ have no common factors, $f(x)$ is divisible by $p_1^{a_1} \cdots p_t^{a_t} = n$, and therefore x is a solution of the congruence $f(x) \equiv 0 \pmod{n}$. Thus, we have seen that the solutions of congruence (1) are the same as the simultaneous solutions of all congruences (2).

Hence, in order to arrive at a complete solution of the congruence $f(x) \equiv 0 \pmod{n}$, we may proceed in two steps:

1. *Solve each of the congruences*

$$f(x) \equiv 0 \pmod{p_i^{a_i}} \qquad (i = 1, \ldots, t)$$

separately.

2. *Determine which solutions are common solutions of all congruences* (2).

To better understand what is involved in step 2, let us assume that part 1 is completely settled and that the solutions of the congruences

$$f(x) \equiv 0 \pmod{p_i^{a_i}} \qquad (i = 1, \ldots, t) \tag{3}$$

are determined (somehow) to be

$$x \equiv b_{i1} \pmod{p_i^{a_i}},\ x \equiv b_{i2} \pmod{p_i^{a_i}}, \ldots. \tag{4}$$

For a given x to be a solution of $f(x) \equiv 0 \pmod{n}$, we have seen that it is necessary and sufficient for x to satisfy each of congruences (3). Therefore, since the solutions of (3) are given by (4), x is a solution of $f(x) \equiv 0 \pmod{n}$ if and only if x is congruent to one of $b_{11}, b_{12}, \ldots \pmod{p_1^{a_1}}$, to one of $b_{21}, b_{22}, \ldots \pmod{p_2^{a_2}}$, and so forth. Thus, we arrive at the following result:

Theorem 1: The integer x is a solution of the congruence $f(x) \equiv 0(\text{mod } n)$ if and only if x satisfies a system of congruences

$$\begin{aligned} x &\equiv b_1(\text{mod } p_1^{a_1}) \\ &\vdots \\ x &\equiv b_t(\text{mod } p_t^{a_t}), \end{aligned}$$

where b_i is a solution of the congruence

$$f(x) \equiv 0(\text{mod } p_i^{a_i}) \qquad (i = 1, \ldots, t).$$

From Theorem 1, we see that, provided part 1 is solved, then part 2 can be solved provided that we can solve systems of simultaneous congruences of the form

$$\begin{aligned} x &\equiv b_1(\text{mod } p_1^{a_1}) \\ &\vdots \\ x &\equiv b_t(\text{mod } p_t^{a_t}), \end{aligned}$$

that is, to find x which satisfies all the congruences simultaneously. Let us now take up this problem, which we shall completely settle. Namely, we shall prove the following result:

Theorem 2 (Chinese Remainder Theorem): Suppose that the positive integers $m_1, m_2, \ldots, m_t$ are relatively prime in pairs; that is, $\gcd(m_i, m_j) = 1$ if $i \neq j$, $1 \leq i, j \leq t$. Further, let $b_1, b_2, \ldots, b_t$ be arbitrary integers. Then the congruences

$$\begin{aligned} x &\equiv b_1(\text{mod } m_1) \\ &\vdots \\ x &\equiv b_t(\text{mod } m_t) \end{aligned}$$

have a simultaneous solution. Moreover, the simultaneous solution is unique modulo $m_1 m_2 \cdots m_t$. That is, if y is another solution, then $x \equiv y(\text{mod } m_1 m_2 \cdots m_t)$.

Before we prove Theorem 2, some remarks are in order.

Remark 1: Since, in the preceding discussion, the integers $p_1^{a_1}, \ldots, p_t^{a_t}$ are relatively prime in pairs (the primes $p_1, p_2, \ldots, p_t$ are all distinct), the system of congruences of Theorem 1 always has solutions, and the solution is unique modulo $p_1^{a_1} p_2^{a_2} \cdots p_t^{a_t} = n$. We shall apply this remark to our original problem of solving polynomial congruences later in this section.

Remark 2: Our proof of Theorem 2 will be constructive, in the sense that the proof will provide a computational technique whereby the solutions of the system of congruences can be calculated in a finite number of steps.

Remark 3: The reason that Theorem 2 is called the Chinese Remainder Theorem is that rudimentary forms of it go back to the ancient Chinese.

Proof of Theorem 2: The idea of the proof is to write x in the form

$$x = y_1 b_1 + \cdots + y_t b_t,$$

where $y_1 \equiv 1(\text{mod } m_1)$ and $y_1 \equiv 0(\text{mod } m_i)$ $(2 \leq i \leq t)$, $y_2 \equiv 1(\text{mod } m_2)$ and $y_2 \equiv 0(\text{mod } m_i)$ $(i = 1, 3, 4, \ldots, t)$, and similarly for $y_3, \ldots, y_t$. To have $y_1 \equiv 0(\text{mod } m_i)$ $(2 \leq i \leq t)$ we must have $m_2 \cdots m_t \mid y_1$, since the m_i are relatively prime in pairs. Thus, in general, set

$$m_i' = \frac{m_1 m_2 \cdots m_t}{m_i}.$$

Then $\gcd(m_i, m_i') = 1$ since $m_1, \ldots, m_t$ are relatively prime in pairs. (Exercise.) Thus, m_i' has an arithmetic inverse $m_i'^*$ mod m_i:

$$m_i'^* m_i' \equiv 1(\text{mod } m_i).$$

Set $y_i = m_i'^* m_i'$, and correspondingly set

$$x = m_1'^* m_1' b_1 + \cdots + m_t'^* m_t' b_t.$$

Let us show that $x \equiv b_1(\text{mod } m_1)$. Indeed, for $2 \leq i \leq t$, we have

$$m_1 \mid m_i',$$

so that $m_i'^* m_i' b_i \equiv 0(\text{mod } m_1)$ for $2 \leq i \leq t$; also $m_1'^* m_1' \equiv 1(\text{mod } m_1)$, so that $m_1'^* m_1' b_1 \equiv b_1(\text{mod } m_1)$. Thus,

$$x \equiv b_1 + 0 + \cdots + 0 \equiv b_1(\text{mod } m_1).$$

It follows similarly that $x \equiv b_i(\text{mod } m_i)$ for all i, $1 \leq i \leq t$.

As to the uniqueness of the solution, let x' be any other simultaneous solution. Then $x \equiv x'(\text{mod } m_i)$ for $i = 1, 2, \ldots, t$. Therefore, since $m_1, m_2, \ldots, m_t$ are relatively prime in pairs and since $x - x'$ is divisible by $m_1, m_2, \ldots, m_t$, we have that $m_1 m_2 \cdots m_t \mid x - x'$. Thus, $x \equiv x'(\text{mod } m_1 m_2 \cdots m_t)$. ■

The above method of proof is completely constructive, in that it gives us a specific method for computing the solution x. Let us first do a simple example. Find x satisfying the congruences

$$x \equiv 1(\text{mod } 3)$$

$$x \equiv 4(\text{mod } 5).$$

Here $m_1 = 3$, $m_2 = 5$, $b_1 = 1$, $b_2 = 4$. Thus, $m_1' = (5 \cdot 3)/3 = 5$ and $m_2' = 3$.

Then since $m_1'^*$ is an arithmetic inverse of $m_1'(\text{mod } m_1)$, $m_1'^*$ must satisfy

$$5m_1'^* \equiv 1(\text{mod } 3),$$

so that $m_1'^* \equiv 2(\text{mod } 3)$. Also $3m_2'^* \equiv 1(\text{mod } 5)$ implies that $m_2'^* \equiv 2(\text{mod } 5)$. Thus,

$$x = 2 \cdot 5 \cdot 1 + 2 \cdot 3 \cdot 4 = 34 \equiv 4(\text{mod } 15).$$

As a more complicated example we find the simultaneous solutions of the congruences

$$x \equiv 7(\text{mod } 8)$$
$$x \equiv 2(\text{mod } 9)$$
$$x \equiv -1(\text{mod } 5).$$

Here

$$b_1 = 7, \qquad b_2 = 2, \qquad b_3 = -1,$$

and

$$m_1 = 8, \qquad m_2 = 9, \qquad m_3 = 5,$$

and thus

$$m_1' = 9 \cdot 5 = 45, \qquad m_2' = 8 \cdot 5 = 40, \qquad m_3' = 8 \cdot 9 = 72.$$

Therefore,

$$45m_1'^* \equiv 1(\text{mod } 8), \qquad 40m_2'^* \equiv 1(\text{mod } 9), \qquad 72m_3'^* \equiv 1(\text{mod } 5),$$

so that

$$m_1'^* \equiv 5(\text{mod } 8), \qquad m_2'^* \equiv -2(\text{mod } 9), \qquad m_3'^* \equiv 3(\text{mod } 5).$$

We therefore set

$$x = 5 \cdot 45 \cdot 7 + (-2) \cdot 40 \cdot 2 + 3 \cdot 72 \cdot (-1) = 1199 \equiv 119(\text{mod } 360)$$

$(360 = 8 \cdot 9 \cdot 5)$. It is easily checked that $x = 119$ satisfies the original congruences.

By combining Theorems 1 and 2, we see that the problem of solving polynomial congruences is reduced completely to the problem of solving polynomial congruences of the form

$$f(x) \equiv 0(\text{mod } p^a),$$

where p is a prime. This may seem like a small simplification, but it is significant for at least two reasons. If n is divisible by small powers of many primes, the congruences modulo the prime powers can be solved by testing all possible cases. This should involve considerably less work than testing all cases modulo n. For example, if $n = 48 = 3 \cdot 2^4$, then the number of residue classes modulo n is 48, whereas the numbers modulo 3 and 2^4 total $3 + 16 = 19$. However, an even more compelling reason for reducing the problem of solving polynomial congruences to the case of congruences modulo a prime power is that many theoretical results can be established about prime power congruences which cannot be proved in general.

Before we go any further, however, let us illustrate the ideas we have introduced thus far by means of the numerical example $x^2 \equiv 1(\text{mod } 56)$. Since this congruence is equivalent to $x^2 - 1 \equiv 0(\text{mod } 56)$, we may set $f(x) = x^2 - 1$. Moreover, since $56 = 2^3 \cdot 7$, the original congruence is equivalent to the pair of congruences

$$x^2 - 1 \equiv 0(\text{mod } 2^3)$$
$$x^2 - 1 \equiv 0(\text{mod } 7).$$

The first congruence has the solutions $x \equiv 1, 3, 5, 7(\text{mod } 2^3)$. (For now, we obtain solutions by just testing all elements of a complete residue system, in this case modulo 8.) Similarly, the second congruence has the solutions $x \equiv 1, 6(\text{mod } 7)$. Therefore, the congruence $x^2 \equiv 1(\text{mod } 56)$ has eight solutions modulo 56, namely the solutions of the following eight pairs of congruences:

$$\begin{cases} x \equiv 1(\text{mod } 2^3) \\ x \equiv 1(\text{mod } 7), \end{cases} \quad \begin{cases} x \equiv 1(\text{mod } 2^3) \\ x \equiv 6(\text{mod } 7), \end{cases} \quad \begin{cases} x \equiv 3(\text{mod } 2^3) \\ x \equiv 1(\text{mod } 7), \end{cases} \quad \begin{cases} x \equiv 3(\text{mod } 2^3) \\ x \equiv 6(\text{mod } 7), \end{cases}$$

$$\begin{cases} x \equiv 5(\text{mod } 2^3) \\ x \equiv 1(\text{mod } 7), \end{cases} \quad \begin{cases} x \equiv 5(\text{mod } 2^3) \\ x \equiv 6(\text{mod } 7), \end{cases} \quad \begin{cases} x \equiv 7(\text{mod } 2^3) \\ x \equiv 1(\text{mod } 7), \end{cases} \quad \begin{cases} x \equiv 7(\text{mod } 2^3) \\ x \equiv 6(\text{mod } 7). \end{cases}$$

The solutions of the pairs of congruences give the solutions to the original congruence. And the solutions of the pairs of congruences can be found using the Chinese Remainder Theorem. They are

$$x \equiv 1(\text{mod } 56), \quad x \equiv 41(\text{mod } 56), \quad x \equiv 43(\text{mod } 56), \quad x \equiv 27(\text{mod } 56),$$
$$x \equiv 29(\text{mod } 56), \quad x \equiv 13(\text{mod } 56), \quad x \equiv 15(\text{mod } 56), \quad x \equiv 55(\text{mod } 56).$$

Let us now make a few comments about congruences of the form

$$f(x) \equiv 0(\text{mod } p^a),$$

where p is a prime. Our first goal will be to show that in order to solve such a congruence, it is actually sufficient to solve the simpler congruence

$$f(x) \equiv 0(\text{mod } p).$$

Actually, what we shall do is prescribe a procedure for calculating the solutions of the congruence

$$f(x) \equiv 0(\text{mod } p^{a+1})$$

from the solutions to the congruence

$$f(x) \equiv 0(\text{mod } p^a).$$

Thus, starting from the solutions of a congruence modulo p, our procedure will allow us to calculate successively the solutions of the same congruence modulo p^2, p^3, and so forth.

Let us suppose that $a \geq 1$ and that the solutions to the congruence

$$f(x) \equiv 0 (\text{mod } p^a)$$

are given by

$$x \equiv b_1 (\text{mod } p^a), \ldots, x \equiv b_r (\text{mod } p^a).$$

Our problem is to determine the solutions of

$$f(x) \equiv 0 (\text{mod } p^{a+1}).$$

First, however, we need a preliminary result about polynomials. If $f(x) = a_0 + a_1 x + \cdots + a_m x^m$, denote by f' the *formal derivative* of f, given by the usual formula

$$f'(x) = a_1 + 2a_2 x + \cdots + m a_m x^{m-1}.$$

The usual rules for differentiating apply to our formal derivative, namely for polynomials f and g we have

$$(f + g)' = f' + g' \tag{5}$$

$$(af)' = af' \qquad \text{for an integer } a. \tag{6}$$

We shall denote the second derivative of f (that is, the derivative of f') by f''. More generally, we shall denote the ith derivative of f by $f^{(i)}$. In what follows, we shall require a particular case of what is called *Taylor's formula* in calculus.

Lemma 3: Let $f(x)$ be any polynomial of degree m. Then

$$f(x + y) = f(x) + \frac{f'(x)}{1} y + \frac{f''(x)}{1 \cdot 2} y^2 + \cdots + \frac{f^{(m)}(x)}{1 \cdot 2 \cdots m} y^m.$$

Moreover, if f has integer coefficients, so do all the polynomials $f'(x)/1$, $f''(x)/1 \cdot 2, \ldots, f^{(m)}(x)/1 \cdot 2 \cdots m$.

Proof: By properties (5) and (6) of the derivative, it suffices to prove our lemma for the polynomial $f(x) = x^m$. In this case, $f(x + y) = (x + y)^m$ is given by the binomial theorem:

$$f(x + y) = x^m + \frac{m}{1} x^{m-1} y + \frac{m(m-1)}{1 \cdot 2} x^{m-2} y^2 + \cdots + y^m,$$

where all the coefficients (the so-called binomial coefficients) are integers. Now

$$\frac{f'(x)}{1} = \frac{m}{1} x^{m-1}, \qquad \frac{f''(x)}{1 \cdot 2} = \frac{m(m-1)}{1 \cdot 2} x^{m-2},$$

and so forth. Therefore,

$$f(x + y) = f(x) + \frac{f'(x)}{1} y + \frac{f''(x)}{1 \cdot 2} y^2 + \cdots + y^m,$$

and the lemma is proved. ■

Let us now return to the congruence $f(x) \equiv 0 (\text{mod } p^{a+1})$. If x is a solution, then x is also a solution of $f(x) \equiv 0 (\text{mod } p^a)$, and therefore $x \equiv$

$b \pmod{p^a}$, where b is one of $b_1, b_2, \ldots, b_r$. Thus, we may write $x = b + kp^a$ for some k. Let us determine the condition which k must satisfy in order for x to be a solution of $f(x) \equiv 0 \pmod{p^{a+1}}$. By Lemma 3, with $x = b$, $y = kp^a$,

$$\begin{aligned} f(x) &= f(b + kp^a) \\ &= f(b) + \frac{f'(b)}{1}kp^a + \frac{f''(b)}{1 \cdot 2}(kp^a)^2 + \cdots + \frac{f^{(m)}(b)}{1 \cdot 2 \cdots m}(kp^a)^m. \end{aligned}$$

By Lemma 3, the polynomials

$$\frac{f'(x)}{1}, \frac{f''(x)}{1 \cdot 2}, \ldots, \frac{f^{(m)}(x)}{1 \cdot 2 \cdots m}$$

have integer coefficients, so that

$$\frac{f'(b)}{1}, \frac{f''(b)}{1 \cdot 2}, \ldots, \frac{f^{(m)}(b)}{1 \cdot 2 \cdots m}$$

are all integers. Moreover, since $a \geq 1$, we have $a + 1 \leq 2a$, so that p^{2a}, $p^{3a}, \ldots, p^{ma}$ are all $\equiv 0 \pmod{p^{a+1}}$. Thus,

$$f(x) \equiv f(b) + \frac{f'(b)}{1}kp^a \pmod{p^{a+1}}.$$

Since $f(b) \equiv 0 \pmod{p^a}$, we may set $f(b) = tp^a$ for some t. Thus,

$$f(x) \equiv p^a(t + f'(b)k) \pmod{p^{a+1}}.$$

Thus, $f(x) \equiv 0 \pmod{p^{a+1}}$ for $x = b + kp^a$ if and only if

$$p^a(t + f'(b)k) \equiv 0 \pmod{p^{a+1}},$$

which is equivalent to

$$t + f'(b)k \equiv 0 \pmod{p} \qquad \text{(Proposition 2.11).}$$

We have thus derived the condition that k must satisfy in order that $x = b + kp^a$ be a solution of $f(x) \equiv 0 \pmod{p^{a+1}}$, namely

$$\boxed{f'(b)k \equiv -\frac{f(b)}{p^a} \pmod{p}.} \tag{7}$$

(Recall that $f(b) \equiv 0 \pmod{p^a}$ by assumption, and so $f(b)/p^a$ is an integer.) Thus, to determine the solutions of $f(x) \equiv 0 \pmod{p^{a+1}}$ knowing the solutions of $f(x) \equiv 0 \pmod{p^a}$, it suffices to solve the *linear* congruence (7) for k, modulo just the prime p. We discussed this subject thoroughly in Section 2. We shall nevertheless carry out the solution explicitly in the present situation. There are two cases to consider.

Case 1: $f'(b) \equiv 0 \pmod{p}$. In this case congruence (7) can only be satisfied provided that

$$\frac{f(b)}{p^a} \equiv 0 \pmod{p},$$

which is equivalent to

$$f(b) \equiv 0 (\text{mod } p^{a+1}).$$

Thus, there are two possibilities. Either $f(b) \equiv 0(\text{mod } p^{a+1})$, in which case $x = b + kp^a$ is a solution of the congruence $f(x) \equiv 0(\text{mod } p^{a+1})$ for arbitrary k. Or $f(b) \not\equiv 0(\text{mod } p^{a+1})$, in which case $x = b + kp^a$ is never a solution of the congruence $f(x) \equiv 0(\text{mod } p^{a+1})$ for any k.

Case 2: $f'(b) \not\equiv 0(\text{mod } p)$. In this case congruence (7) will have exactly one solution, namely

$$k \equiv -f'(b)^* \frac{f(b)}{p^a} (\text{mod } p),$$

where $f'(b)^*$ is an arithmetic inverse of $f'(b)(\text{mod } p)$. Thus, in Case 2, there is a unique solution of the congruence $f(x) \equiv 0(\text{mod } p^{a+1})$ of the form $x \equiv b + kp^a(\text{mod } p^{a+1})$, namely

$$x \equiv b - f'(b)^* \frac{f(b)}{p^a} p^a \, (\text{mod } p^{a+1}).$$

Thus, we have seen that in all cases we can determine the solutions modulo p^{a+1} once we know the solutions modulo p^a. We shall conclude this section by giving some examples.

Example 4: Find all solutions of the congruence

$$f(x) = x^3 - 2x + 1 \equiv 0(\text{mod } 3^2).$$

We start by determining all the solutions of $f(x) \equiv 0(\text{mod } 3)$ by inspection. There are only three possibilities, namely $x \equiv 0, \pm 1(\text{mod } 3)$. We readily check that $x \equiv 1(\text{mod } 3)$ is the only solution.

We now solve $f(x) \equiv 0(\text{mod } 3^2)$. There is one solution mod 3, so $b = b_1 = 1$. Also, $f(1) = 0$ and $f'(x) = 3x^2 - 2$, so that $f'(1) = 1$. Thus, by (7), we must solve the congruence

$$1 \cdot k \equiv \tfrac{0}{3}(\text{mod } 3),$$

which has $k \equiv 0(\text{mod } 3)$ as the unique solution. Thus,

$$x \equiv b + k \cdot 3 \equiv 1(\text{mod } 9)$$

is the unique solution mod 9. (In this case it would not be too difficult to check this result directly by checking all the nine possible cases modulo 9.)

Example 5: We shall find all the solutions of the congruence

$$f(x) = x^2 + x + 1 \equiv 0(\text{mod } 49).$$

We must again start by solving the congruence $f(x) \equiv 0(\text{mod } 7)$. Checking the seven possibilities directly we see that $f(x) \equiv 0(\text{mod } 7)$ if and only if

$$x \equiv 2 \quad \text{or} \quad -3(\text{mod } 7).$$

We now solve $f(x) \equiv 0 \pmod{7^2}$. Here $b_1 = 2$ and $b_2 = -3$. We compute

$$f(2) = 7 \quad \text{and} \quad f(-3) = 7.$$

Also $f'(x) = 2x + 1$, and so

$$f'(2) = 5 \quad \text{and} \quad f'(-3) = -5.$$

To find the solutions of the form $x = 2 + 7k$ we must solve (7), in this case

$$5k \equiv -\tfrac{7}{7} \pmod 7,$$

so $k \equiv -3 \pmod 7$ and $x \equiv -19 \pmod{49}$. Similarly, if $x = -3 + 7k$, we must solve

$$-5k \equiv -\tfrac{7}{7} \pmod 7$$

so $k \equiv 3 \pmod 7$ and $x \equiv 18 \pmod{49}$. Thus, the solutions of $f(x) \equiv 0 \pmod{49}$ are $x \equiv 30, 18 \pmod{49}$.

Both Examples 4 and 5 fall under Case 2 above. Let us give an example where Case 1 occurs.

Example 6: Find all the solutions of

$$f(x) = x^2 + x + 7 \equiv 0 \pmod{27}.$$

We see that $x \equiv 1 \pmod 3$ is the only solution of $f(x) \equiv 0 \pmod 3$. We now solve $f(x) \equiv 0 \pmod{3^2}$. Here $b_1 = 1$, $f(1) = 9$. Since $f'(x) = 2x + 1$, we have $f'(1) = 3$. Thus, the solutions are of the form $x = 1 + 3k$, where k satisfies

$$3k \equiv -\tfrac{9}{3} \pmod 3$$

or

$$3k \equiv -3 \pmod 3.$$

It is clear that any k satisfies this congruence, so that $k \equiv 0, 1, 2 \pmod 3$ yield solutions $x = 1 + 3k$ and so

$$x \equiv 1, 4, \text{ or } 7 \pmod 9$$

are the solutions of $f(x) \equiv 0 \pmod 9$.

We now, finally, solve $f(x) \equiv 0 \pmod{3^3}$. Here $b_1 = 1$, $b_2 = 4$, $b_3 = 7$. Thus,

$$f(1) = 9, \qquad f(4) = 27, \qquad f(7) = 63,$$

and

$$f'(1) = 3, \qquad f'(4) = 9, \qquad f'(7) = 15.$$

The solutions are of the form $x = b_i + 9k$ for $i = 1, 2, 3$. For $i = 1$, k must satisfy

$$3k \equiv -\tfrac{9}{9} \pmod 3$$

and there are clearly no solutions. For $i = 2$, k must satisfy

$$9k \equiv -\tfrac{27}{9} \pmod 3,$$

and so any k works mod 3 yielding solutions $x \equiv 4 + 9 \cdot 0 \equiv 4 \pmod{27}$, $x \equiv 4 + 9 \cdot 1 \equiv 13 \pmod{27}$, and $x \equiv 4 + 9 \cdot 2 \equiv 22 \pmod{27}$. For $i = 3$, k must satisfy

$$15k \equiv -\tfrac{63}{9} \equiv -7 \pmod{3},$$

and so there are no solutions. Thus, $f(x) \equiv 0 \pmod{27}$ if and only if

$$x \equiv 4,\ 13,\ \text{or } 22 \pmod{27}.$$

3.4 Exercises

1. Find all simultaneous solutions of the congruences
   (a) $x \equiv 3 \pmod{7}$, $x \equiv 2 \pmod{6}$.
   (b) $x \equiv 5 \pmod{2}$, $x \equiv 1 \pmod{3}$, and $x \equiv 2 \pmod{5}$.
   (c) $x \equiv 1 \pmod{9}$, $x \equiv 5 \pmod{7}$, and $x \equiv 3 \pmod{5}$.
2. Find all integers n which leave the remainder 4 when divided by 8, leave the remainder 6 when divided by 7, and leave the remainder 1 when divided by 5.
3. Show that for any integer $k \geq 1$, there are k successive integers which are divisible by squares larger than 1. (*Hint:* Use the Chinese Remainder Theorem.)
4. Solve the following congruences using the Chinese Remainder Theorem:
   (a) $x^2 + 1 \equiv 0 \pmod{65}$.
   (b) $5x^2 + 7x - 3 \equiv 0 \pmod{35}$.
   (c) $11x + 1 \equiv 0 \pmod{210}$.
5. Determine whether the following sets of simultaneous congruences have solutions. If they have solutions, find them all.
   (a) $x \equiv 5 \pmod{6}$ and $x \equiv 7 \pmod{10}$.
   (b) $x \equiv 1 \pmod{6}$ and $x \equiv 8 \pmod{15}$.

*6. Let m and n be positive integers. Let $d = \gcd(m, n)$. Prove that the simultaneous congruences

$$x \equiv a \pmod{m} \quad \text{and} \quad x \equiv b \pmod{n}$$

are solvable if and only if $a \equiv b \pmod{d}$.

7. Find all simultaneous solutions of the congruences
   (a) $3x \equiv 1 \pmod{10}$ and $4x \equiv 2 \pmod{7}$.
   (b) $3x \equiv 2 \pmod{4}$, $2x \equiv 7 \pmod{15}$, and $4x \equiv -1 \pmod{7}$.
8. Let $m_1, m_2, \ldots, m_k$ be positive integers which are relatively prime in pairs. Let $b_1, b_2, \ldots, b_k$ and $a_1, a_2, \ldots, a_k$ be given integers such that

$\gcd(b_i, m_i) = 1$ for $i = 1, 2, \ldots, k$. Prove that the simultaneous congruences

$$b_1 x \equiv a_1 (\text{mod } m_1)$$
$$b_2 x \equiv a_2 (\text{mod } m_2)$$
$$\vdots$$
$$b_k x \equiv a_k (\text{mod } m_k)$$

are always solvable. Give a formula like the one given in the Chinese Remainder Theorem for x in this case.

9. Using the method developed in this section solve the following congruences:
   (a) $x^4 + 2x + 4 \equiv 0(\text{mod } 27)$.
   (b) $x^2 \equiv -1(\text{mod } 125)$.
   (c) $x^2 + 3x - 3 \equiv 0(\text{mod } 49)$.
   (d) $x^2 + 3x - 10 \equiv 0(\text{mod } 49)$.

10. Find all solutions of the following congruence:

$$4x^4 + 9x^3 - 5x^2 - 21x + 61 \equiv 0(\text{mod } 1125).$$

(*Note:* This problem has been set up to illustrate all the concepts and cases discussed in this section. The answers are $x \equiv 323, -52, -427(\text{mod } 1125)$.)

11. Let p be a prime larger than 2 and suppose that $p \nmid a$. Assume that $x^2 - a \equiv 0(\text{mod } p)$ is solvable. Show that for all $n \geq 1$

$$x^2 - a \equiv 0(\text{mod } p^n)$$

has precisely two solutions.

*12. Let $p > 2$ be a prime. Analyze the solvability of $x^2 + a \equiv 0(p^n)$ for all $n \geq 1$ when $p \mid a$. (There will be many cases depending on the highest power m such that $p^m \mid a$ and the resulting quotient a/p^m).

13. Let m and n be positive integers such that $\gcd(m, n) = 1$. Let $r_1, \ldots, r_{\phi(m)}$ be a reduced residue system modulo m and let $s_1, \ldots, s_{\phi(n)}$ be a reduced residue system modulo n.
   (a) Show that if $\gcd(x, mn) = 1$, then there is a unique pair (i, j) with $1 \leq i \leq \phi(m)$ and $1 \leq j \leq \phi(n)$ such that $x \equiv r_i(\text{mod } m)$ and $x \equiv s_j(\text{mod } n)$.
   (b) Conversely show that given any pair (i, j) there is a unique x modulo mn such that $x \equiv r_i(\text{mod } m)$ and $x \equiv s_j(\text{mod } n)$ and $\gcd(x, mn) = 1$.

14. Use Exercise 13 to prove that if $\gcd(m, n) = 1$, then $\phi(mn) = \phi(m)\phi(n)$.

15. Prove that if $n = p_1^{k_1}p_2^{k_2}\cdots p_r^{k_r}$ is the canonical factorization of n as a product of primes, then

$$\phi(n) = p_1^{k_1-1}(p_1 - 1)p_2^{k_2-1}(p_2 - 1)\cdots p_r^{k_r-1}(p_r - 1).$$

(Use Exercise 14 above and Exercise 6 in Section 3.2.)

16. Prove directly from the definition of formal derivative of a polynomial given in the text that for polynomials f and g and integer a, we have
    (a) $(af)' = af'$.
    (b) $(f + g)' = f' + g'$.
    (c) $(fg)' = f'g + fg'$.

3.5 Solving Polynomial Congruences, II

In this section we shall develop further the theory of polynomial congruences $f(x) \equiv 0(\text{mod } n)$. Our primary goal is to study the congruence $f(x) \equiv 0(\text{mod } p)$ where p is a prime. The main effort of the last section was in showing that if we could solve polynomial congruences modulo primes, then we could pass fairly routinely to the case of a general modulus n. Thus, all the difficulties must already occur in the prime case. Unfortunately, all we shall do in this section is prove some general facts about solving polynomial congruences modulo primes. We shall not give a procedure for actually solving them or even give a procedure for determining whether they have a solution. Indeed, no such procedures are known. Moreover, all of Chapter 4 deals with this problem just for *quadratic* polynomials.

Let us begin by making a few general observations. Since we are attempting to solve

$$f(x) \equiv 0(\text{mod } n) \tag{1}$$

for a polynomial $f(x) = a_m x^m + \cdots + a_1 x + a_0$, with integer coefficients, it certainly does not matter if we change one (or more) of the a_i to another integer congruent to it modulo n. We would have precisely the same solutions. Indeed, if we are working modulo n, it is very natural to consider two polynomials, all of whose coefficients differ by multiples of n, as the "same" modulo n. We are thus motivated to give the following definition:

Definition 1: Suppose that $f(x) = a_0 + a_1 x + \cdots$ and $g(x) = b_0 + b_1 x + \cdots$ are polynomials with integer coefficients. We say that $f(x)$ *is congruent to* $g(x)$ *modulo* n and write

$$f(x) \equiv_x g(x)(\text{mod } n)$$

provided that $a_i \equiv b_i(\text{mod } n)$ for all i.

Example 2:

(i) $x^3 + x + 1 \equiv_x x^3 + x - 2 \pmod 3$.
(ii) $x^3 + x + 1 \equiv_x 4x^3 + x + 7 \pmod 3$.
(iii) $x^3 + x + 1 \equiv_x 3x^7 + x^3 + x + 1 \pmod 3$.
(iv) $58x^7 + 89x^5 + 18 \equiv_x 3x^7 + 4x^5 + 3 \pmod 5$.

It is clear that, in general, we may replace congruence (1) by one in which all the coefficients of the polynomial are between 0 and $n - 1$ (or in any complete residue system). This helps considerably in computations.

Now if $f(x) \equiv_x g(x) \pmod n$, there is a polynomial $t(x)$ with integer coefficients such that

$$f(x) - g(x) = nt(x)$$

(since the coefficients of $f(x) - g(x)$ are the $a_i - b_i$ in Definition 1). Thus, if $f(a) \equiv 0 \pmod n$, we see that

$$g(a) = f(a) - nt(a) \equiv 0 \pmod n.$$

That is, congruence (1) has precisely the same solutions as the congruence $g(x) \equiv 0 \pmod n$. We have proved the following result:

Proposition 3: Suppose that $f(x) \equiv_x g(x) \pmod n$. Then the congruence $f(x) \equiv 0 \pmod n$ has precisely the same solutions as the congruence $g(x) \equiv 0 \pmod n$. Moreover, it always suffices to consider polynomial congruences where the polynomials have coefficients between 0 and $n - 1$.

Now let us return to the idea in part (iii) of Example 2. All the coefficients of $f(x)$ in Eq. (1) which are divisible by n may be replaced by 0. In particular we might be able to lower the degree of $f(x)$. This motivates the following definition:

Definition 4: Let $f(x) = a_0 + a_1x + \cdots + a_mx^m$ be a polynomial with integer coefficients. Assume that not all the a_i are divisible by n. Then, by the *degree of f modulo n*, written $\deg_n f(x)$, we mean the largest integer i such that $a_i \not\equiv 0 \pmod n$.

Thus, for example, if $f(x) = 12x^4 + 3x^2 + x + 1$, then $\deg_5 f = 4$, $\deg_{12} f = 2$, $\deg_4 f = 2$, and $\deg_3 f = 1$.

Now let us begin our theory concerning the actual solutions to Eq. (1). Let $f(x)$ and $g(x)$ be polynomials, $g(x) \neq 0$. Then the process of long division of polynomials you learned in high school produces two polynomials $q(x)$, the quotient, and $r(x)$, the remainder, such that

$$\frac{f(x)}{g(x)} = q(x) + \frac{r(x)}{g(x)}$$

or

$$f(x) = g(x)q(x) + r(x), \tag{$*$}$$

where the degree of $r(x)$ is less than the degree of $g(x)$. It is not true in general that $q(x)$ and $r(x)$ have integer coefficients even if both $f(x)$ and $g(x)$ have integer coefficients.

We shall not need this general situation. We shall only need long division in the case where $g(x) = x - a$. We shall prove the result $(*)$ for this particular case completely. The proof will involve a slick application of induction. However, if you just set up and do the long division of $x - a$ into $f(x) = a_m x^m + \cdots + a_0$, you will see that the induction is simply a way of saying "and so forth" in the long division process:

$$
\begin{array}{r|l}
 & a_m x^{m-1} + \cdots \\
\hline
x - a & a_m x^m + a_{m-1}x^{m-1} + a_{m-2}x^{m-2} + \cdots + a_0 \\
 & a_m x^m - a a_m x^{m-1} \\
\hline
 & (a_{m-1} + a a_m)x^{m-1} + a_{m-2}x^{m-2} + \cdots + a_0 \\
 & \cdot \\
 & \cdot \\
 & \cdot
\end{array}
$$

Lemma 5: Let $f(x) = a_m x^m + \cdots + a_0$ be a polynomial with integer coefficients and let a be an integer. Then there is a polynomial $q(x)$ with integer coefficients such that

$$f(x) = (x - a)q(x) + f(a).$$

Proof: We shall prove the result by induction on the degree of $f(x)$. If $f(x)$ is the constant polynomial $f(x) = a_0$, then we may take $q(x) = 0$, as we may readily check. Thus, we have taken care of the case where the degree of $f(x)$ is zero.

Thus, we may assume the truth of Lemma 5 for all polynomials of degree $\leq m - 1$. Set

$$
\begin{aligned}
f_1(x) &= f(x) - a_m x^{m-1}(x - a) \\
&= (a_{m-1} + a a_m)x^{m-1} + a_{m-2}x^{m-2} + \cdots + a_0.
\end{aligned}
$$

The degree of $f_1(x) \leq m - 1$, and we may apply the induction hypothesis. Thus, there is a polynomial $q_1(x)$ with integer coefficients such that

$$f_1(x) = (x - a)q_1(x) + f_1(a).$$

Thus,

$$
\begin{aligned}
f(x) &= f_1(x) + a_m x^{m-1}(x - a) \\
&= (x - a)(q_1(x) + a_m x^{m-1}) + f_1(a).
\end{aligned}
$$

Let $q(x) = q_1(x) + a_m x^{m-1}$. Then $q(x)$ has integer coefficients, and

$$f(x) = (x - a)q(x) + f_1(a).$$

Finally, plugging $x = a$ into this last result we see that $f(a) = f_1(a)$. Thus, Lemma 5 is completely proved. ■

We obtain immediately from Lemma 5 that if a is a solution to (1) then we may "factor out" $x - a$ modulo n. Namely,

Proposition 6: Let $f(x)$ be a polynomial with integer coefficients. Let a be an integer. Then $f(a) \equiv 0 \pmod{n}$ if and only if there is a polynomial $q(x)$ with integer coefficients such that

$$f(x) \equiv_x (x - a)q(x) \pmod{n}. \tag{2}$$

(Equation (2) is a congruence of polynomials as in Definition 1).

Proof: If $f(a) \equiv 0 \pmod{n}$, we obtain from Lemma 5 that $f(x) = (x - a)q(x) + f(a)$, where $n \mid f(a)$. Thus, with $q(x)$ as in Lemma 5 we indeed have (2). Conversely, applying Proposition 3, we see that since a is a solution of the congruence $(x - a)q(x) \equiv 0 \pmod{n}$, it must be a solution of $f(x) \equiv 0 \pmod{n}$ also. ■

Corollary 7: The polynomial $q(x)$ in Proposition 6 can always be chosen so that its degree is at most $\deg(f(x)) - 1$.

Proof: The $q(x)$ from Proposition 6 came directly from Lemma 5. Examining the proof of Lemma 5, we see that

$$q(x) = a_m x^{m-1} + (a_{m-1} + aa_m)x^{m-2} + \cdots,$$

and so the result is clear. ■

We now must specialize to the case where we solve

$$f(x) \equiv 0 \pmod{p},$$

that is, where $n = p$ is a prime. In this case we can limit the number of solutions by the degree of the equation. First we prove

Theorem 8: Let $f(x)$ be a polynomial with integer coefficients. Let $b_1, b_2, \ldots, b_t$ be t incongruent solutions of $f(x) \equiv 0 \pmod{p}$. Then there is a polynomial $q(x)$ with integer coefficients such that

$$f(x) \equiv_x (x - b_1)(x - b_2) \cdots (x - b_t)q(x) \pmod{p}.$$

Moreover, $\deg_p q(x) \le \deg_p f(x) - t$.

Proof: By replacing $f(x)$ by a polynomial congruent to it modulo p we may assume that $f(x) = a_m x^m + \cdots + a_0$, where $a_m \not\equiv 0 \pmod{p}$. That is, $\deg_p f(x) = m$. By Proposition 6 and Corollary 7 there is a polynomial $q_1(x)$ such that

$$f(x) \equiv_x (x - b_1)q_1(x) \pmod{p}$$

with $\deg_p q_1(x) \le m - 1 = \deg_p f(x) - 1$.

Plug $x = b_2$ into this last result, and we obtain

$$f(b_2) \equiv (b_2 - b_1)q_1(b_2) \pmod{p}.$$

But $f(b_2) \equiv 0 \pmod{p}$ implies that $p \mid (b_2 - b_1)q_1(b_2)$, and so $p \mid b_2 - b_1$ or

$p \mid q_1(b_2)$ (by Lemma 2.4.3). By assumption, $b_2 \not\equiv b_1 \pmod p$ and thus $p \mid q_1(b_2)$, so that $q_1(b_2) \equiv 0 \pmod p$. Again applying Proposition 6 and Corollary 7 we see there is a polynomial $q_2(x)$ such that

$$q_1(x) \equiv_x (x - b_2)q_2(x) \pmod p$$

with $\deg_p q_2(x) \leq \deg_p q_1(x) - 1$. Therefore, a quick resort to definitions implies that

$$f(x) \equiv_x (x - b_1)(x - b_2)q_2(x) \pmod p$$

with $\deg_p q_2(x) \leq \deg_p q_1(x) - 1 \leq \deg_p f(x) - 1 - 1 = \deg_p f(x) - 2$.

Now plug $x = b_3$ into this last result. Proceed similarly and obtain

$$f(x) \equiv_x (x - b_1)(x - b_2)(x - b_3)q_3(x) \pmod p$$

with $\deg_p q_3(x) \leq \deg_p f(x) - 3$.

Continuing in this way we arrive at Theorem 8. ■

Corollary 9: Let $f(x)$ be a polynomial with integer coefficients not all divisible by p. Then the number of incongruent solutions of $f(x) \equiv 0 \pmod p$ is at most the degree of $f(x)$ modulo p.

Proof: If $b_1, \ldots, b_t$ are t incongruent solutions, then we have from Theorem 8 that

$$f(x) \equiv_x (x - b_1) \cdots (x - b_t)q(x) \pmod p,$$

where $\deg_p q(x) \leq \deg_p f(x) - t$. Now $q(x)$ must have some coefficient not divisible by p or else all the coefficients of $f(x)$ will be divisible by p. Thus,

$$0 \leq \deg_p q(x) \leq \deg_p f(x) - t$$

or $t \leq \deg_p f(x)$, as asserted. ■

Example 10: The assertion of Corollary 9 is false if we are not working modulo a prime. For example,

$$x^2 - 1 \equiv 0 \pmod 8$$

has the four solutions $x \equiv 1, 3, 5, 7 \pmod 8$, whereas $\deg_8(x^2 - 1) = 2$.

But of course, $x^2 - 1 \equiv 0 \pmod p$, p a prime, can only have the solutions $x \equiv \pm 1 \pmod p$ because we know from Corollary 9 there are at most two solutions, and $x \equiv \pm 1 \pmod p$ are obviously solutions.

Example 11: By Fermat's theorem (Theorem 3.1) we have that if $p \nmid a$, then $a^{p-1} \equiv 1 \pmod p$, and so $a^p \equiv a \pmod p$. Thus, a is a solution of

$$x^p - x \equiv 0 \pmod p. \tag{3}$$

It is clear that $x = 0$ also satisfies this congruence. Thus, *every integer satisfies congruence* (3). Applying Theorem 8 to (3) and the solutions $0, 1, \ldots, p - 1$ we see that

$$x^p - x \equiv_x x(x - 1) \cdots (x - (p - 1))q(x) \pmod p, \tag{4}$$

where $\deg_p q(x) \leq p - p = 0$. Thus, $q(x)$ is a constant, say $q(x) = b$, where b is an integer. Then the coefficient of x^p on the left-hand side of (4) is 1 and on the right-hand side of (4) is b. Thus, $b \equiv 1 \pmod p$, and

$$x^p - x \equiv_x x(x-1)(x-2) \cdots (x-(p-1)) \pmod p.$$

For example, $x^3 - x \equiv_x x(x-1)(x-2) \pmod 3$. (Multiply it out and check it directly.)

It is possible to reduce further the degree of the polynomial in a congruence modulo a prime by using the results in Example 11. Since $x^p \equiv x \pmod p$ is satisfied for any value of x, we may replace x^p in a polynomial by x without altering the set of solutions. For example,

$$x^7 + x^2 + 5 \equiv 0 \pmod 7$$

has precisely the same solutions as

$$x + x^2 + 5 \equiv 0 \pmod 7$$

since $x^7 \equiv x \pmod 7$ for any value of x. However, $x^7 + x^2 + 5 \not\equiv_x x + x^2 + 5 \pmod 7$.

As a more elaborate example, consider

$$x^{35} - x^{10} + x - 3 \equiv 0 \pmod 5. \tag{5}$$

Since $x^{35} = (x^5)^7$, we have

$$x^{35} \equiv x^7 \pmod 5$$

for all values of x. Since $x^7 = x^5x^2$, we have

$$x^7 \equiv x \cdot x^2 = x^3 \pmod 5$$

for all x or

$$x^{35} \equiv x^3 \pmod 5$$

for every integer x. Similarly,

$$x^{10} \equiv x^2 \pmod 5$$

for all x. Thus, congruence (5) has precisely the same solutions as the congruence

$$x^3 - x^2 + x - 3 \equiv 0 \pmod 5.$$

We started with a polynomial of degree 35 and found it sufficed to solve one of degree 3.

It should be observed that by replacing x^p by x in a congruence, possibly repeating the operation many times, we can always end up with a polynomial of degree $\leq p - 1$. We record this in

Theorem 12: Let p be a prime. Then the congruence

$$f(x) \equiv 0 \pmod p$$

can be transformed into a congruence

$$g(x) \equiv 0 \pmod p$$

having precisely the same solutions and where the degree of $g(x)$ is at most $p - 1$ or $g(x)$ is the zero polynomial.

Thus, modulo 5, we need never consider polynomials of degree larger than 4.

3.5 Exercises

1. Test whether the following pairs of polynomials are congruent modulo 7:
   (a) $x^3 + 2x + 1$ and $8x^3 - 5x + 1$.
   (b) $x^3 + 2x + 1$ and $8x^3 - 6x + 1$.
   (c) $3x^5 + 2x^2 + x$ and $10x^5 - 12x^2 + x + 7$.
   (d) $7x^2 + 2$ and $7x + 2$.
2. Let $f(x) = 35x^4 + 7x^2 + 2x + 1$. What are $\deg_2 f$, $\deg_3 f$, $\deg_5 f$, $\deg_7 f$, and $\deg_{11} f$?
3. Let f be a polynomial. Show that for all primes p, $\deg_p f \leq \deg f$ and that there are only finitely many primes p such that $\deg_p f < \deg f$.
4. For the following polynomials $f(x)$ and integers a notice that $f(a) \equiv 0 \pmod{11}$. In each of these cases find a polynomial $g(x)$ such that
$$f(x) \equiv_x (x - a)g(x) \pmod{11}.$$
   (a) $f(x) = x^2 + 10x + 3$, $a = 6$.
   (b) $f(x) = x^3 - x^2 + x + 10$, $a = 1$.
   (c) $f(x) = x^3 - 6x^2 - 2x + 20$, $a = -3$.
5. The solutions of $f(x) \equiv 0 \pmod{13}$, where $f(x) = x^4 - 6x^3 - 3x^2 - 7x + 2$, are $x \equiv \pm 1 \pmod{13}$. Find a polynomial $g(x)$ such that
$$f(x) \equiv_x (x - 1)(x + 1)g(x) \pmod{13}.$$
6. Show that the polynomial $f(x) = x^3 + 3x^2 + 2x + 2$ cannot be factored modulo 5. (That is, we cannot find polynomials $g(x)$ and $h(x)$ such that $\deg_5 g(x) < 3$ and $\deg_5 h(x) < 3$ such that $f(x) \equiv_x h(x)g(x) \pmod{5}$.)
7. Convert the following polynomial congruences into congruences with the same solutions having degree less than 5:
   (a) $2x^{17} + 3x^2 + 1 \equiv 0 \pmod{5}$.
   (b) $x^{10} + 2x^5 + 1 \equiv 0 \pmod{5}$.
   (c) $3x^{23} + 2x^{20} + 4x^{17} - x^6 + x^5 - 3x^3 + 2x + 1 \equiv 0 \pmod{5}$.
8. Let p be a prime. Let $f(x) = a_n x^n + \cdots + a_1 x + a_0$ be a polynomial. Assume that $a_0 \not\equiv 0 \pmod{p}$. Show that one can find a polynomial $g(x)$ of degree $\leq p - 2$ such that the congruences $f(x) \equiv 0 \pmod{p}$ and $g(x) \equiv 0 \pmod{p}$ have the same *nonzero* solutions.

9. Find all solutions of the congruences in Exercise 7.

10. Let p be a prime. Let a and b be integers such that $a \geq 1$. Find all solutions of the congruence
$$x^{p^a} \equiv b \pmod{p}.$$

11. Prove the following polynomial congruences for primes p:
 (a) $x^{p-1} - 1 \equiv_x (x-1)(x-2)\cdots(x-(p-1)) \pmod{p}$.
 (b) $x^{p-2} + x^{p-3} + \cdots + x + 1 \equiv_x (x-2)\cdots(x-(p-1)) \pmod{p}$.

12. By comparing coefficients of the two sides of the polynomial congruence in Exercise 11(a), prove the following for primes p:
 (a) Wilson's theorem (look at the constant term).
 (b) For $p \geq 3$, $1 + 2 + \cdots + (p-1) \equiv 0 \pmod{p}$.
 (c) For $p \geq 5$, $1\cdot 2 + 1\cdot 3 + \cdots + 1\cdot(p-1) + 2\cdot 3 + \cdots + 2\cdot(p-1) + 3\cdot 4 + \cdots + 3\cdot(p-1) + \cdots + (p-2)(p-1) \equiv 0 \pmod{p}$.
 (d) For $p \geq 5$, $1\cdot 2 \cdots (p-2) + 1\cdot 2\cdots(p-3)(p-1) + 1\cdot 2\cdots(p-4)(p-2)(p-1) + \cdots + 1\cdot 3\cdots(p-1) + 2\cdot 3\cdots(p-1) \equiv 0 \pmod{p}$.

**13. Let p be a prime and let $a \geq 1$ be an integer. Let $r_1, \ldots, r_{\phi(p^a)}$ be a reduced residue system mod p^a. Prove the following polynomial congruence:
$$(x^{p-1} - 1)^{p^{a-1}} \equiv_x (x - r_1)(x - r_2)\cdots(x - r_{\phi(p^a)}) \pmod{p^a}.$$

3.6 Primitive Roots

In this section we shall give an extremely useful application of the theory of polynomial congruences modulo a prime p. We shall explore the congruence properties of the various powers of an integer a modulo p.

Let us begin with some experimentation. Suppose that $p = 7$. In Table 3-1, we have tabulated the powers a^k modulo 7 for $a = 0, 1, 2, \ldots, 6$ (a complete residue system) and $k = 0, 1, 2, \ldots, 6$, with respect to the complete residue system $0, 1, 2, \ldots, 6$ modulo 7.

Why did we quit with $k = 6$? We quit because we know by Fermat's Little Theorem (Theorem 3.1) that if $7 \nmid a$, then $a^6 \equiv 1 \pmod{7}$, and so $a^7 = a^6 a \equiv a \pmod{7}$, $a^8 = a^6 a^2 \equiv a^2 \pmod{7}$, $a^9 = a^6 a^3 \equiv a^3 \pmod{7}$, and so forth. Thus, the powers repeat themselves when the exponent changes modulo 6.

What do we observe from the table about the powers of a given integer a (read across the rows of the table)? There are two integers a in the list, $a = 3, 5$, which have all the integers 1, 2, 3, 4, 5, 6 (i.e., 1 through $p - 1$) as powers of them. In other words, every integer b such that $b \not\equiv 0 \pmod{p}$ is congruent to a power of 3 (or 5) modulo 7. Does this phenomenon persist for all primes p? If it did, we would be able to reduce the study of arithmetic

TABLE 3-1

Powers a^k of $a \pmod 7$

a \ k	0	1	2	3	4	5	6
0	0	0	0	0	0	0	0
1	1	1	1	1	1	1	1
2	1	2	4	1	2	4	1
3	1	3	2	6	4	5	1
4	1	4	2	1	4	2	1
5	1	5	4	6	2	3	1
6	1	6	1	6	1	6	1

modulo p to the study of the powers of a single element (together with zero). It is, in fact, true that for every prime p there are such integers, and they are called *primitive roots*. The proof of their existence and their uses comprise the subject of this section. Let us first formally define a primitive root.

Definition 1: Let p be a prime. By a *primitive root* modulo p we mean an integer g such that

$$g^0 = 1, g, g^2, \ldots, g^{p-2} \tag{1}$$

is a reduced residue system modulo p. That is, the integers (1) are a rearrangement, modulo p, of $1, 2, \ldots, p-1$.

Although Definition 1 tells us what primitive roots are all about, it is not convenient for determining them, and so let us go back to Table 3-1 and see what else we can discover. We notice in the rows corresponding to integers other than 3 and 5 (excluding $a = 0$, of course) that the entries repeat themselves. For example, for $a = 2$ we have 1, 2, 4 repeated. This happens since $2^3 \equiv 1 \pmod 7$, and so $2^4 = 2^3 \cdot 2 \equiv 2 \pmod 7$, $2^5 = 2^3 \cdot 2^2 \equiv 4 \pmod 7$, and so forth. That is, if $a^k \equiv 1 \pmod 7$ then $a^{k+1} = a^k a \equiv a \pmod 7$, and so forth. Thus, we see that the important point is that there is a lowest power of a that yields 1 modulo 7. Thereafter the powers of a repeat themselves. In particular, a can be a primitive root modulo 7 if and only if the first power of a that gives $1 \pmod 7$ is 6. (This power can be no larger than 6 by Fermat's Little Theorem.) We are thus led to the following definition:

Definition 2: Let p be a prime and a an integer such that $a \not\equiv 0 \pmod p$. Then by the *order of a mod p* we mean the least integer $k \geq 1$ such that

$$a^k \equiv 1 (\text{mod } p).$$

We denote this integer by $\text{ord}_p\, a$.

For example, $\text{ord}_7\, 2 = 3$, $\text{ord}_7\, 3 = 6$, $\text{ord}_7\, 4 = 3$, $\text{ord}_7\, 5 = 6$, and $\text{ord}_7\, 6 = 2$.

By Fermat's Little Theorem, we see that $\text{ord}_p\, a$ is an integer such that

$$1 \leq \text{ord}_p\, a \leq p - 1.$$

Also from our above discussion we are led to the following result.

Theorem 3: Let p be a prime and let g be an integer such that $g \not\equiv 0 (\text{mod } p)$. Then g is a primitive root mod p if and only if $\text{ord}_p\, g = p - 1$.

Proof: We first suppose that g is a primitive root mod p. Then, by Definition 1, $g^0, g^1, g^2, \ldots, g^{p-2}$ is a reduced residue system modulo p. In particular, no two of these numbers are congruent. Since $g^0 = 1$, we have $g^k \not\equiv 1 (\text{mod } p)$ for $1 \leq k \leq p - 2$. Since $g^{p-1} \equiv 1 (\text{mod } p)$ (Fermat's theorem again), we have immediately from the definition of $\text{ord}_p\, g$ that $\text{ord}_p\, g = p - 1$.

Conversely, suppose that $\text{ord}_p\, g = p - 1$. Since there are $p - 1$ numbers $g^0, g^1, \ldots, g^{p-2}$ and for each k, $p \nmid g^k$, it suffices to show that no two of these numbers are congruent modulo p in order to show that g is a primitive root modulo p.

Suppose then that this statement is false. Then we can find i, j such that $0 \leq i < j \leq p - 2$ and

$$g^i \equiv g^j (\text{mod } p).$$

Since $g^j = g^i g^{j-i}$ and $p \nmid g^i$, we may cancel g^i (Proposition 2.10) and obtain

$$1 \equiv g^{j-i} (\text{mod } p).$$

Since $1 \leq j - i < p - 1$ we have now contradicted the hypothesis that $\text{ord}_p\, g = p - 1$. ■

It should now be clear from Theorem 3 that we must make a careful study of $\text{ord}_p\, a$. We shall accumulate the results in

Proposition 4: Let p be a prime and let a be an integer such that $p \nmid a$. Then

(i) $\text{ord}_p\, a$ divides $p - 1$.

(ii) *If* $a^v \equiv 1 (\text{mod } p)$, then $\text{ord}_p\, a$ divides v.

(iii) $\text{ord}_p(a^u) = \text{ord}_p\, a / \gcd(u, \text{ord}_p\, a)$.

Proof: Since, by Fermat's Little Theorem, we have $a^{p-1} \equiv 1 (\text{mod } p)$, we see that part (i) follows immediately from part (ii).

Part (ii) is an immediate application of the general principle noted following the proof of Theorem 2.2.3. Namely, writing $k = \text{ord}_p\, a$ (just shorthand) we can find q and r such that

$$v = kq + r, \qquad 0 \leq r < k.$$

Then

$$1 \equiv a^v = a^{kq+r} = (a^k)^q a^r \equiv 1^q a^r = a^r (\operatorname{mod} p).$$

But k is the least integer $t \geq 1$ such that $a^t \equiv 1(\operatorname{mod} p)$, and thus $a^r \equiv 1(\operatorname{mod} p)$ and $0 \leq r < k$ imply that $r = 0$. Thus, $k \mid v$, as we asserted.

For part (iii), let us again write $k = \operatorname{ord}_p a$; also write $m = \gcd(u, \operatorname{ord}_p a) = \gcd(u, k)$. Then by part (ii), we have

$$(a^u)^t = a^{ut} \equiv 1(\operatorname{mod} p)$$

if and only if $k \mid ut$. But this is equivalent to saying

$$\frac{k}{m} \,\Big|\, \frac{ut}{m}.$$

Since $\gcd(k/m, u/m) = 1$ (Theorem 2.3.6) this last statement is equivalent to the condition

$$\frac{k}{m} \,\Big|\, t$$

(Theorem 2.3.6).

Thus, we have shown that $(a^u)^t \equiv 1(\operatorname{mod} p)$ if and only if $(k/m) \mid t$. Thus, k/m is the least integer t such that $(a^u)^t \equiv 1(\operatorname{mod} p)$, and so $\operatorname{ord}_p(a^u) = k/m$, as we wished to prove. ∎

We need one more result before we can prove that primitive roots exist.

Proposition 5: Let p be a prime and let a_1, a_2 be such that $p \nmid a_1$ and $p \nmid a_2$. Further, suppose that $\operatorname{ord}_p a_1 = k_1$, $\operatorname{ord}_p a_2 = k_2$ with $\gcd(k_1, k_2) = 1$. Then $\operatorname{ord}_p(a_1 a_2) = k_1 k_2$.

Proof: Since

$$(a_1 a_2)^{k_1 k_2} = (a_1^{k_1})^{k_2} (a_2^{k_2})^{k_1} \equiv 1^{k_2} 1^{k_1} = 1(\operatorname{mod} p),$$

we have by Proposition 4, part (ii), that $\operatorname{ord}_p(a_1 a_2)$ divides $k_1 k_2$.

Now suppose that

$$(a_1 a_2)^t \equiv 1(\operatorname{mod} p).$$

Then $a_1^t a_2^t \equiv 1(\operatorname{mod} p)$, and a_2^t is an arithmetic inverse of a_1^t.

The following lemma is then relevant:

Lemma 6: If $p \nmid a$ and a^* is an arithmetic inverse of a, then $\operatorname{ord}_p a = \operatorname{ord}_p a^*$.

Proof: Let $k = \operatorname{ord}_p a$. Since $a^k \equiv 1(\operatorname{mod} p)$, we see that

$$a^{*k} = 1 \cdot a^{*k} \equiv a^k a^{*k} = (aa^*)^k \equiv 1^k = 1(\operatorname{mod} p),$$

and so by Proposition 4, part (ii), $\operatorname{ord}_p a^*$ divides k. Now suppose that $\operatorname{ord}_p a^* = v$. Then, exactly as above,

$$a^v \equiv a^{*v} a^v \equiv 1^v = 1 \pmod{p},$$

and so $\operatorname{ord}_p a = k$ divides $v = \operatorname{ord}_p a^*$. That is $k \mid v$ and $v \mid k$, and so $v = k$. ■

Proof of Proposition 5 Continued: From what we had before, we deduce immediately from Lemma 6 that

$$\operatorname{ord}_p a_1^t = \operatorname{ord}_p a_2^t.$$

Thus, we conclude from Proposition 4, part (iii), that

$$\frac{k_1}{\gcd(t, k_1)} = \frac{k_2}{\gcd(t, k_2)}.$$

Since $\gcd(k_1, k_2) = 1$, k_1 and k_2 have no common factors greater than 1; the last relation implies that

$$\frac{k_1}{\gcd(t, k_1)} = \frac{k_2}{\gcd(t, k_2)} = 1.$$

Then $k_1 = \gcd(t, k_1)$ implies that $k_1 \mid t$. Similarly, $k_2 \mid t$. Again, since $\gcd(k_1, k_2) = 1$, we may conclude that $k_1 k_2 \mid t$.

This all shows, in particular, that $k_1 k_2 \mid \operatorname{ord}_p(a_1 a_2)$. Since we also had that $\operatorname{ord}_p(a_1 a_2) \mid k_1 k_2$, we have $k_1 k_2 = \operatorname{ord}_p(a_1 a_2)$. ■

Corollary 7: Let p be a prime and let $a_1, \ldots, a_r$ be integers such that $p \nmid a_1 a_2 \cdots a_r$. Assume that $\operatorname{ord}_p a_i = k_i$ and that the k_i are relatively prime in pairs. Then

$$\operatorname{ord}_p(a_1 a_2 \cdots a_r) = k_1 k_2 \cdots k_r.$$

Proof: Exercise (Ex. 6). ■

Let us now prove the fundamental result of this section.

Theorem 8: Let p be a prime. Then there exists a primitive root modulo p.

Proof: If $p = 2$, then any odd number will do, so assume that p is odd. Then $p - 1 > 1$, and so we may write

$$p - 1 = p_1^{a_1} p_2^{a_2} \cdots p_r^{a_r},$$

where $p_1, p_2, \ldots, p_r$ are distinct primes. It suffices to find integers $g_1, g_2, \ldots, g_r$ such that

$$\operatorname{ord}_p g_i = p_i^{a_i}, \qquad 1 \le i \le r, \tag{$*$}$$

since then, by Corollary 7, setting $g = g_1 g_2 \cdots g_r$, we would have

$$\operatorname{ord}_p g = p_1^{a_1} p_2^{a_2} \cdots p_r^{a_r} = p - 1,$$

and so g would be a primitive root modulo p by Theorem 3.

The proof of the existence of the g_i is based on

Lemma 9: Let $k \mid p-1$. Then the congruence $x^k - 1 \equiv 0 \pmod{p}$ has precisely k solutions.

Proof: Write $p - 1 = kt$. We use the following polynomial identity familiar from high school:

$$x^{p-1} - 1 = (x^k - 1)(x^{k(t-1)} + x^{k(t-2)} + \cdots + 1).$$

Let x be one of the integers $1, 2, \ldots, p-1$. Since $x^{p-1} - 1 \equiv 0 \pmod{p}$ (Fermat's Little Theorem), we have that

$$(x^k - 1)(x^{k(t-1)} + x^{k(t-2)} + \cdots + 1) \equiv 0 \pmod{p}.$$

Thus, $p \mid (x^k - 1)(x^{k(t-1)} + \cdots + 1)$, and so by Euclid's lemma (Lemma 2.4.3) we see that

$$p \mid x^k - 1 \quad \text{or} \quad p \mid x^{k(t-1)} + \cdots + 1.$$

In other words, every x among $1, 2, \ldots, p-1$ is a solution of one of the congruences

$$x^k - 1 \equiv 0 \pmod{p} \tag{2}$$

or

$$x^{k(t-1)} + \cdots + 1 \equiv 0 \pmod{p}. \tag{3}$$

Since $x = 0$ satisfies neither congruence (2) nor (3), the pair of congruences (2) and (3) have a total of $p - 1$ solutions. By Corollary 5.9 the first must have $\leq k$ solutions, and the second must have $\leq k(t-1)$ solutions. Thus, the two of them must have $\leq k + k(t-1) = kt = p - 1$ solutions. The only way for this to happen is for (2) to have k solutions and (3) to have $k(t-1)$ solutions. The assertion for (2) is the statement of Lemma 9. ■

Proof of Theorem 8 Continued: By (*), we see that it is enough to show that there is a g_1 such that $\text{ord}_p\, g_1 = p_1^{a_1}$. What are the conditions that g_1 must satisfy? First,

$$g_1^{p_1^{a_1}} \equiv 1 \pmod{p}. \tag{4}$$

That is, g_1 must be a solution of the congruence

$$x^{p_1^{a_1}} - 1 \equiv 0 \pmod{p}. \tag{5}$$

Let us assume that we have any solution g_1 of (5). What other condition must g_1 satisfy in order to guarantee that $\text{ord}_p\, g_1 = p_1^{a_1}$? Let $b = \text{ord}_p\, g_1$. Then, since g_1 satisfies (4), we see that $b \mid p_1^{a_1}$, so that $b = p_1^v$ for some $v \leq a_1$. Now $\text{ord}_p\, g_1 \neq p_1^{a_1}$ if and only if $b \neq p_1^{a_1}$, which is equivalent to saying $v \leq a_1 - 1$. Now if $v \leq a_1 - 1$, we have

$$g_1^{p_1^{a_1-1}} = (g_1^b)^{p_1^{a_1-1-v}} \equiv 1 \pmod{p},$$

and so g_1 satisfies the congruence

$$x^{p_1^{a_1-1}} - 1 \equiv 0 \pmod{p}. \tag{6}$$

Conversely, if g_1 satisfies (6), then $\text{ord}_p\, g_1 \mid p_1^{a_1-1}$, so that $v \leq a_1 - 1$. Thus, we see that $\text{ord}_p\, g_1 \neq p_1^{a_1}$ if and only if g_1 satisfies congruence (6). Thus, $\text{ord}_p\, g_1 = p_1^{a_1}$ if and only if g_1 satisfies (5) but does not satisfy (6). By Lemma 9, (5) has $p_1^{a_1}$ solutions and (6) has $p_1^{a_1-1}$ solutions. Since $p_1^{a_1-1} < p_1^{a_1}$, we can find a solution of (5) which is not a solution of (6). Thus, we have proved the existence of g_1.

In a similar way, we can construct $g_2, \ldots, g_r$. ■

We observe that superficially the above proof of the existence of primitive roots is constructive, that is, gives a method of finding primitive roots. However, it is of little practical value. For in the notation of the proof of Theorem 8, we must find a solution of (5) which is not a solution of (6). We have no method of solving (5) other than plugging in specific values for x and determining if they yield a solution. This is no better a computational technique than determining the order of specific integers directly by computing their powers and determining the first power congruent to 1 modulo p. However, it should be observed that the ideas used in the proof of Theorem 8 together with the results of Proposition 4 and Corollary 7 can often simplify the work. We shall give an example.

Example 10: Let $p = 23$. Then $p - 1 = 22 = 2 \cdot 11$. Thus, for integers a such that $23 \nmid a$, $\text{ord}_{23}\, a = 1, 2, 11$, or 22 (Proposition 4(i)). We begin by computing $\text{ord}_{23}\, 2$. By an easy computation we see that $2^2 \not\equiv 1 \pmod{23}$ and that $2^{11} \equiv 1 \pmod{23}$, and thus $\text{ord}_{23}\, 2 = 11$. Now let us observe that $\text{ord}_{23}(-1) = 2$. Thus, applying Corollary 7, we see that $\text{ord}_{23}(-2) = 2 \cdot 11 = 22$. That is, -2 is a primitive root mod 23.

Let us see how this example fits into the scheme of the proof of Theorem 8. Since $p - 1 = 2 \cdot 11$, we wished to find a solution to $x^2 - 1 \equiv 0 \pmod{23}$ which was not a solution of $x - 1 \equiv 0(23)$ and a solution of $x^{11} - 1 \equiv 0(23)$ which was not a solution of $x - 1 \equiv 0(23)$. We showed that 2 satisfied the last condition and observed that -1 satisfied the first condition and concluded that $2 \cdot (-1) = -2$ was a primitive root.

Let us go into more detail now on how we use primitive roots. Let g be a primitive root modulo a prime p. Then $g^0, g^1, \ldots, g^{p-2}$ is a reduced residue system mod p, and so no two of them can be congruent modulo p. In general, when can g^i and g^j be the same mod p? Well, if $i < j$ and

$$g^i \equiv g^j \pmod{p},$$

then we have $g^{j-i} \equiv 1 \pmod{p}$. Thus, $\text{ord}_p\, g = p - 1$ implies that $p - 1 \mid j - i$ (Proposition 4 (ii)). Conversely if $p - 1 \mid j - i$, then $j = i + k(p - 1)$, and so

$$g^j = g^{i+k(p-1)} = g^i(g^{p-1})^k \equiv g^i 1^k = g^i \pmod{p}.$$

We have shown

Proposition 11: Let g be a primitive root modulo p. Then $g^i \equiv g^j \pmod{p}$ if and only if $i \equiv j \pmod{p-1}$.

In this way, *multiplicative problems* modulo p may be reduced to additive problems modulo $p - 1$. (Note the analogy with logarithms.) We shall illustrate this by studying the following example in some detail. We wish to solve the congruence

$$x^n \equiv a \pmod{p}, \tag{7}$$

where $n > 0$ and a are given integers. Integers a for which (7) has a solution x are called nth *power residues mod p*.

If $p \mid a$, then the answer is easy, for then $a \equiv 0 \pmod{p}$, and so our congruence (7) is $x^n \equiv 0 \pmod{p}$. If x is a solution, then $p \mid x^n$. But then $p \mid x$ by Euclid's lemma (Lemma 2.4.3) and so $x \equiv 0 \pmod{p}$. Conversely $x \equiv 0 \pmod{p}$ is clearly a solution. Thus, the solutions when $p \mid a$ are precisely the x such that $x \equiv 0 \pmod{p}$. We may, from now on, assume that $p \nmid a$. In this case, we see that $x \equiv 0 \pmod{p}$ cannot be a solution.

Congruence (7) need not have any solutions. For example,

$$x^2 \equiv 2 \pmod{5}$$

has no solutions. We see this simply by checking the four possible cases $x \equiv 1, 2, 3, 4 \pmod{5}$, obtaining $x^2 \equiv 1, 4, 4, 1 \pmod{5}$, respectively. In fact, congruence (7) is quite subtle, its theory is very involved, and there remain many unsolved problems. Let us make an initial attempt at developing a criterion for determining whether (7) is solvable.

Let us fix a primitive root g mod p. Then $g^0 = 1, g, g^2, \ldots, g^{p-2}$ is a reduced residue system mod p. Since $p \nmid a$, there must be an integer b such that

$$a \equiv g^b \pmod{p}.$$

Moreover, any solution x of (7) cannot be divisible by p and thus must be of the form $x \equiv g^y \pmod{p}$. Congruence (7) is then the same as the congruence

$$g^{ny} \equiv g^b \pmod{p}, \tag{8}$$

and we must find solutions y. By Proposition 11, (8) is equivalent to

$$ny \equiv b \pmod{p-1}, \tag{9}$$

our old friend the linear congruence in one variable. We know that (9) is solvable for y if and only if $\gcd(n, p-1) \mid b$. Thus, we have proved

Theorem 12: Let p be a prime and a be an integer such that $p \nmid a$. Let g be a primitive root mod p and let $a \equiv g^b \pmod{p}$. Then the congruence $x^n \equiv a \pmod{p}$ is solvable if and only if $\gcd(n, p-1) \mid b$.

Example 13: Let us consider the case where $p = 23$. We observed in Example 10 that $g = -2 \equiv 21 \pmod{23}$ is a primitive root mod 23. Moreover, we know from Theorem 12 that if $a \equiv (-2)^b \pmod{23}$, then $x^n \equiv a \pmod{23}$ is solvable if and only if $\gcd(n, 22) \mid b$. Thus, for example, if $n = 2$, then b must be even. If $n = 11$, then b must be divisible by 11. Also if $2 \nmid n$ and $11 \nmid n$, any b works, and hence there is a solution for any a.

How do we actually compute solutions? To do this we must actually solve congruence (9). But before we can do this we must know what b is. Now b is determined by the condition that $a \equiv g^b \pmod{p}$, where a is given to us in advance. To do this for explicit a's we must then tabulate the powers of g. Using the primitive root $g = -2$ of Example 10 we tabulate in Table 3-2 the powers g^b for $0 \leq b \leq p - 2 = 21$, using the reduced residue system $1, 2, \ldots, p - 1$.

It is easy to solve equations of the form $x^n \equiv a \pmod{23}$ using Table 3-2.

TABLE 3-2

Powers of the Primitive Root -2 Modulo 23

b	0	1	2	3	4	5	6	7	8	9	10	11	12	13	14	15	16	17	18	19	20	21
g^b	1	21	4	15	16	14	18	10	3	17	12	22	2	19	8	7	9	5	13	20	6	11

For example, we solve

$$x^7 \equiv 17 \pmod{23}.$$

Looking at Table 3-2, we see that $17 \equiv (-2)^9 \pmod{23}$. We write $x \equiv (-2)^y \pmod{23}$. Then, $x^7 \equiv 17 \pmod{23}$ is the same as

$$(-2)^{7y} \equiv (-2)^9 \pmod{23}$$

or

$$7y \equiv 9 \pmod{22}.$$

We check easily that $7^* \equiv 19 \pmod{22}$ (noting that $3 \cdot 7 \equiv -1 \pmod{22}$), and so

$$y \equiv 19 \cdot 9 \equiv -27 \equiv 17 \pmod{22}.$$

Thus, $x \equiv (-2)^{17} \pmod{23}$ is a solution to the original congruence. Again using Table 3-2, we see that $x \equiv 5 \pmod{23}$ is a solution. It is the only solution. (Why?)

Theorem 12 has the disadvantage that one must first find a primitive root g mod p to calculate b. However, we may deduce from Theorem 12 the following simple criterion which does not suffer from this defect:

Theorem 14 (Euler's criterion): Let p be a prime and let a be an integer such that $p \nmid a$ and let n be positive. Set $s = \gcd(n, p - 1)$. Then the congruence $x^n \equiv a \pmod{p}$ is solvable if and only if $a^{(p-1)/s} \equiv 1 \pmod{p}$.

Proof: Let g be a primitive root mod p and write $a \equiv g^b \pmod{p}$.
First suppose that $x^n \equiv a \pmod{p}$ has a solution x. Then

$$a^{(p-1)/s} \equiv (x^n)^{(p-1)/s} = (x^{p-1})^{n/s} \pmod{p}.$$

(Recall that $s \mid n$, so that n/s is an integer.) By Fermat's Little Theorem, $x^{p-1} \equiv 1 \pmod{p}$, and so

$$a^{(p-1)/s} \equiv 1^{n/s} = 1 \pmod{p}.$$

Conversely, suppose that $a^{(p-1)/s} \equiv 1 \pmod{p}$. Then $a \equiv g^b \pmod{p}$ implies that

$$1 \equiv a^{(p-1/s} \equiv g^{b((p-1)/s)} \pmod{p}.$$

Thus, since $\text{ord}_p\, g = p - 1$ (Theorem 3), we conclude from Proposition 4(ii), that

$$p - 1 \mid b\frac{p-1}{s},$$

and thus b/s is an integer. That is, $s \mid b$, and so by Theorem 12, $x^n \equiv a \pmod{p}$ is solvable. ■

Let us specialize Theorem 14 to the case $n = 3$. If $p \equiv 2 \pmod{3}$, then $p - 1 \equiv 1 \pmod{3}$, so that $3 \nmid p - 1$ and $\gcd(3, p - 1) = 1$. If $p \equiv 1 \pmod{3}$, then $3 \mid p - 1$, so that $\gcd(3, p - 1) = 3$. Thus, we derive the following result:

Corollary 15: Let p be an odd prime, $p \neq 3$, and suppose that $3 \nmid a$. Then the congruence

$$x^3 \equiv a \pmod{p}$$

is always solvable if $p \equiv 2 \pmod{3}$. If $p \equiv 1 \pmod{3}$, then the congruence is solvable if and only if

$$a^{(p-1)/3} \equiv 1 \pmod{p}.$$

Proof: Use Theorem 14. If $p \equiv 2 \pmod{3}$, the congruence is solvable if and only if $a^{p-1} \equiv 1 \pmod{p}$, which holds by Fermat's Little Theorem. ■

3.6 Exercises

1. Let n be a positive integer (not necessarily a prime), and a an integer such that $\gcd(a, n) = 1$. Define the *order of a modulo n* to be the least positive integer k such that $a^k \equiv 1 \pmod{n}$ (write $k = \text{ord}_n\, a$). Prove the following analogues of the statements in Proposition 4:
   (a) $\text{ord}_n\, a$ exists.
   (b) $\text{ord}_n\, a$ divides $\phi(n)$.
   (c) If $a^v \equiv 1 \pmod{n}$, then $\text{ord}_n\, a \mid v$.
   (d) $\text{ord}_n(a^u) = (\text{ord}_n\, a)/\gcd(u, \text{ord}_n\, a)$.

2. Determine $\text{ord}_n\, a$ where
   (a) $n = 11,\ 1 \leq a \leq 10$.
   (b) $n = 13,\ 1 \leq a \leq 12$.
   (c) $n = 9,\ a = 1, 2, 4, 5, 7, 8$.
   (d) $n = 12,\ a = 1, 5, 7, 11$.
   (e) $n = 15,\ a = 1, 2, 4, 7, 8, 11, 13, 14$.
   Recall that Proposition 4 or Exercise 1 simplifies your work.
3. Let p be a prime. Prove that $\text{ord}_p\, a = 2$ if and only if $a \equiv -1 (\text{mod } p)$. Is this result true if p is not a prime?
4. Do Exercise 8 of Section 3 using the concept of order. That is, prove that $\text{ord}_n\, a = n - 1$ implies that n is a prime.
5. Prove that part (ii) follows directly from part (iii) of Proposition 4.
6. Prove Corollary 7.
7. Determine primitive roots for the following primes: $p = 11, 13, 19, 23, 29$.
8. Let p be a prime and let g be a primitive root mod p. For integers $n \geq 1$, show that g^n is a primitive root mod p if and only if $\gcd(n, p - 1) = 1$.
9. Let p be an odd prime, $p \nmid a$. Show that $a^{(p-1)/2} \equiv \pm 1 (\text{mod } p)$.
10. Let p be a prime. Let a be an integer such that $p \nmid a$. Call a a quadratic residue mod p if there is a solution to $x^2 \equiv a (\text{mod } p)$. Otherwise call a a quadratic nonresidue. Use Theorem 14 (Euler's criterion) and Exercise 9 to show that the product of two quadratic residues or of two quadratic nonresidues mod p is a quadratic residue mod p, whereas the product of a quadratic residue and quadratic nonresidue is a quadratic nonresidue mod p. Make up some numerical examples illustrating this result.
11. Use the existence of primitive roots to prove Wilson's theorem.
12. Let p be a prime and let a be an integer such that $p \nmid a$.
    (a) Show that if $\text{ord}_p\, a = nm$ and $\gcd(n, m) = 1$, then $a \equiv bc (\text{mod } p)$, where $\text{ord}_p\, b = n$ and $\text{ord}_p\, c = m$ (Use the note on p. 21.)
    (b) Show that if $\text{ord}_p\, a = n_1 n_2 \cdots n_k$ and the n_i are pairwise relatively prime, then $a \equiv b_1 b_2 \cdots b_k (\text{mod } p)$, where $\text{ord}_p\, b_i = n_i$. (*Hint:* Use induction.)
13. Let p be a prime. Show there are $\phi(p - 1)$ primitive roots mod p. (*Hint:* Use Exercise 8).
14. Let p be a prime and let n be a positive integer such that $n \mid p - 1$. Show that the number of integers a mod p such that $\text{ord}_p\, a = n$ is $\phi((p - 1)/n)$. (Look at Exercise 13).
15. (Indices) Let p be a prime and let g be a primitive root mod p. Then if a is any integer such that $p \nmid a$, we know that $a \equiv g^i (\text{mod } p)$ for some

integer i, $0 \leq i \leq p-1$. Call i the *index of a with respect to g* mod p. We write $i = \text{ind } a$ when p and g are understood. Prove

(a) $a \equiv b \pmod{p}$ if and only if $\text{ind } a \equiv \text{ind } b \pmod{p-1}$.
(b) $\text{ind } ab \equiv \text{ind } a + \text{ind } b \pmod{p-1}$.
(c) $\text{ind } a^* \equiv -\text{ind } a \pmod{p-1}$, where a^* denotes the arithmetic inverse of a mod p.

Note the analogy between indices and logarithms.

16. Compute tables of indices for $a = 1, \ldots, p-1$ for the primitive roots mod p you computed in Exercise 7.

17. Use the tables computed in Exercise 16 to compute the values mod p of the following numbers (your answer should lie between 1 and $p-1$):
    (a) $5 \cdot 6 \cdot 7 \cdot 8 \equiv ? \pmod{11}$.
    (b) $17^5 \cdot 14^9 \cdot 9^{21} \cdot 25 \equiv ? \pmod{29}$.
    (c) $15^4 \cdot (-8)^{10} \cdot 4^{11} \cdot 21^6 \equiv ? \pmod{13}$.

18. Find all solutions of the following congruences:
    (a) $x^5 \equiv 13 \pmod{23}$.
    (b) $x^{17} \equiv 50 \pmod{23}$.
    (c) $5x^9 \equiv 43 \pmod{23}$.
    (d) $x^{93} \equiv 100 \pmod{23}$.
    (e) $x^{10} \equiv 8 \pmod{23}$.
    (f) $x^{26} \equiv 10 \pmod{23}$.
    (*Note:* Use Table 3-2.)

19. Let p be a prime. An integer a such that $p \nmid a$ is called a fifth power residue if and only if the congruence $x^5 \equiv a \pmod{p}$ is solvable. Show that
    (a) If $\gcd(5, p-1) = 1$, then every integer is a fifth power residue.
    (b) If $5 \mid p-1$, then there are precisely $(p-1)/5$ fifth power residues mod p.

20. Let p be a prime and let $n \geq 1$, a be integers. Give a formula for the number of solutions of $x^n \equiv a \pmod{p}$ (assuming there are solutions).

In Exercises 21–29 we shall determine precisely which moduli n have primitive roots. By a primitive root mod n we mean an integer a such that $\gcd(a, n) = 1$ and $\text{ord}_n\, a = \phi(n)$. (See Ex. 1.) You should assume the formula for $\phi(n)$ given in Exercise 15 in Section 3.4. In these problems p denotes an *odd* prime.

21. Deduce the following congruence from the binomial theorem:

$$(a + p^k b)^{np^\ell} \equiv a^{np^\ell} + np^\ell a^{np^\ell - 1} p^k b \pmod{p^{2k+\ell}},\ k \geq 1.$$

22. (a) Let g be a primitive root mod p. Let $r = g + pt$. Show that there is a t such that

$$r^{p-1} = 1 + ps,$$

where $p \nmid s$. Conclude that r is a primitive root mod p^2.

(b) Conversely, show that if r is a primitive root modulo p^2, then $r^{p-1} = 1 + ps$ for some s such that $p \nmid s$.

**23. Show that the integer r of Exercise 22 is a primitive root mod p^k for all integers $k \geq 1$. Do this in the following steps:

(a) Let $m = \text{ord}_{p^k} r$. Show that $m = p^{\ell}d$, where $\ell \leq k - 1$ and $d | p - 1$.

(b) Looking at $(r^d)^{p^\ell}$ mod p, show that $d = p - 1$.

(c) Using the formula $r^{p-1} = 1 + ps$ $(p \nmid s)$, show that $\ell = k - 1$.

(d) Notice by parts (a)–(c) and Exercise 22 that, if r is a primitive root modulo p^2, then r is also a primitive root modulo p^k $(k \geq 1)$.

24. Determine primitive roots mod 9, 27, 81, 243, 25, 125.

25. Let p be an *odd* prime and let r be a primitive root mod p^k. If r is odd, let $s = r$, and if r is even, let $s = r + p^k$. Show that s is a primitive root mod $2p^k$.

26. Find a primitive root mod 50 and 98.

27. Show that 2 and 4 have primitive roots.

28. Use Exercise 3.11 to conclude that if $k \geq 3$, then there is no primitive root mod 2^k.

29. Now let $n = p_1^{a_1} \cdots p_t^{a_t}$. Let $M =$ the least common multiple of $\phi(p_1^{a_1}), \ldots, \phi(p_t^{a_t})$.

(a) Show that $a^M \equiv 1 (\text{mod } n)$ for all a such that $\gcd(a, n) = 1$.

(b) Show that in order for n to have a primitive root we must have $\phi(n) | M$, so that $n = 2p^k, p^k, 2$, or 4.

30. Using the primitive root obtained in Exercise 24, solve the following congruences:

(a) $x^{17} \equiv 50 (\text{mod } 81)$.

(b) $7x^{25} \equiv 10 (\text{mod } 81)$.

31. Here is a method for computing a table of the powers of $a (\text{mod } p)$. First make a table:

1	2	$\cdots$	$p - 1$
a	$2a$	$\cdots$	$(p - 1)a$ (mod p)

If $a^r \equiv j (\text{mod } p)$, $0 \leq j < p - 1$, we can find $a^{r+1} \equiv aj (\text{mod } p)$ from this table. Use this method to compute the powers of 13 modulo 23.

3.7 Congruences—Some Historical Notes

We have gone through this chapter without previously discussing to any serious extent the historical development of the theory of congruences. Let us rectify this now.

The essential idea behind congruences was already used in the seventeenth and eighteenth centuries, and many special facts concerning congruences were noted. For example, the congruences of Fermat, Euler, and Wilson date from this period. During the eighteenth century, further contributions were made by Lagrange and Legendre. However, the real birth of congruences as a coherent theory occurred in 1799, with the publication of Gauss' *Disquisitiones Arithmeticae*. It was Gauss who first systematically studied congruences for their own sake and who introduced the convenient notation still in use today. It also was Gauss who first posed the problem of solving the general polynomial congruence

$$a_0 + a_1x + a_2x^2 + \cdots + a_mx^m \equiv 0 \pmod{n}.$$

In his *Disquisitiones* and in later works, Gauss made a study of the general congruence as far as he could carry it, and then he launched into a deep study of congruences of the first, second, third, and fourth degrees. To solve congruences of the second degree, Gauss discovered and proved* the law of quadratic reciprocity, which will be the topic of the next chapter.

Cubic congruences were studied further by Gauss' star pupil Eisenstein, who in the 1840's proved a law of cubic reciprocity. Gauss, himself, proved a law of quartic reciprocity, having to do with the solution of congruences of the fourth degree.

Gauss' work provided the direction which much of number theory took for the entire nineteenth century. In an effort to reformulate and better understand Gauss' *Disquisitiones*, Lejeune Dirichlet wrote the definitive treatise in 1863 entitled *Vörlesungen über Zahlentheorie*. Over the years, Dirichlet added supplements to various editions of his book in order to reflect current research. In a very famous supplement, Dirichlet's successor Richard Dedekind reinterpreted Gauss' theory of congruences in terms of ideals, which had been introduced by Ernst Kummer in connection with his work on Fermat's Last Theorem. Dedekind's so-called *Twelfth Supplement* provides the first organized treatment of the subject known today as *algebraic number theory*. Thus, we see that Gauss' work led in a quite direct manner to the development of a whole new field of number theory.

---

*Actually parts of it were guessed by Euler, and the full law was conjectured by Legendre, although Gauss was unaware of their work.

Over the past fifty years, extensive research has been done on polynomial congruences, especially congruences in several variables. This work is very technical. Often simpleminded arithmetic questions about congruences in several variables become inseparable from rather highbrow mathematics such as algebraic geometry. Suffice it to say that congruences provide for extensive research opportunities, even today.

4

The Law of Quadratic Reciprocity

4.1 Introduction

In Chapter 3, we devoted a considerable amount of time to a discussion of the polynomial congruence

$$f(x) \equiv 0(\text{mod } m),$$

where $f(x)$ is a polynomial in one variable and m is a positive integer. We showed that if we could solve the congruence

$$f(x) \equiv 0(\text{mod } p)$$

for every prime p dividing m, then we could solve the original congruence modulo m. We then went on to discuss polynomial congruences modulo primes and discovered a number of properties which greatly facilitate their solution. However, when all was said and done, we gave no general method for solving congruences modulo a prime or even for determining whether or not a solution exists. (Of course, trial and error is a method which involves checking only p cases for a congruence modulo p, but we have in mind a somewhat more enlightening method than just checking a complete residue system.) The reason we gave no method for solving general polynomial congruences is that no such method is currently known. Indeed, as we mentioned in the preceding chapter, finding a general method for solving polynomial congruences modulo a prime is one of the most important unsolved problems in the theory of numbers. In general, the problem is not nearly solved. However, if we restrict our attention to a special class of congruences, it is sometimes possible to arrive at a satisfactory solution.

For example, we prescribed a method for solving linear congruences in the last chapter. Let us recall our results. In the case of linear congruences, $f(x) = ax + b$, where we may assume, without loss of generality, that $p \nmid a$. (Otherwise, our congruence is equivalent to the trivial congruence $b \equiv 0 \pmod{p}$.) Then the linear congruence $ax + b \equiv 0 \pmod{p}$ is always solvable and there is precisely one solution modulo p. This solution is given explicitly by

$$x \equiv -ba^{p-2} \pmod{p}.$$

(See the remark after Theorem 3.3.2). Thus, in the case of linear congruences, we have the best of all possible worlds: A solution always exists, and we can write down an explicit formula for the solution.

In this chapter, we shall discuss the next case in order of difficulty, namely the one in which $f(x)$ is a quadratic polynomial, say $f(x) = ax^2 + bx + c$. It will turn out that the theory of the congruence

$$ax^2 + bx + c \equiv 0 \pmod{p} \tag{1}$$

is much more complicated than the theory of linear congruences. First, no solution may exist. Second, even when it is known that a solution exists, it is not easy to compute it explicitly. However, there are some affirmative statements we can make. We shall describe a procedure for determining when congruence (1) has solutions and when it does not. This procedure, which is the main result of this chapter, makes use of Gauss' Law of quadratic reciprocity. We must emphasize, however, that although we shall always be able to determine, via the reciprocity law, whether (1) is solvable, our procedure will not give us any means better than trial and error for finding the solutions when they exist.

The law of quadratic reciprocity is one of the most celebrated and important results in all of number theory. In addition to its role in determining whether congruence (1) is solvable, it turns out to be the key tool, often unexpectedly, for solving many number-theoretic problems. We shall meet some of these applications in Chapter 6 and in the second half of this book.

Let us begin, then, by considering the general quadratic congruence

$$ax^2 + bx + c \equiv 0 \pmod{p}, \tag{2}$$

where p is a prime. Without loss of generality, assume that $p \nmid a$, for if $p \mid a$, then congruence (2) is equivalent to the linear congruence $bx + c \equiv 0 \pmod{p}$, which can be handled as we described above. If $p = 2$, then it is easy to solve (2) by trial and error. We leave the results to the exercises. Henceforth, let us assume that $p \neq 2$. (Unfortunately, the discussion which follows is not valid for $p = 2$.) Let us now try to solve (2) by imitating the procedure used in high school algebra for solving quadratic equations, namely, completing the square. Let us then recall the procedure for solving the equation

$$ax^2 + bx + c = 0,$$

where a, b, c are any real numbers, $a \neq 0$. We write

$$ax^2 + bx + c = a\left(x^2 + \frac{b}{a}x + \frac{c}{a}\right)$$

$$= a\left(\left(x + \frac{b}{2a}\right)^2 + \frac{c}{a} - \frac{b^2}{4a^2}\right).$$

Thus, $ax^2 + bx + c = 0$ if and only if

$$\left(x + \frac{b}{2a}\right)^2 = \frac{b^2}{4a^2} - \frac{c}{a} = \frac{b^2 - 4ac}{4a^2}.$$

Thus, in order to find x, we must look for the square root of $(b^2 - 4ac)/4a^2$. In terms of the square root, x may be given as

$$x = \frac{-b \pm \sqrt{b^2 - 4ac}}{2a}.$$

The last formula is the so-called *quadratic formula* of high school algebra. From the quadratic formula, we see that the main difficulty in solving the quadratic equation is in extracting the square root of $b^2 - 4ac$. If $b^2 - 4ac \geq 0$, then there exists a nonnegative real number α whose square is $b^2 - 4ac$, so that we may set $\alpha = \sqrt{b^2 - 4ac}$. However, if $b^2 - 4ac < 0$, then there is no real number α whose square is $b^2 - 4ac$. To extract $\sqrt{b^2 - 4ac}$ in the latter case, it is necessary to use complex numbers. However, if $b^2 - 4ac < 0$ and if we insist on solutions x which are real, then our original equations has no solutions. The situation for quadratic congruences will be similar to that for quadratic equations.

Let us now return to the quadratic congruence

$$ax^2 + bx + c \equiv 0 \pmod p,$$

where a, b, c are integers and $p \nmid a$ and $p \neq 2$. Let us mimic the method for solving quadratic equations. The first step was to factor out a. Instead of $1/a$, we need the arithmetic inverse a^* of a modulo p. We can find such an inverse since $p \nmid a$ (Proposition 3.2.8). Then a^* has the property $aa^* \equiv 1 \pmod p$, so that

$$ax^2 + bx + c \equiv_x a(x^2 + a^*bx + a^*c) \pmod p. \tag{3}$$

The next step is to complete the square inside the parentheses of (3). To do this, we need an inverse 2^* for 2 modulo p. We can find such an inverse since $p \neq 2$. In fact, we may take $2^* = (p + 1)/2$, since $2 \cdot 2^* = p + 1 \equiv 1 \pmod p$. Then

$$x^2 + a^*bx + a^*c \equiv_x (x + 2^*a^*b)^2 + (a^*c - 2^{*2}a^{*2}b^2) \pmod p. \tag{4}$$

(Multiply this out and check it.) Therefore, since $p \nmid a$, Euclid's lemma of Chapter 2 together with (3) and (4) imply that $ax^2 + bx + c \equiv 0 \pmod p$ if and only if

$$(x + 2^*a^*b)^2 \equiv 2^{*2}a^{*2}b^2 - a^*c \pmod p). \tag{5}$$

Set $y = x + 2^*a^*b$, $d = 2^{*2}a^{*2}b^2 - a^*c$. Then we can solve our original congruence for x if and only if we can solve the congruence

$$y^2 \equiv d(\text{mod } p)$$

for y.

Example 1: As an illustration of the above technique, let us solve

$$5x^2 + 9x + 11 \equiv 0(\text{mod } 13). \tag{6}$$

Note that since $8 \cdot 5 \equiv 1(\text{mod } 13)$, an inverse of 5 modulo 13 is 8; i.e., we may choose $5^* = 8$. Therefore, (6) is equivalent to

$$8 \cdot 5x^2 + 8 \cdot 9x + 8 \cdot 11 \equiv 0(\text{mod } 13)$$

or

$$x^2 + 7x + 10 \equiv 0(\text{mod } 13). \tag{7}$$

Then an inverse of 2 modulo 13 is 7, so that (7) is equivalent to

$$(x + 7 \cdot 7)^2 + 10 - (7 \cdot 7)^2 \equiv 0(\text{mod } 13)$$

or

$$(x + 10)^2 \equiv -1(\text{mod } 13).$$

Setting $y = x + 10$, we see that the problem of solving (6) is reduced to solving

$$y^2 \equiv -1(\text{mod } 13). \tag{8}$$

Therefore, we need to find the square roots of -1 modulo 13, if any exists. We can resort to trial and error. (Or, we can observe that since $13 \equiv 1 \pmod 4$, Theorem 3.3.5 implies that $((13 - 1)/2)! = 6!$ has the property $6!^2 \equiv -1(\text{mod } 13)$.) In any case, we observe that $y \equiv \pm 5(\text{mod } 13)$ are two solutions of (8), and by Corollary 3.5.9, (8) can have at most two solutions, so $y \equiv \pm 5(\text{mod } 13)$ are the only solutions of (8). Thus,

$$x = y - 10 \equiv -5, -2(\text{mod } 13)$$

are the solutions of the congruence (6). Indeed,

$$5 \cdot (-5)^2 + 9 \cdot (-5) + 11 = 125 - 45 + 11 = 91 \equiv 0(\text{mod } 13).$$

The reader should check that $x \equiv -2(\text{mod } 13)$ also is a solution.

Note that if we replace (6) by

$$5x^2 + 9x + 9 \equiv 0(\text{mod } 13),$$

then there would be no solutions since (8) would be replaced by

$$y^2 \equiv 2(\text{mod } 13),$$

which has no solutions. (Check this.)

The point of our discussion is that in order to solve the congruence

$$ax^2 + bx + c \equiv 0(\text{mod } p), \qquad p \nmid a, p \neq 2,$$

it always suffices to solve a congruence of the type

$$y^2 \equiv d \pmod{p}. \tag{9}$$

The remainder of this chapter will be devoted to a discussion of congruence (9). The point is that (9) is much simpler to deal with than the general quadratic congruence and, moreover, that the Law of Quadratic Reciprocity will yield a very simple, computationally viable procedure for determining whether or not (9) has a solution. We shall delay stating the Law of Quadratic Reciprocity for now, and we shall also postpone a discussion of its long history until we have developed a little more background.

4.1 Exercises

1. Reduce each of the following congruences to a congruence of the form $x^2 \equiv a \pmod{p}$:
   (a) $2x^2 + x + 3 \equiv 0 \pmod{5}$.
   (b) $8x^2 + 5x + 2 \equiv 0 \pmod{7}$.
   (c) $4x^2 + 2 \equiv 0 \pmod{17}$.
   (d) $14x^2 + 8x + 7 \equiv 0 \pmod{23}$.
   (e) $3x^2 + x + 9 \equiv 0 \pmod{11}$.
2. Solve parts (a)–(e) of Exercise 1 by using the results of Exercise 1 and by testing a complete residue system. (Proper choice of the complete residue system and a few observations should considerably lighten the computational burden.)
3. Let n be any odd positive integer (not necessarily prime). Show that the congruence $ax^2 + bx + c \equiv 0 \pmod{n}$ can be reduced to a congruence of the form $x^2 \equiv d \pmod{n}$ provided that $\gcd(a,n) = 1$.
4. Solve the following congruences:
   (a) $x^2 \equiv 2 \pmod{15}$.
   (b) $3x^2 + 2x + 5 \equiv 0 \pmod{22}$.
5. Show that the general cubic congruence $ax^3 + bx^2 + cx + d \equiv 0 \pmod{p}$ can be reduced to a congruence of the form $x^3 + rx + q \equiv 0 \pmod{p}$ provided that $p \neq 3$ and $p \nmid a$.
6. Let a be an odd integer. Completely determine all solutions of the congruence $ax^2 + bx + c \equiv 0 \pmod{2}$.

4.2 Basic Properties of Quadratic Residues

Let us now initiate a study of the congruence

$$x^2 \equiv a \pmod{p}, \tag{1}$$

where p is any odd prime and a is any integer. As we have seen in Section 1, all quadratic congruences modulo p can be reduced to congruence (1).

If $a \equiv 0(\text{mod } p)$, then we have seen that the only solution to (1) is $x \equiv 0(\text{mod } p)$. Therefore, let us henceforth assume that $p \nmid a$. For some values of a, (1) will have a solution, whereas for some other values of a, (1) will have no solution. Let us distinguish between these two sorts of a.

Definition 1: Let p be a prime, and let a be any integer such that $p \nmid a$. We say that a is a *quadratic residue modulo p* provided that

$$x^2 \equiv a(\text{mod } p)$$

has a solution. Otherwise, we say that a is a *quadratic nonresidue modulo p*.

Suppose that p is given. Let us consider the problem of determining all quadratic residues modulo p. If a is a quadratic residue modulo p, then $p \nmid a$ and $a \equiv x^2(\text{mod } p)$ for some x. However, since any integer is congruent to one of $0, 1, \ldots, p-1(\text{mod } p)$, we see that a must be congruent to one of

$$1^2, 2^2, \ldots, (p-1)^2(\text{mod } p).$$

If p is not too large, then this procedure can actually be used for computation.

Example 2: Let $p = 13$. Then a is a quadratic residue modulo 13 if and only if a is congruent to one of $1^2, 2^2, \ldots, 12^2(\text{mod } 13)$; that is, a is a quadratic residue modulo 13 if and only if $a \equiv 1, 4, 9, 3, 12, 10, 10, 12, 3, 9, 4$, or $1(\text{mod } 13)$. Thus, the quadratic residues modulo 13 are 1, 3, 4, 9, 10, 12. Hence, the quadratic nonresidues modulo 13 are 2, 5, 6, 7, 8, 11.

Note that in Example 2 the initial list of quadratic residues we obtain is symmetric, with each element of the list appearing exactly twice. This is a general phenomenon. Indeed, we have $p - x \equiv -x(\text{mod } p)$, so that $(p-x)^2 \equiv (-x)^2(\text{mod } p)$, and thus

$$(p-x)^2 \equiv x^2(\text{mod } p). \tag{2}$$

Therefore, if a is a quadratic residue modulo p, then a is congruent to one of

$$1^2, 2^2, \ldots, \left(\frac{p-1}{2}\right)^2(\text{mod } p).$$

Lemma 3: Suppose that $p > 2$ is a prime and a is a quadratic residue modulo p, $p \nmid a$. Then the congruence $x^2 \equiv a(\text{mod } p)$ has exactly two distinct solutions.

Proof: By the general theory of polynomial congruences the congruence $x^2 \equiv a(\text{mod } p)$ has at most two solutions (Corollary 3.5.9). Since a is a quadratic residue modulo p, there exists x_0 such that $x_0^2 \equiv a(\text{mod } p)$. Moreover, since $p \nmid a$, we have $p \nmid x_0$. Now $(-x_0)^2 \equiv x_0^2(\text{mod } p) \equiv a(\text{mod } p)$, so that

$-x_0$ is also a solution of the congruence $x^2 \equiv a \pmod{p}$. Moreover, $x_0 \not\equiv -x_0 \pmod{p}$, since $p > 2$ and $p \nmid x_0$. Therefore, the solutions x_0 and $-x_0$ are distinct, so that there are at least two solutions. We have proved that there are at most two solutions, so there are exactly two distinct solutions. ■

We proved above that every quadratic residue modulo p is congruent to one of

$$1^2, 2^2, \ldots, \left(\frac{p-1}{2}\right)^2 \pmod{p}.$$

Actually no two of these numbers can be congruent modulo p, for if $x_0^2 \equiv y_0^2 \pmod{p}$ for $1 \leq x_0 < y_0 \leq (p-1)/2$, the congruence $x^2 \equiv a \pmod{p}$ with $a = y_0^2$ has the solutions $x = x_0, y_0, p - x_0$ (Eq. (2)). Moreover, these solutions are distinct (exercise). However, the congruence cannot have three distinct solutions by Lemma 3. Therefore, $x_0^2 \not\equiv y_0^2 \pmod{p}$, and no two of the integers

$$1^2, 2^2, \ldots, \left(\frac{p-1}{2}\right)^2$$

can be congruent modulo p. As a consequence of what we have just proved, we note that a quadratic residue modulo p is congruent to one and only one of the numbers

$$1^2, 2^2, \ldots, \left(\frac{p-1}{2}\right)^2.$$

Let us summarize our observations in a proposition.

Proposition 4: Let p be an odd prime, $p \nmid a$. Then a is a quadratic residue modulo p if and only if a is congruent to one of

$$1^2, 2^2, \ldots, \left(\frac{p-1}{2}\right)^2 \tag{3}$$

modulo p. No two of the integers (3) are congruent modulo p. Hence, among the integers $1, 2, \ldots, p-1$, precisely $(p-1)/2$ are quadratic residues modulo p and precisely $(p-1)/2$ are quadratic nonresidues modulo p.

We shall find the following notation very convenient:

Definition 5: Let p be an odd prime and a an integer such that $p \nmid a$. Let us define the *Legendre symbol* $\left(\frac{a}{p}\right)$ as follows: Set

$$\left(\frac{a}{p}\right) = \begin{cases} +1 & \text{if } a \text{ is a quadratic residue modulo } p, \\ -1 & \text{if } a \text{ is a quadratic nonresidue modulo } p. \end{cases}$$

The reader should not confuse the Legendre symbol $\left(\frac{a}{p}\right)$ with the fraction a/p. From our above calculation of the quadratic residues modulo 13

(Example 2), we see that

$$\left(\tfrac{2}{13}\right) = -1, \qquad \left(\tfrac{3}{13}\right) = 1, \qquad \left(\tfrac{4}{13}\right) = 1, \qquad \left(\tfrac{5}{13}\right) = -1.$$

Moreover, since $18 \equiv 5 \pmod{13}$ and 5 is a quadratic nonresidue modulo 13, so is 18, and thus

$$\left(\tfrac{18}{13}\right) = -1.$$

The Legendre symbol was first introduced by the French mathematician Legendre in the eighteenth century in order to facilitate computations with quadratic residues. In fact, we shall state the quadratic reciprocity law in terms of the properties of the Legendre symbol. Let us first establish some elementary properties of the Legendre symbol.

Proposition 6: Let p be an odd prime and let a and b be integers such that $p \nmid a$ and $p \nmid b$. Then the following results hold:

(i) $\left(\dfrac{a^2}{p}\right) = 1.$

(ii) $\left(\dfrac{1}{p}\right) = 1.$

(iii) If $a \equiv b \pmod p$, then $\left(\dfrac{a}{p}\right) = \left(\dfrac{b}{p}\right)$.

Proof:

(i) The conguence $x^2 \equiv a^2 \pmod p$ has as a solution $x = a$.

(ii) Set $a = 1$ in result (i).

(iii) If $a \equiv b \pmod p$, then the solutions of $x^2 \equiv a \pmod p$ are the same as the solutions of $x^2 \equiv b \pmod p$. Therefore, the first congruence has solutions if and only if the second does. Thus, $\left(\dfrac{a}{p}\right) = \left(\dfrac{b}{p}\right)$. ■

The properties of the Legendre symbol given in Proposition 6 are very elementary. However, a property of the symbol which is by no means obvious is the following result:

Theorem 7 (Euler's criterion): Let p be an odd prime and let a be an integer such that $p \nmid a$. Then

$$\left(\frac{a}{p}\right) \equiv a^{(p-1)/2} \pmod p.$$

Proof: By Fermat's Little Theorem (Theorem 3.3.1), we have $(a^{(p-1)/2})^2 = a^{p-1} \equiv 1 \pmod p$. Thus, if $h = a^{(p-1)/2}$, then $h^2 \equiv 1 \pmod p$, and so $p \mid (h-1)(h+1)$. Therefore, $p \mid h - 1$ or $p \mid h + 1$, and hence $h = a^{(p-1)/2} \equiv \pm 1 \pmod p$. Now, p is odd, and thus Theorem 3.6.14 implies that $\left(\dfrac{a}{p}\right)$

$= +1$ if and only if $a^{(p-1)/2} \equiv 1 \pmod p$. Consequently, $\left(\frac{a}{p}\right) = \pm 1$ if and only if $a^{(p-1)/2} \equiv \pm 1 \pmod p$, respectively. ■

Theorem 7 is due to the Swiss mathematician Leonhard Euler, who proved it in 1755. Euler's criterion will be an extremely useful technical device in proving various properties of the Legendre symbol. However, as far as yielding a simple test for determining the solvability of $x^2 \equiv a \pmod p$, Euler's criterion leaves much to be desired, for to use Euler's criterion, we would need to compute $a^{(p-1)/2} \pmod p$ in order to determine $\left(\frac{a}{p}\right)$. However, let us now demonstrate two easy consequences of Euler's criterion, which should suffice to demonstrate its usefulness.

Corollary 8: Let p be an odd prime, and let a and b be integers such that $p \nmid a, p \nmid b$. Then

$$\left(\frac{ab}{p}\right) = \left(\frac{a}{p}\right)\left(\frac{b}{p}\right).$$

Proof: By Euler's criterion

$$\left(\frac{ab}{p}\right) \equiv (ab)^{(p-1)/2} = a^{(p-1)/2} b^{(p-1)/2} \equiv \left(\frac{a}{p}\right)\left(\frac{b}{p}\right) \pmod p.$$

Since the numbers $\left(\frac{ab}{p}\right)$ and $\left(\frac{a}{p}\right)\left(\frac{b}{p}\right)$ can be equal only to 1 or -1 and since they are congruent modulo p with $p > 2$, we see that

$$\left(\frac{ab}{p}\right) = \left(\frac{a}{p}\right)\left(\frac{b}{p}\right),$$

as desired. ■

It is an immediate consequence of Corollary 8 that (i) the product of two quadratic residues modulo p is a quadratic residue modulo p, (ii) the product of two quadratic nonresidues modulo p is a quadratic residue modulo p, and (iii) the product of a quadratic residue and a quadratic nonresidue is a quadratic nonresidue.* These results are at once striking and also very useful. Let us check these facts with a numerical example modulo 13. (See Example 2.) By our previous calculations, 3 and 12 are quadratic residues modulo 13, and, indeed, $3 \cdot 12 = 36 \equiv 6^2 \pmod{13}$ is a quadratic residue. However, 2 and 5 are quadratic nonresidues, and $2 \cdot 5 = 10 \equiv 6^2 \pmod{13}$ is a quadratic residue. Finally, 7 is a nonresidue, and 10 is a residue, and indeed, $7 \cdot 10 = 70 \equiv 5 \pmod{13}$ is a nonresidue.

---

*Note that (i) and (iii) are easy to prove directly. However, (ii) uses the full strength of Corollary 8.

Another consequence of Euler's criterion is a result which we proved in Chapter 3, although we did not state it then in terms of quadratic residues or Legendre symbols. Namely we have the following result:

Corollary 9: Let p be an odd prime. Then

$$\left(\frac{-1}{p}\right) = (-1)^{(p-1)/2}.$$

In other words,

$$\left(\frac{-1}{p}\right) = \begin{cases} +1 & \text{if } p \equiv 1 (\text{mod } 4), \\ -1 & \text{if } p \equiv 3 (\text{mod } 4). \end{cases}$$

Proof: By Euler's criterion,

$$\left(\frac{-1}{p}\right) \equiv (-1)^{(p-1)/2} (\text{mod } p).$$

Therefore, since $\left(\frac{-1}{p}\right) = \pm 1$ and since $p > 2$, we have the desired result. ■

Note that Corollary 9 is equivalent to Theorem 3.3.5. Fermat knew of Corollary 9 in the early seventeenth century. However, it was first proved, only with great difficulty, by Euler in 1749. Since Euler did not discover his simple criterion until 1755, he had to devise other methods to prove Corollary 9. These other methods were very laborious. It was not until 1773 that Lagrange observed that one could write down explicitly the square root of $-1 (\text{mod } p)$ for $p \equiv 1 (\text{mod } 4)$, namely $((p - 1)/2)!$, using Wilson's theorem. Thus, our earlier proof of Corollary 9 is due to Lagrange.

Let us conclude this section with a few examples of the properties of the Legendre symbol we have proved, as well as a few comments.

Example 10: By Proposition 6, part (i), Corollary 8, and Corollary 9, we have

$$\left(\frac{-a^2}{p}\right) = \left(\frac{-1}{p}\right)\left(\frac{a^2}{p}\right) = \left(\frac{-1}{p}\right) = (-1)^{(p-1)/2}.$$

Therefore, the congruence $x^2 \equiv -a^2 (\text{mod } p)$ is solvable if and only if $p \equiv 1 (\text{mod } 4)$.

Example 11: Can we solve the congruence $x^2 \equiv 19 (\text{mod } 23)$? Well, since $19 \equiv -4 (\text{mod } 23)$, we have

$$\left(\frac{19}{23}\right) = \left(\frac{-4}{23}\right) = \left(\frac{-1}{23}\right)\left(\frac{2}{23}\right)^2 = -1,$$

since $23 \equiv 3 (\text{mod } 4)$. Thus, $x^2 \equiv 19 (\text{mod } 23)$ is not solvable.

Example 12: How do we go about computing $\left(\frac{a}{p}\right)$ for $p \nmid a$? Suppose that $a = \pm p_1^{a_1} \cdots p_t^{a_t}$, where $p_1, \ldots, p_t$ are distinct primes. Since $p \nmid a$, we see that $p \neq p_i$. Then by Corollary 7, we have

$$\left(\frac{a}{p}\right) = \left(\frac{\pm 1}{p}\right)\left(\frac{p_1}{p}\right)^{a_1} \cdots \left(\frac{p_t}{p}\right)^{a_t}. \tag{4}$$

Thus, for example, if $p = 5$ and $a = -24$, then

$$\left(\frac{-24}{5}\right) = \left(\frac{-1}{5}\right)\left(\frac{2}{5}\right)^3\left(\frac{3}{5}\right)$$
$$= 1 \cdot (-1)^3 \cdot (-1) = 1.$$

The point of the representation (4) is that it shows that in order to compute $\left(\frac{a}{p}\right)$ it suffices to be able to compute $\left(\frac{q}{p}\right)$ where p and q are distinct primes. This is precisely what the law of quadratic reciprocity will allow us to do.

4.2 Exercises

1. Show that the congruence $x^2 \equiv 0 \pmod{p}$ has only one solution mod p.
2. Determine all the quadratic residues and quadratic nonresidues modulo p for
   (a) $p = 5$.
   (b) $p = 7$.
   (c) $p = 11$.
   (d) $p = 17$.
3. Compute the following:
   (a) $\left(\frac{2}{3}\right)$.
   (b) $\left(\frac{7}{5}\right)$.
   (c) $\left(\frac{11}{7}\right)$.
   (d) $\left(\frac{9}{17}\right)$.
   (e) $\left(\frac{-1}{11}\right)$.
   (f) $\left(\frac{2}{17}\right)$.
   (g) $\left(\frac{2}{11}\right)$.
   (h) $\left(\frac{16}{11}\right)$.
4. Use the results of Exercise 2 to verify that
   (a) $\left(\frac{2}{5}\right)\left(\frac{3}{5}\right) = \left(\frac{6}{5}\right)$.
   (b) $\left(\frac{7}{17}\right)\left(\frac{5}{17}\right) = \left(\frac{35}{17}\right)$.
5. Check numerically that
   $$\left(\frac{6}{17}\right) \equiv 6^{(17-1)/2} \pmod{17}.$$
6. Prove that a primitive root modulo p cannot be a quadratic residue modulo p.

7. Show that the congruence $x^2 \equiv a(\text{mod } p)$ $(p \nmid a)$ has either two solutions or none by using the existence of a primitive root modulo p.
8. Solve the congruence $x^2 \equiv 38(\text{mod } 47)$. (*Hint:* $38 \equiv -9(\text{mod } 47)$.)
9. Suppose that n is a positive integer and that $n = m^2k$, where k is not divisible by any perfect square. (k is called the *square free* part of n.) Show that if $p \nmid n$, then $p \nmid k$ and

$$\left(\frac{n}{p}\right) = \left(\frac{k}{p}\right).$$

10. Let p be an odd prime. Define quadratic residues and quadratic nonresidues modulo p^a in the obvious way. Show that the product of two quadratic residues modulo p^a is a quadratic residue, the product of two quadratic nonresidues is a quadratic residue, and the product of a quadratic residue and a quadratic nonresidue is a quadratic nonresidue. (*Hint:* By Exercise 3.6.23 there is a primitive root mod p^a.)
11. Show that the results of Exercise 10 are not necessarily true if p^a is replaced by an arbitrary integer n. (*Hint:* Look at modulo 15.)
12. Show that the sum of the quadratic residues modulo p is divisible by p.
13. Show that

$$\left(\frac{1}{p}\right) + \left(\frac{2}{p}\right) + \cdots + \left(\frac{p-1}{p}\right) = 0.$$

14. Prove Proposition 6, part (iii), using the Euler criterion.

For Exercises 15 and 16,* note the following: Let p be an odd prime. We showed that there are $(p - 1)/2$ quadratic nonresidues mod p and there are $\phi(p - 1)$ primitive roots between 0 and $p - 1$. Also, by Exercise 6, every primitive root is a quadratic nonresidue.

15. (a) Suppose that $p = 2q + 1$, where q is a prime. Then there are q quadratic nonresidues and $q - 1$ primitive roots. Therefore, the primitive roots are exactly the quadratic nonresidues with one exception. Show that $2q \equiv -1(\text{mod } p)$ is the exception.
    (b) Use part (a) to calculate the primitive roots modulo 7, 11, 23, 47.
16. Let p be an odd prime. Show that the number of quadratic residues modulo p^n (i.e., integers a for which the congruence $x^2 \equiv a(\text{mod } p^n)$ is solvable) equals

$$\frac{p^{n+1} - p}{2(p+1)} + 1, \qquad n \text{ even}$$

$$\frac{p^{n+1} - 1}{2(p+1)} + 1, \qquad n \text{ odd}.$$

---

*The authors would like to thank Dr. John Hemperly for pointing out these exercises.

4.3 The Gauss Lemma

Let us now give a criterion for a given integer a to be a quadratic residue modulo a prime p. The criterion we shall establish is a rather remarkable one, due to Gauss. At first it will look very strange. Indeed, it took Gauss 10 years to discover this property of quadratic residues. It is by no means obvious that it ought even be true. However, not only is it correct, but it will provide us with the key to an elementary and relatively simple proof of the quadratic reciprocity law. Let us state Gauss' criterion.

Theorem 1 (Gauss' lemma): Let p be an odd prime and let a be an integer such that $p \nmid a$. Look at the list of integers $a, 2a, 3a, \ldots, ((p-1)/2)a$. Replace each integer in the list by the one congruent to it modulo p which lies between $-(p-1)/2$ and $(p-1)/2$. Let ν be the number of negative integers in the resulting list. Then

$$\left(\frac{a}{p}\right) = (-1)^{\nu}.$$

Before we proceed with the simple proof, let us look at an example.

Example 2: Let us work again with $p = 13$, so that $(p-1)/2 = 6$. Let $a = 5$. Then we look at the list

$$5, \quad 10, \quad 15, \quad 20, \quad 25, \quad 30.$$

By subtracting an appropriate multiple of 13 from each element in the list, we ensure that all remainders lie between -6 and 6 and replace the list by

$$5, \quad -3, \quad 2, \quad -6, \quad -1, \quad 4.$$

Then $\nu = 3$ and $(\frac{5}{13}) = (-1)^3 = -1$. If $a = 3$, then we look at

$$3, \quad 6, \quad 9, \quad 12, \quad 15, \quad 18,$$

which are replaced by

$$3, \quad 6, \quad -4, \quad -1, \quad 2, \quad 5,$$

and so $\nu = 2$ and $(\frac{3}{13}) = (-1)^2 = 1$. These results coincide with what we observed in Example 2.2.

Proof of Theorem 1: First observe that if $k \neq k'$ and $1 \leq k, k' \leq (p-1)/2$, then

$$ka \not\equiv k'a (\text{mod } p)$$

and

$$ka \not\equiv -k'a (\text{mod } p).$$

Indeed, if $ka \equiv k'a (\text{mod } p)$, then $k \equiv k' (\text{mod } p)$. But then $k = k'$, since $1 \leq k, k' \leq (p-1)/2$, which is a contradiction. On the other hand, if $ka \equiv -k'a \pmod p$, then $k + k' \equiv 0 \pmod p$. However, since $1 \leq k, k' \leq (p-1)/2$, we have $1 \leq k + k' < p$, so that $k + k' \not\equiv 0 \pmod p$.

For each k such that $1 \leq k \leq (p-1)/2$, let r_k be the replacement of $ka \pmod p$. That is, choose r_k such that

$$-\frac{p-1}{2} \leq r_k \leq \frac{p-1}{2}$$

$$r_k \equiv ka \pmod p.$$

Then the replaced list referred to in Gauss' lemma is just

$$r_1, r_2, \ldots, r_{(p-1)/2}. \tag{*}$$

Assume that $k \neq k'$. From our reasoning above, we have $r_k \not\equiv r_{k'} \pmod p$ and $r_k \not\equiv -r_{k'} \pmod p$, so that $r_k \neq \pm r_{k'}$. In other words, $|r_k| \neq |r_{k'}|$. Since there are $(p-1)/2$ numbers

$$|r_1|, \ldots, |r_{(p-1)/2}|$$

lying between 1 and $(p-1)/2$ and they are all different, they must consist of the numbers $1, 2, \ldots, (p-1)/2$, in some order. Therefore, since the list $(*)$ has ν negative entries, we have

$$r_1 r_2 \cdots r_{(p-1)/2} = (-1)^\nu 1 \cdot 2 \cdots \frac{p-1}{2}. \tag{1}$$

On the other hand, since $r_k \equiv ka \pmod p$, we have

$$r_1 \cdots r_{(p-1)/2} \equiv a(2a) \cdots \left(\frac{p-1}{2}a\right) \equiv a^{(p-1)/2} \cdot 1 \cdot 2 \cdots \frac{p-1}{2} \pmod p. \tag{2}$$

Comparing (1) and (2), we have

$$(-1)^\nu \cdot 1 \cdot 2 \cdots \frac{p-1}{2} \equiv a^{(p-1)/2} \cdot 1 \cdot 2 \cdots \frac{p-1}{2} \pmod p,$$

so that, cancelling the factor $1 \cdot 2 \cdots (p-1)/2$, we have

$$a^{(p-1)/2} \equiv (-1)^\nu \pmod p. \tag{3}$$

Now by Euler's criterion (Theorem 2.7) and (3), we see that

$$\left(\frac{a}{p}\right) \equiv (-1)^\nu \pmod p.$$

Thus,

$$\left(\frac{a}{p}\right) = (-1)^\nu,$$

since $p > 2$. ■

Let us now give examples of how the Gauss lemma may be used to calculate $\left(\frac{a}{p}\right)$.

Example 3: Let us begin with a simple example. Let us apply Gauss' lemma to show (for the third time) that

$$\left(\frac{-1}{p}\right) = (-1)^{(p-1)/2}.$$

Here $a = -1$, and we look at the list

$$1\cdot(-1), 2\cdot(-1), \ldots, \frac{p-1}{2}\cdot(-1).$$

All these numbers are negative and lie between $-(p-1)/2$ and $(p-1)/2$. Therefore, $\nu = (p-1)/2$, as desired.

Example 4: Let us now do the less trivial example $\left(\frac{2}{p}\right)$. Here $a = 2$, and we look at the list

$$1\cdot 2, 2\cdot 2, 3\cdot 2, \ldots, \frac{p-1}{2}\cdot 2.$$

All these numbers lie between 1 and p. Moreover, it is clear that those that become negative when replaced are those lying between $p/2$ and p. (Carry out explicit calculations for $p = 13, 17$ if you do not see this.) Therefore, ν is the number of integers k such that

$$\frac{p}{2} \leq 2k \leq p,$$

or equivalently,

$$\frac{p}{4} \leq k \leq \frac{p}{2}.$$

Write $p = 8m + r$, where $0 \leq r < 8$. Since p is odd, we must have $r = 1, 3, 5$, or 7. Thus, we must count the number of integers k such that

$$2m + \frac{r}{4} \leq k \leq 4m + \frac{r}{2}.$$

We merely check the four possibilities for r. When $r = 1$, we want to count the k such that

$$2m + \tfrac{1}{4} \leq k \leq 4m + \tfrac{1}{2},$$

and these integers are $2m + 1, 2m + 2, \ldots, 4m$. Thus, there are $2m$ of them, and $\nu = 2m$. In particular, we see that when $p \equiv 1 \pmod 8$, then $\nu = 2m$ is even, and so

$$\left(\frac{2}{p}\right) = (-1)^{\nu} = (-1)^{2m} = 1.$$

If $r = 3$, then the range for k is

$$2m + \tfrac{3}{4} \leq k \leq 4m + \tfrac{3}{2},$$

and thus the relevant values of k are $2m+1, \ldots, 4m, 4m+1$. Therefore, $\nu = 2m+1$ is odd, and $\left(\frac{2}{p}\right) = (-1)^\nu = -1$. For $r = 5$, the range is

$$2m + \tfrac{5}{4} \leq k \leq 4m + \tfrac{5}{2},$$

giving $k = 2m+2, \ldots, 4m+2$, and so $\nu = 2m+1$. Therefore, if $p \equiv 5 \pmod 8$, then $\left(\frac{2}{p}\right) = (-1)^{2m+1} = -1$. Finally, $r = 7$ gives $2m + \frac{7}{4} \leq k \leq 4m + \frac{7}{2}$ or $k = 2m+2, \ldots, 4m+2, 4m+3$, or $\nu = 2m+2$. Thus, $p \equiv 7 \pmod 8$ implies that $\left(\frac{2}{p}\right) = 1$. Thus, we have completely determined the value of $\left(\frac{2}{p}\right)$ for all odd primes p. Let us summarize our result in a separate theorem, since our present result is usually stated as part of the Law of Quadratic Reciprocity but will not follow from our reasoning to be given in Section 4.

Theorem 5: Let p be an odd prime. Then

$$\left(\frac{2}{p}\right) = +1 \qquad \text{if and only if } p \equiv \pm 1 \pmod 8$$

$$\left(\frac{2}{p}\right) = -1 \qquad \text{if and only if } p \equiv \pm 5 \pmod 8.$$

Thus, for example, $\left(\frac{2}{13}\right) = -1$ and $\left(\frac{2}{17}\right) = 1$ as $13 \equiv 5 \pmod 8$ and $17 \equiv 1 \pmod 8$.

If $p \equiv \pm 1 \pmod 8$, then $(p^2 - 1)/8$ is even, whereas if $p \equiv \pm 5 \pmod 8$, then $(p^2 - 1)/8$ is odd. Therefore, Theorem 5 may be rewritten in the form

Corollary 6: Let p be an odd prime. Then

$$\left(\frac{2}{p}\right) = (-1)^{(p^2-1)/8}.$$

The net goal of this chapter is to give a criterion for determining whether or not the congruence $x^2 \equiv a \pmod p$ is solvable. And such a criterion just amounts to specifying a method for calculating the Legendre symbol $\left(\frac{a}{p}\right)$. We shall demonstrate with two rather lengthy examples that the Gauss lemma provides such a method. These examples are worthwhile even though the Quadratic Reciprocity Law will supersede them, since Gauss' lemma is the crucial ingredient in the proof of that law. We hope that the examples will give further insight into the behavior of the Legendre symbol.

Example 7: Let $a = 3$, and let p be an odd prime such that $p \nmid a$. Then p must be greater than 3. We shall determine $\left(\frac{3}{p}\right)$. We saw in Example 4 (for

$a = 2$) that the value of $\left(\frac{2}{p}\right)$ depends on the residue class of p modulo 8. Similarly, the value of $\left(\frac{3}{p}\right)$ depends on the residue class of p modulo $4a = 12$. In the Gauss lemma, we look at the list

$$1\cdot 3, 2\cdot 3, 3\cdot 3, \ldots, \frac{p-1}{2}\cdot 3.$$

All these numbers lie between 1 and $3p/2$. When the list is replaced, it is clear that the integers between 1 and $p/2$ will be replaced by positive integers, those between $p/2$ and p will be replaced by negative integers, and those between p and $3p/2$ will be replaced by positive integers. Therefore, $\nu =$ the number of integers k such that

$$\frac{p}{2} \leq 3k \leq p$$

or

$$\frac{p}{6} \leq k \leq \frac{p}{3}.$$

Write $p = 12m + r$ where* $r = 1, 5, 7$, or 11. Then ν is the number of integers k such that

$$2m + \frac{r}{6} \leq k \leq 4m + \frac{r}{3}.$$

If we check the values $r = 1, 5, 7, 11$ explicitly and reason as in Example 4, we derive the results in Table 4-1. Thus, for $p > 3$,

$$\left(\frac{3}{p}\right) = +1 \quad \text{if and only if} \quad p \equiv \pm 1 (\text{mod } 12).$$

TABLE 4-1

Value of $\left(\frac{3}{p}\right)$, $p > 3$

r	Range of k	ν	Parity of ν	$\left(\frac{3}{p}\right)$
1	$2m + 1$ to $4m$	$2m$	Even	$+1$
5	$2m + 1$ to $4m + 1$	$2m + 1$	Odd	-1
7	$2m + 2$ to $4m + 2$	$2m + 1$	Odd	-1
11	$2m + 2$ to $4m + 3$	$2m + 2$	Even	$+1$

*If $r = 0, 2, 3, 4, 6, 8, 9, 10$, then $p = 12m + r$ would not be prime for $m \geq 1$, and for $m = 0$ could be prime only for $r = 3$, but $p = 3$ has been excluded.

Before we do a more elaborate example, let us make a few general observations. First, observe that we do not need to determine ν explicitly. We need only know whether ν is even or odd. That is, we must determine the parity of ν.

Next, we observe that a positive integer n with $p \nmid n$ will be replaced by a negative number when it is reduced modulo p to lie between $-(p-1)/2$ and $(p-1)/2$ if and only if

$$\frac{p}{2} \leq n \leq p$$

or

$$\tfrac{3}{2}p \leq n \leq 2p$$

or

$$\tfrac{5}{2}p \leq n \leq 3p,$$

and so forth. To put it another way, n is replaced by a negative integer if and only if there is an integer $t \geq 1$ such that

$$\frac{2t-1}{2}p \leq n \leq tp.$$

Since $2t-1$ and p are odd, we see that $(2t-1)p/2$ is not an integer, and thus it will be necessary to count the number of integers k such that $\alpha \leq k \leq \beta$, where α and β are given real numbers and α is *not* an integer. We shall now prove a lemma (Lemma 9), collecting all the facts we need to facilitate the determination of the parity of the number of such integers.

Definition 8: Let α be a real number. Let us denote by $[\alpha]$ the largest integer $\leq \alpha$. Therefore, $[3] = 3$, $[\pi] = 3$, $[\frac{27}{4}] = 6$. Also, given any real number α, let $\alpha_1 = \alpha - [\alpha]$, so that $0 \leq \alpha_1 < 1$. Then α_1 is called the *fractional part* of α.

Lemma 9: Let α and β be real numbers, α not an integer, $\alpha \leq \beta$. Then

(i) The number of integers k such that $\alpha \leq k \leq \beta$ is $[\beta] - [\alpha]$.

(ii) If n is an integer, then $[n + \beta] = n + [\beta]$.

(iii) If $n_1 \leq n_2$ are integers, then the number of integers k satisfying $2n_1 + \alpha \leq k \leq 2n_2 + \beta$ and the number satisfying $\alpha \leq k \leq \beta$ have the same parity.

Proof:

(i) These integers k are precisely $[\alpha] + 1, [\alpha] + 2, \ldots, [\beta]$ (since α is not an integer), and there are $[\beta] - [\alpha]$ of them.

(ii) Let $\beta_1 = \beta - [\beta]$ be the fractional part of β. Then $n + \beta = (n + [\beta]) + \beta_1$ and $n + [\beta]$ is an integer, so the assertion is clear.

(iii) By parts (i) and (ii), the two numbers of part (iii) are $[2n_2 + \beta] - [2n_1 + \alpha] = 2n_2 + [\beta] - 2n_1 - [\alpha]$ and $[\beta] - [\alpha]$, which clearly have the same parity. ∎

Let us close with a fairly complicated example.

Example 10: Let us compute $\left(\frac{7}{p}\right)$ for all primes $p \neq 2, 7$. Analogous to Examples 3, 4, 7, the answer will depend on the residue of p modulo $4 \cdot 7 = 28$. In Gauss' lemma, we look at the integers

$$1 \cdot 7, 2 \cdot 7, \ldots, \frac{p-1}{2} \cdot 7.$$

We must determine the parity of the total number of integers in the intervals

$$\frac{p}{2} \leq 7k \leq p,$$

$$\tfrac{3}{2}p \leq 7k \leq 2p,$$

$$\tfrac{5}{2}p \leq 7k \leq 3p.$$

These are the only intervals to be considered since $1 \leq k \leq p/2$ implies that $7k \leq 7p/2$, and the next interval beyond those given would be

$$\tfrac{7}{2}p \leq 7k \leq 4p,$$

which lies outside our range for $7k$. Thus, the three intervals for k are

$$\frac{p}{14} \leq k \leq \frac{p}{7},$$

$$\tfrac{3}{14}p \leq k \leq \tfrac{2}{7}p,$$

$$\tfrac{5}{14}p \leq k \leq \tfrac{3}{7}p.$$

Write $p = 28m + r$, where r is one of the 12 numbers 1, 3, 5, 9, 11, 13, 15, 17, 19, 23, 25, 27. Then our intervals for k are

$$2m + \frac{r}{14} \leq k \leq 4m + \frac{r}{7},$$

$$6m + \frac{3r}{14} \leq k \leq 8m + \frac{2r}{7},$$

$$10m + \frac{5r}{14} \leq k \leq 12m + \frac{3r}{7}.$$

By Lemma 9, part (iii), we get the same parity if we count the total number of integers in the three intervals

$$\frac{r}{14} \leq k \leq \frac{r}{7},$$

$$\frac{3r}{14} \leq k \leq \frac{2r}{7},$$

$$\frac{5r}{14} \leq k \leq \frac{3r}{7}.$$

Now we simply check the 12 values of r. For example, if $r = 1$, then there are no k's in any of the intervals, and thus $\nu \equiv 0 \pmod 2$ is even. Or if $r = 13$, there is precisely one k in each of the three intervals ($k = 1, 3, 5$, respectively), and thus $\nu \equiv 3 \pmod 2$ is odd. The results are tabulated in Table 4-2.

TABLE 4-2

Value of $\left(\frac{7}{p}\right)$

$p \equiv r \pmod{28}$	Number of k's	ν	$\left(\frac{7}{p}\right)$
1	0	Even	+1
3	0	Even	+1
5	1	Odd	−1
9	2	Even	+1
11	3	Odd	−1
13	3	Odd	−1
15	3	Odd	−1
17	3	Odd	−1
19	4	Even	+1
23	5	Odd	−1
25	6	Even	+1
27	6	Even	+1

4.3 Exercises

1. Use Gauss' lemma to compute
   (a) $(\frac{11}{17})$.
   (b) $(\frac{12}{23})$.
   (c) $(\frac{5}{11})$.
   (d) $(\frac{6}{31})$.

2. Show that $\left(\frac{-3}{p}\right) = 1$ if and only if $p \equiv 1 \pmod 3$.

3. Prove Corollary 6.

4. Write a computer program to determine $\left(\frac{a}{p}\right)$ using the Gauss lemma.

5. Use the Chinese Remainder Theorem and the results of Sections 3.4 and 4.3 to determine all positive integers n such that $x^2 \equiv 2 \pmod n$ has a solution.

6. Use the Gauss lemma to determine
   (a) $\left(\frac{5}{p}\right)$.
   (b) $\left(\frac{11}{p}\right)$.

7. Use the facts in Section 4.3 as well as Exercise 6 to show the following facts:

(a) $$\left(\frac{5}{p}\right) = \left(\frac{p}{5}\right) \qquad (p \neq 2, 5).$$

(b) $$\left(\frac{3}{p}\right) = (-1)^{(p-1)/2}\left(\frac{p}{3}\right) \qquad (p \neq 2, 3).$$

(c) $$\left(\frac{7}{p}\right) = (-1)^{(p-1)/2}\left(\frac{p}{7}\right) \qquad (p \neq 2, 7).$$

(d) $$\left(\frac{11}{p}\right) = (-1)^{(p-1)/2}\left(\frac{p}{11}\right) \qquad (p \neq 2, 11).$$

8. (a) Prove that if p and q are primes such that $p \equiv q \pmod{20}$, then
$$\left(\frac{5}{p}\right) = \left(\frac{5}{q}\right).$$
(b) Prove that if p and q are primes such that $p \equiv q \pmod{44}$, then
$$\left(\frac{11}{p}\right) = \left(\frac{11}{q}\right).$$

9. Suppose that p, q are primes and that a is one of 3, 5, 7, 11. Show that if $p \equiv -q \pmod{4a}$, then
$$\left(\frac{a}{p}\right) = \left(\frac{a}{q}\right).$$

10. Evaluate $[7.1]$, $[5/2]$, $[22/13]$, $[-1.8]$, $[e]$, $[\sqrt{5}]$, $[\pi^3]$.

11. Show that the number of integers n such that $\alpha \leq n \leq \beta$, where α and β are real numbers and α is an integer, equals $[\beta] - [\alpha] + 1 = [\beta] - \alpha + 1$.

*12. Show that there are an infinite number of primes $p \equiv +1 \pmod 3$ and an infinite number congruent to $-1 \pmod 3$. (*Hint:* For $p \equiv -1 \pmod 3$, generalize Euclid's proof of the infinitude of primes as follows: Suppose that there are only a finite number, say $p_1, \ldots, p_t$. If t is odd, look at $p_1 \cdots p_t + 3$. If t is even, look at $p_1 \cdots p_t + 1$. For $p \equiv 1 \pmod 3$, reason in the same way, except look at $(p_1 \cdots p_t)^2 + 3$ and use Exercise 2.)

13. Let x be any positive number and let n be a positive integer. Show that
$$\left[\frac{[x]}{n}\right] = \left[\frac{x}{n}\right].$$

14. Let n be a positive integer. Show that the exact power of the prime p dividing $n!$ is
$$\sum_{m=1}^{\infty}\left[\frac{n}{p^m}\right].$$
(Note that after a certain point all terms of the series are zero.)

15. Show that for a given positive integer n all the binomial coefficients $\binom{n}{j}$ $(0 \leq j \leq n)$ are odd if and only if n is of the form $2^k - 1$.

*16. (Eisenstein) Let m, n be odd positive integers, $m \neq 1$, $n \neq 1$. Show that

$$\sum_{x=1}^{(m-1)/2} \left[\frac{nx}{m}\right] + \sum_{y=1}^{(n-1)/2} \left[\frac{my}{n}\right] = \frac{m-1}{2} \cdot \frac{n-1}{2}.$$

(*Hint:* Draw an $m \times n$ square in the first quadrant of the plane, with one vertex at $(0, 0)$ and the axes as sides. Draw the diagonal and count points with integer coordinates below the diagonal.)

17. Show that the number of points (x, y), with x, y positive integers, which lie on or below the hyperbola $xy = n$ is just $2\sum_{0<x\leq\sqrt{n}}[n/x] - [\sqrt{n}\,]^2$.

18. Compute the number of points (x, y), with x, y integers, which are within or on the circle $x^2 + y^2 = n$ in terms of the function [ ].

19. (Eisenstein) Let x, n be positive, n an integer. Show that

$$[x] + \left[x + \frac{1}{n}\right] + \left[x + \frac{2}{n}\right] + \cdots + \left[x + \frac{n-1}{n}\right] = [nx].$$

*20. Let a, b, m be integers and assume that $\gcd(a, m) = d$. Show that

$$\left[\frac{b}{m}\right] + \left[\frac{a+b}{m}\right] + \left[\frac{2a+b}{m}\right] + \cdots + \left[\frac{(m-1)a+b}{m}\right]$$
$$= \frac{(a-1)(m-1)}{2} + \frac{d-1}{2} + b - r,$$

where $b = qd + r$, $0 \leq r < d$.

21. (a) Show that if h, k, ℓ are positive integers, then $[2h/k] + [2\ell/k] \geq [h/k] + [\ell/k] + [(h+\ell)/k]$.
    (b) Show that
    $$\frac{(2m)!(2n)!}{m!n!(m+n)!}$$
    is an integer.

22. Let p be a prime of the form $4q + 1$, q a prime: Examples are $p = 13$, 29, 53. Show that 2 is a primitive root modulo p by the following argument: By direct checking, we may assume that $p > 16$.
    (a) Show that $\left(\frac{2}{p}\right) = -1$.
    (b) Show that $2^{2q} \equiv -1 \pmod{p}$. (*Hint:* Use the Euler criterion.)
    (c) Show that $\left(\frac{2^q}{p}\right) = -1$.
    (d) Conclude that 2 is a primitive root modulo p. (*Hint:* What are the possibilities for $\operatorname{ord}_p 2$?).

4.4 The Law of Quadratic Reciprocity

We now come to the main topic of this chapter, the Law of Quadratic Reciprocity. As we have already stated, the Law of Quadratic Reciprocity will give us a method for calculating the Legendre symbol $\left(\frac{a}{p}\right)$ and thereby for determining whether the congruence $x^2 \equiv a(\text{mod } p)$ can be solved. Euler, after extensive numerical calculation, was able to conjecture the law by computing $\left(\frac{a}{p}\right)$ for many specific values of a and p. In the examples of the last section we computed many special cases of $\left(\frac{a}{p}\right)$. Indeed we have more data there than Euler did since we computed $\left(\frac{a}{p}\right)$ for $a = -1, 2, 3, 7$, and all primes p. (Of course, Euler did not have the Gauss lemma at his disposal.)

Let us examine our data for $\left(\frac{2}{p}\right)$, $\left(\frac{3}{p}\right)$ and $\left(\frac{7}{p}\right)$, and let us retrace Euler's reasoning. (Look at Theorem 3.5 and Examples 3.7 and 3.10, respectively.)

First, we note that in all cases we computed, the precise value of p is not needed to compute $\left(\frac{a}{p}\right)$, but rather only the remainder left when p is divided by $4a$. In other words, if p and q are primes and $p \equiv q(\text{mod } 4a)$, then $\left(\frac{a}{p}\right) = \left(\frac{a}{q}\right)$.

Second, we observe that Tables 4-1 and 4-2 are symmetric about their respective centers. That is, $\left(\frac{a}{p}\right)$ has the same value for the remainders r and $4a - r$. To put it another way, we observe that if p and q are primes and $p \equiv -q(\text{mod } 4a)$, then $\left(\frac{a}{p}\right) = \left(\frac{a}{q}\right)$.

Do the properties we have observed on the basis of our calculations persist as general properties of the Legendre symbol $\left(\frac{a}{p}\right)$? Euler conjectured that they do, and this is precisely the content of the Law of Quadratic Reciprocity, which we shall now state.

Theorem 1 (Law of Quadratic Reciprocity—first form): Let $a > 1$ be a fixed integer. For a prime p such that $p \nmid 4a$, write $p = 4am + r$, $0 < r < 4a$. Then $\left(\frac{a}{p}\right)$ depends only on the remainder r and not on p. Moreover, $\left(\frac{a}{p}\right)$ assumes the same value for the remainders r and $4a - r$.

Euler was not able to prove his conjecture, however. Gauss, working without any knowledge of Euler's work, rediscovered Euler's conjecture and supplied the first proof. Gauss' work was done by the time he was 19 and first apepared in his famous and extremely influential book *Disquisitiones Arithmeticae*, published in 1799. (An English translation has recently appeared in paperback.*) Gauss' original proof of the Law of Quadratic Reciprocity was very complicated and used a double induction. However, Gauss was so fascinated by the result that he continued his investigations and eventually provided seven proofs of the Law of Quadratic Reciprocity. The theorem has continued to fascinate number theorists, and by now there are several hundred proofs. The proof of Theorem 1 which we shall give is one of Gauss' seven proofs and relies on Gauss' lemma.

Proof of Theorem 1: From Gauss' lemma (Theorem 3.1) and our ensuing discussion in Section 3, we must determine the parity of the total number ν of integers k such that $1 \leq k \leq p/2$ and such that k lies in one of the intervals

$$\tfrac{1}{2}p \leq ka \leq p,$$
$$\tfrac{3}{2}p \leq ka \leq 2p,$$
$$\tfrac{5}{2}p \leq ka \leq 3p,$$

and so forth. Dividing through by a, we see that we must determine the parity of the total number ν of integers k such that $1 \leq k \leq p/2$ and such that k lies in one of the intervals

$$\frac{p}{2a} \leq k \leq \frac{p}{a},$$
$$\frac{3p}{2a} \leq k \leq \frac{2p}{a},$$
$$\frac{5p}{2a} \leq k \leq \frac{3p}{a},$$

and so forth. What is the last interval we must consider? The intervals are all of the form

$$\frac{2s-1}{2a}p \leq k \leq \frac{sp}{a}$$

for integers $s \geq 1$. Therefore, since $1 \leq k \leq p/2$, we certainly need not consider any interval beyond the one containing $p/2$, and for this interval we have

$$\frac{2t-1}{2a}p \leq \frac{p}{2} \leq \frac{tp}{a}$$

**Disquisitiones Arithmeticae*, A. A. Clark, S. J. (trans.), Yale University Press, New Haven, Conn., 1966.

or, equivalently,

$$\frac{2t-1}{2} \le \frac{a}{2} \le t. \tag{1}$$

From (1), we see that since a and t are integers, we must have

$$t = \begin{cases} \dfrac{a}{2} & \text{if } a \text{ is even,} \\[2ex] \dfrac{a}{2} + \dfrac{1}{2} & \text{if } a \text{ is odd.} \end{cases}$$

Therefore, we need to determine the parity of the total number ν of integers k such that $1 \le k \le p/2$ and such that k lies in one of the intervals

$$\begin{gathered} \frac{p}{2a} \le k \le \frac{p}{a}, \\ \frac{3p}{2a} \le k \le \frac{2p}{a}, \\ \vdots \\ \frac{2t-1}{2a}p \le k \le \frac{tp}{a}. \end{gathered} \tag{2}$$

If a is even, then $tp/a = p/2$, so that $1 \le k \le p/2$ is automatically implied by the set of inequalities (2), so that the condition $1 \le k \le p/2$ may be omitted. If a is odd, then $t = a/2 + 1/2$ and the last interval of (2) is just

$$\frac{p}{2} \le k \le \frac{p}{2} + \frac{p}{2a},$$

so that we may omit the condition $1 \le k \le p/2$ if we omit the last interval of (2). Thus, let us set $u = t$ if a is even, and $u = t - 1$ if a is odd. Then we must determine the parity of the total number ν of integers k lying in one of the intervals

$$\begin{gathered} \frac{p}{2a} \le k \le \frac{p}{a}, \\ \frac{3p}{2a} \le k \le \frac{2p}{a}, \\ \vdots \\ \frac{2u-1}{2a}p \le k \le \frac{up}{a}. \end{gathered}$$

Thus, we see the important point that the last interval we must consider depends only on a and *not* on p, since $u = a/2$ if a is even and $u = (a/2) - \frac{1}{2}$ if a is odd.

As in the preceding section, set $p = 4am + r$, where $0 < r < 4a$. (Note that $r \neq 0$ since $p \nmid 4a$.) Then we need to determine the parity of the total number ν of integers k lying in one of the intervals

$$2m + \frac{r}{2a} \leq k \leq 4m + \frac{r}{a},$$

$$6m + \frac{3r}{2a} \leq k \leq 8m + \frac{2r}{a},$$

$$10m + \frac{5r}{2a} \leq k \leq 12m + \frac{3r}{a}$$

and so forth, up to the interval

$$2(2u - 1)m + \frac{2u - 1}{2a} r \leq k \leq 4um + \frac{ur}{a}.$$

By using Lemma 3.9, part (iii), we see that the parity of ν above is the same as the parity of the total number of integers k lying in one of the intervals

$$\begin{gathered} \frac{r}{2a} \leq k \leq \frac{r}{a}, \\ \frac{3r}{2a} \leq k \leq \frac{2r}{a}, \\ \vdots \\ \frac{2u-1}{2a} r \leq k \leq \frac{ur}{a}. \end{gathered} \tag{3}$$

However, this parity clearly depends only on r and not on p. Thus, we have proved that $\left(\frac{a}{p}\right)$ with $p = 4am + r$ depends only on r and not on p. This proves the first half of Theorem 1.

Suppose that we now replace r by $4a - r$ in (3). Then the intervals (3) become

$$\frac{4a - r}{2a} \leq k \leq \frac{4a - r}{a},$$

$$\frac{3(4a - r)}{2a} \leq k \leq \frac{2(4a - r)}{a},$$

$$\frac{5(4a - r)}{2a} \leq k \leq \frac{3(4a - r)}{a},$$

$$\vdots$$

$$\frac{(2u - 1)(4a - r)}{2a} \leq k \leq \frac{u(4a - r)}{a},$$

or, equivalently,

$$
\begin{aligned}
2 - \frac{r}{2a} &\le k \le 4 - \frac{r}{a}, \\
6 - \frac{3r}{2a} &\le k \le 8 - \frac{2r}{a}, \\
10 - \frac{5r}{2a} &\le k \le 12 - \frac{3r}{a}, \qquad (4)\\
&\vdots \\
4u - 2 - \frac{r(2u-1)}{2a} &\le k \le 4u - \frac{ru}{a}.
\end{aligned}
$$

By Lemma 3.9, parts (i) and (ii), the number of integers in the first interval of (4) is

$$
\begin{aligned}
\left[4 - \frac{r}{a}\right] - \left[2 - \frac{r}{2a}\right] &= 4 + \left[-\frac{r}{a}\right] - \left(2 + \left[\frac{-r}{2a}\right]\right) \\
&= 2 + \left[\frac{r}{2a}\right] - \left[\frac{r}{a}\right]
\end{aligned}
$$

since $r/2a$ and r/a are not integers and $[-\alpha] = -[\alpha] - 1$ for any real α which is not an integer (exercise). Therefore, the number of integers in the first interval of (4) is

$$2 + \left[\frac{r}{2a}\right] - \left[\frac{r}{a}\right] \equiv \left[\frac{r}{2a}\right] - \left[\frac{r}{a}\right] \pmod 2.$$

Thus, the parity of the number of integers in the first interval of (4) is the same as the parity of

$$\left[\frac{r}{2a}\right] - \left[\frac{r}{a}\right],$$

which is just the number of integers in the first interval of (3). Reasoning in the same way with each interval in (4), we see that the parity of the total number ν' of integers lying in one of the intervals (4) is the same as the parity of the total number ν of integers lying in one of the intervals (3). Thus, by Gauss' lemma, if p and q are primes, $p \nmid 4a$, $q \nmid 4a$, with $p = 4am + r$, $q = 4am' + (4a - r)$, then

$$\left(\frac{a}{p}\right) = (-1)^{\nu} = (-1)^{\nu'} = \left(\frac{a}{q}\right).$$

Thus, the Legendre symbol $\left(\frac{a}{p}\right)$ is the same for the remainders r and $4a - r$. This is the second assertion of Theorem 1. ∎

We may reword Theorem 1 in the following equivalent way:

Corollary 2: Let a be an integer greater than 1, and let p, q be primes not dividing $4a$. If $p \equiv \pm q \pmod{4a}$, then $\left(\frac{a}{p}\right) = \left(\frac{a}{q}\right)$.

It is possible to reformulate the Law of Quadratic Reciprocity in a more elegant, if somewhat more mysterious, manner. This reformulation was first stated by Legendre in 1785 as a conjecture.

Theorem 3 (Law of Quadratic Reciprocity—second form): Let p and q be distinct odd primes. Then

$$\left(\frac{p}{q}\right)\left(\frac{q}{p}\right) = (-1)^{((p-1)/2)((q-1)/2)}.$$

Thus, $\left(\frac{p}{q}\right) = \left(\frac{q}{p}\right)$, unless $p \equiv q \equiv -1 \pmod 4$, in which case $\left(\frac{p}{q}\right) = -\left(\frac{q}{p}\right)$.

Theorem 3 is very striking and is quite unexpected, for it says that there is a relationship between the solvability of the congruence $x^2 \equiv p \pmod q$ and the congruence $x^2 \equiv q \pmod p$. Moreover, as we shall see below, Theorem 3 provides us with a simple algorithm for calculating $\left(\frac{p}{q}\right)$.

Proof of Theorem 3: First we assume that $p \equiv q \pmod 4$. Without loss of generality, assume that $p > q$. Write $p = q + 4a$. Then, by Corollaries 2.8 and 2.9 and Proposition 2.6, we have

$$\left(\frac{p}{q}\right) = \left(\frac{q+4a}{q}\right) = \left(\frac{4a}{q}\right) = \left(\frac{4}{q}\right)\left(\frac{a}{q}\right) = \left(\frac{a}{q}\right)$$

and

$$\left(\frac{q}{p}\right) = \left(\frac{p-4a}{p}\right) = \left(\frac{-4a}{p}\right) = \left(\frac{-1}{p}\right)\left(\frac{4}{p}\right)\left(\frac{a}{p}\right) = (-1)^{(p-1)/2}\left(\frac{a}{p}\right).$$

Since $p \equiv q \pmod{4a}$, we have by Corollary 2 that $\left(\frac{a}{p}\right) = \left(\frac{a}{q}\right)$ so that

$$\left(\frac{p}{q}\right)\left(\frac{q}{p}\right) = (-1)^{(p-1)/2}.$$

Finally $(p-1)/2 = (q-1)/2 + 2a$, and so $(p-1)/2$ is even if and only if $(q-1)/2$ is even, and thus

$$(-1)^{(p-1)/2} = (-1)^{((p-1)/2)((q-1)/2)},$$

so that the theorem is proved in this case. If $p \not\equiv q \pmod 4$, then since p and q are odd, we must have either $p \equiv 1 \pmod 4$, $q \equiv 3 \pmod 4$ or $p \equiv 3 \pmod 4$, $q \equiv 1 \pmod 4$. In any case, if $p \not\equiv q \pmod 4$, then $p \equiv -q \pmod 4$. Then we may write $p = -q + 4a$. Again from Corollary 2.8 and Proposition 2.6, we have

$$\left(\frac{p}{q}\right) = \left(\frac{-q+4a}{q}\right) = \left(\frac{4a}{q}\right) = \left(\frac{4}{q}\right)\left(\frac{a}{q}\right) = \left(\frac{a}{q}\right)$$

and

$$\left(\frac{q}{p}\right) = \left(\frac{-p+4a}{p}\right) = \left(\frac{4a}{p}\right) = \left(\frac{4}{p}\right)\left(\frac{a}{p}\right) = \left(\frac{a}{p}\right).$$

Since $p \equiv -q \pmod{4a}$, Corollary 2 implies that $\left(\frac{a}{p}\right) = \left(\frac{a}{q}\right)$, so that $\left(\frac{p}{q}\right) = \left(\frac{q}{p}\right)$ and $\left(\frac{p}{q}\right)\left(\frac{q}{p}\right) = 1$. However, in this case,

$$\frac{p-1}{2} + \frac{q-1}{2} = 2a - 1$$

is odd, so that one of $(p-1)/2$, $(q-1)/2$ must be even. Therefore, $(-1)^{((p-1)/2)((q-1)/2)} = 1$, and the theorem is completely proved. ■

Using the Law of Quadratic Reciprocity in the form stated in Theorem 3, together with Proposition 2.6, Corollaries 2.8, 2.9 and Theorem 3.5, we can completely reduce the question of the solvability of the congruence $x^2 \equiv a \pmod{p}$ to a simple calculation. (We emphasize again, however, that our procedure gives no information about what the solutions are, if there are any.) Let us give a few examples.

Example 4: Can the congruence $x^2 \equiv 3 \pmod{43}$ be solved? Well, we have

$$\left(\frac{3}{43}\right) = \left(\frac{43}{3}\right)(-1)^{((3-1)/2)((43-1)/2)} = -\left(\frac{1}{3}\right) = -1,$$

so that the congruence cannot be solved. Since $3 \equiv -40 \pmod{43}$, we know from Proposition 2.6 that

$$\left(\frac{-40}{43}\right) = \left(\frac{3}{43}\right) = -1.$$

However, as a further example, let us compute $\left(\frac{-40}{43}\right)$ directly:

$$\begin{aligned}
\left(\frac{-40}{43}\right) &= \left(\frac{(-1)2^3\cdot 5}{43}\right) \\
&= \left(\frac{-1}{43}\right)\left(\frac{2}{43}\right)^3\left(\frac{5}{43}\right) \\
&= (-1)^{(43-1)/2}\left(\frac{2}{43}\right)\left(\frac{5}{43}\right) && \text{(Example 3.3)} \\
&= -(-1)\left(\frac{5}{43}\right) && \text{(since } 43\equiv 3 \pmod 8\text{), using Theorem 3.5)} \\
&= \left(\frac{5}{43}\right).
\end{aligned}$$

Therefore, using the Law of Quadratic Reciprocity, we see that

$$\left(\frac{-40}{43}\right) = \left(\frac{5}{43}\right) = \left(\frac{43}{5}\right)(-1)^{((43-1)/2)((5-1)/2)}$$
$$= \left(\frac{3}{5}\right) = \left(\frac{5}{3}\right)(-1)^{((5-1)/2)((3-1)/2)}$$
$$= \left(\frac{2}{3}\right) = -1.$$

To give you some idea of the efficiency of the above procedure, let us compute an example involving larger numbers.

Example 5: Can the congruence

$$x^2 \equiv 20964(\text{mod } 1987)$$

be solved? Note that 1987 is a prime. (You can check that it is not divisible by any prime less than $\sqrt{1987}$.) Therefore, it suffices to compute $\left(\frac{20964}{1987}\right)$. Since $20964 \equiv 1094(\text{mod } 1987)$ and since $1094 = 2 \cdot 547$, we see that

$$\left(\frac{20964}{1987}\right) = \left(\frac{1094}{1987}\right) = \left(\frac{2}{1987}\right)\left(\frac{547}{1987}\right) = -\left(\frac{547}{1987}\right)$$

since $1987 \equiv 3(\text{mod } 8)$ implies that $\left(\frac{2}{1987}\right) = -1$ (Theorem 3.5). Now 547 is a prime and $547 \equiv 1987 \equiv -1(\text{mod } 4)$, so that by the Law of Quadratic Reciprocity, we have

$$\left(\frac{20964}{1987}\right) = -\left(\frac{547}{1987}\right) = \left(\frac{1987}{547}\right) = \left(\frac{346}{547}\right) = \left(\frac{2}{547}\right)\left(\frac{173}{547}\right) = -\left(\frac{173}{547}\right)$$

since $547 \equiv 3(\text{mod } 8)$. Using the reciprocity law again (173 is a prime), we have

$$\left(\frac{20964}{1987}\right) = -\left(\frac{173}{547}\right) = -\left(\frac{547}{173}\right) = -\left(\frac{28}{173}\right) = -\left(\frac{4}{173}\right)\left(\frac{7}{173}\right) = -\left(\frac{7}{173}\right)$$
$$= -\left(\frac{173}{7}\right) = -\left(\frac{5}{7}\right) = -\left(\frac{7}{5}\right) = -\left(\frac{2}{5}\right) = 1.$$

Thus, our original congruence is solvable.

Let us close this section with one more example, similar to the examples we worked out using Gauss' lemma.

Example 6: Let us completely determine $\left(\frac{5}{p}\right)$ for $p > 5$. We write $p = 5a + r$, $r = 1, 2, 3,$ or 4. Since $5 \equiv 1(\text{mod } 4)$, we have from the reciprocity law that

$$\left(\frac{5}{p}\right) = \left(\frac{p}{5}\right)(-1)^{((p-1)/2)((5-1)/2)} = \left(\frac{p}{5}\right) = \left(\frac{5a+r}{5}\right) = \left(\frac{r}{5}\right).$$

Since $\left(\frac{r}{5}\right) = 1$ if and only if $r = 1$ or 4, we see that

$$\left(\frac{5}{p}\right) = \begin{cases} 1 & \text{if } p \equiv \pm 1 (\text{mod } 5), \\ -1 & \text{if } p \equiv \pm 2 (\text{mod } 5). \end{cases}$$

Notice that we ended up with a congruence condition modulo 5 instead of a condition modulo $4 \cdot 5 = 20$ as we know we must by Theorem 1. But, of course, any congruence condition modulo 5 could be interpreted as one modulo 20 since $5 \mid 20$.

4.4 Exercises

1. Use the quadratic reciprocity law to determine whether the following congruences are solvable:
   (a) $x^2 \equiv 15 (\text{mod } 31)$.
   (b) $x^2 \equiv 48 (\text{mod } 89)$.
   (c) $x^2 \equiv -17 (\text{mod } 89)$.
   (d) $x^2 \equiv 17 (\text{mod } 31)$.
   (e) $x^2 \equiv 23 (\text{mod } 59)$.
   (f) $x^2 \equiv 264 (\text{mod } 173)$.
   (g) $x^2 \equiv 4977 (\text{mod } 1987)$.

2. Use the quadratic reciprocity law to completely determine for all odd primes p
   (a) $\left(\frac{3}{p}\right)$, $(p \neq 3)$.
   (b) $\left(\frac{7}{p}\right)$, $(p \neq 7)$.
   (c) $\left(\frac{31}{p}\right)$, $(p \neq 31)$.
   (d) $\left(\frac{6}{p}\right)$, $(p \neq 3)$.
   (e) $\left(\frac{-5}{p}\right)$, $(p \neq 5)$.

   (You may check parts (a) and (b) by the examples in Section 3).

3. Determine whether the following congruences are solvable for integers x and y with $x \not\equiv 0 (\text{mod } 17)$ and $y \not\equiv 0 (\text{mod } 17)$:
   (a) $x^2 + 15y^2 \equiv 0 (\text{mod } 17)$.
   (b) $x^2 + 11y^2 \equiv 0 (\text{mod } 17)$.

4. Determine whether the following congruences can be solved:
   (a) $x^2 \equiv 23 (\text{mod } 177)$, $(177 = 59 \cdot 3)$.
   (b) $x^2 \equiv 5 (\text{mod } 1102)$, $(1102 = 2 \cdot 19 \cdot 29)$.

5. Write a computer program to evaluate $\left(\frac{a}{p}\right)$ for odd primes p.

*6. Show that Theorem 3 implies Theorem 1. (*Hint:* Write a as a product of primes.)

7. Show that if p is a prime and $p \equiv \pm 1 \pmod{4a}$, then $\left(\frac{a}{p}\right) = 1$. (*Hint:* Write a as a product of primes and use the quadratic reciprocity law.)

8. Show that if p is a prime and $p \equiv 2a - 1 \pmod{4a}$ and $a \equiv 1 \pmod 4$, then $\left(\frac{a}{p}\right) = 1$.

The rest of these exercises extend the definition and properties of the Legendre symbol $\left(\frac{a}{p}\right)$.

Definition: If n is an odd integer greater than 1 and $\gcd(a,n) = 1$, write

$$n = p_1 p_2 \cdots p_t$$

as a product of primes. Define the *Jacobi symbol* $\left(\frac{a}{n}\right)$ by

$$\left(\frac{a}{n}\right) = \left(\frac{a}{p_1}\right)\left(\frac{a}{p_2}\right) \cdots \left(\frac{a}{p_t}\right).$$

Note that this definition agrees with the old definition when n is a prime. Its main use is that it simplifies the work in computing the Legendre symbol. Throughout these exercises n denotes an odd integer greater than 1 and a denotes an integer such that $\gcd(a,n) = 1$.

9. Show that if $a \equiv a' \pmod n$, then $\left(\frac{a}{n}\right) = \left(\frac{a'}{n}\right)$.

10. Show that $\left(\frac{aa'}{n}\right) = \left(\frac{a}{n}\right)\left(\frac{a'}{n}\right)$.

11. Show that $\left(\frac{a}{n}\right)\left(\frac{a}{m}\right) = \left(\frac{a}{nm}\right)$. Thus, $\left(\frac{a}{n^2}\right) = 1$.

12. Show that if $p_1, p_2, \ldots, p_t$ are odd primes (not necessarily distinct), then

$$\frac{p_1 - 1}{2} + \frac{p_2 - 1}{2} + \cdots + \frac{p_t - 1}{2} \equiv \frac{p_1 p_2 \cdots p_t - 1}{2} \pmod 2.$$

13. Show that $\left(\frac{-1}{n}\right) = (-1)^{(n-1)/2}$.

14. Show that if $p_1, p_2, \ldots, p_t$ are odd primes, then

$$\frac{p_1^2 - 1}{8} + \frac{p_2^2 - 1}{8} + \cdots + \frac{p_t^2 - 1}{8} \equiv \frac{(p_1 p_2 \cdots p_t)^2 - 1}{8} \pmod 2.$$

15. Show that $\left(\frac{2}{n}\right) = (-1)^{(n^2-1)/8}$. Thus, $\left(\frac{2}{n}\right) = 1$ if and only if $n \equiv \pm 1 \pmod 8$.

16. Show that if $p_1, p_2, \ldots, p_t$ and $q_1, q_2, \ldots, q_s$ are odd primes, then $\sum_{i=1}^{t} \sum_{j=1}^{s} \frac{p_i - 1}{2} \frac{q_j - 1}{2} \equiv \frac{p_1 p_2 \cdots p_t - 1}{2} \frac{q_1 q_2 \cdots q_s - 1}{2} \pmod 2$.

17. (Reciprocity Law) Show that if n and m are odd integers greater than 1 and $\gcd(m, n) = 1$, then

$$\left(\frac{n}{m}\right)\left(\frac{m}{n}\right) = (-1)^{((n-1)/2)((m-1)/2)}.$$

18. Decide whether the following congruences are solvable using the above results on the Jacobi symbol:
(a) $x^2 \equiv 264 \pmod{173}$.
(b) $x^2 \equiv 4977 \pmod{1987}$.
(c) $x^2 \equiv 187 \pmod{389}$.
Note that the advantage of the above method over the method of just using the Legendre symbol is that it is no longer necessary to factor the numerator before inverting the symbol.

19. Write a computer program to evaluate $\left(\frac{a}{n}\right)$ for odd integers n. (Throw away your computer program from Exercise 5.)

20. (a) Show that if $x^2 \equiv a \pmod n$ is solvable, then $\left(\frac{a}{n}\right) = 1$.
(b) Show that the converse of part (a) is *false;* i.e., there are integers a, n with n odd and $\gcd(a,n) = 1$ such that $\left(\frac{a}{n}\right) = 1$ and $x^2 \equiv a \pmod n$ is *not* solvable.

4.5 Applications to Diophantine Equations

Let us now give a few applications of the theory of the preceding sections to Diophantine equations.

As a first example, let us consider the Diophantine equation

$$ax^2 + by + c = 0, \tag{1}$$

where a, b, c are given integers and x and y are integers to be determined. If $b = 0$, Eq. (1) has a solution if and only if $-c/a$ is an integral square. Thus, we assume that $b > 0$. The equation is equivalent to the congruence

$$ax^2 \equiv -c \pmod b. \tag{2}$$

Let $d = \gcd(a, b)$. Then it is clear that if (1) can be solved, we must have $d \mid c$.

If $d \mid c$ and $c = dc'$, $a = da'$, $b = db'$, then $\gcd(a', b') = 1$ and Eq. (1) is equivalent to the equation

$$a'x^2 + b'y + c' = 0,$$

an equation of the same type as the first, except that in addition we have $\gcd(a', b') = 1$. Thus, it suffices to consider Eq. (1) for which a and b are relatively prime. From Proposition 3.2.8 we can find a^* such that

$$aa^* \equiv 1 (\text{mod } b).$$

Then congruence (2) is equivalent to

$$x^2 \equiv -a^*c (\text{mod } b). \tag{3}$$

Let $b = p_1^{a_1} \cdots p_t^{a_t}$ be the decomposition of b into a product of distinct prime powers. We know from the Chinese Remainder theorem and Theorem 3.4.1 that (3) is solvable if and only if

$$x^2 \equiv -a^*c (\text{mod } p_i^{a_i}) \tag{4}$$

is solvable for $1 \leq i \leq t$. Moreover, from Exercise 3.4.11, we know that if p_i is odd, (4) is solvable if and only if we can solve

$$x^2 \equiv -a^*c (\text{mod } p_i), \tag{5}$$

and this is solvable if and only if $\left(\frac{-a^*c}{p_i}\right) = 1$. Thus, we see that the Diophantine equation (1) with $\gcd(a, b) = 1$ and b odd is solvable if and only if

$$\left(\frac{-a^*c}{p}\right) = 1$$

for all primes p such that $p \mid b$. Thus, for example, if p is an odd prime, then

$$x^2 + py + c = 0$$

has a solution if and only if $\left(\frac{-c}{p}\right) = 1$.

Therefore, by using the calculations of Examples 4.4, 4.5, and 4.6, we see that

$$x^2 + 43y - 3 = 0 \text{ is not solvable}$$
$$x^2 + 43y + 40 = 0 \text{ is not solvable}$$
$$x^2 + 1987y - 20964 = 0 \text{ is solvable}$$

and that

$x^2 + py - 5 = 0$ is solvable, for an odd prime p, if and only if $p \equiv \pm 1 (\text{mod } 5)$.

As a second example, let us consider the equation

$$y^2 = x^3 + 45, \tag{6}$$

and let us show how our knowledge of $\left(\frac{2}{p}\right)$ can be used to show that there are no solutions. Note that Eq. (6) is a special case of the equation $y^2 = x^3 + k$. The case $k = 23$ was discussed in Section 3.3, and we shall give a similar treatment in the present case. We have included a second example of this type to show how the knowledge of the Legendre symbol (i.e., the reciprocity law) can often come into a problem in a nontrivial and unexpected way.

Let us now show that (6) has no solutions. The plan of the proof is to show that each of the eight possible residue classes mod 8 are impossible for x. We first eliminate all the even residues 0, 2, 4, 6. This is the same as eliminating the possibility that x is even. If $x \equiv 0 \pmod 2$, then $x^3 \equiv 0 \pmod 8$ implies (from (6)) that $y^2 \equiv 45 \equiv 5 \pmod 8$, but this is impossible since a perfect square is congruent to 0, 1, or $4 \pmod 8$. Thus, x cannot be even. Next, we show that $x \not\equiv 1$ or $5 \pmod 8$. For if $x \equiv 1$ or $5 \pmod 8$, then $x \equiv 1 \pmod 4$, so that $x^3 \equiv 1 \pmod 4$, so that from (6) we deduce that $y^2 \equiv 46 \equiv 2 \pmod 4$, which is again impossible. Thus, the only possibilities left for x are $x \equiv 3 \pmod 8$ or $x \equiv 7 \pmod 8$. These cases require a clever trick. First consider the case where $x \equiv 7 \equiv -1 \pmod 8$. Write (6) in the form

$$y^2 - 2 \cdot 3^2 = x^3 + 27 = (x + 3)(x^2 - 3x + 9). \tag{7}$$

Then $x \equiv -1 \pmod 8$ implies that $x^2 - 3x + 9 \equiv 1 + 3 + 9 \equiv -3 \pmod 8$. It then follows that there is a prime p dividing $x^2 - 3x + 9$ such that $p \equiv \pm 3 \pmod 8$, for otherwise $x^2 - 3x + 9 = p_1 p_2 \cdots p_r$, where all p_i are prime and $p_i \equiv \pm 1 \pmod 8$ for all i. (Note that all p_i must be odd since $x^2 - 3x + 9 \equiv -3 \pmod 8$.) But then $x^2 - 3x + 9 = p_1 p_2 \cdots p_r \equiv (\pm 1) \cdots (\pm 1) \equiv \pm 1 \pmod 8$, which is a contradiction. Then, for the prime $p \equiv \pm 3 \pmod 8$ whose existence we have just established, we have $y^2 - 2 \cdot 3^2 \equiv 0 \pmod p$. Thus, $\left(\frac{2 \cdot 3^2}{p}\right) = 1$. But

$$1 = \left(\frac{2 \cdot 3^2}{p}\right) = \left(\frac{2}{p}\right)\left(\frac{3^2}{p}\right) = \left(\frac{2}{p}\right).$$

However, from Theorem 3.5, we see that $\left(\frac{2}{p}\right) = 1$ if and only if $p \equiv \pm 1 \pmod 8$, which contradicts the choice of p. Thus, x cannot be congruent to $7 \pmod 8$. Finally, we consider the case $x \equiv 3 \pmod 8$. Let us write (6) in the form

$$y^2 - 2 \cdot 6^2 = x^3 - 27 = (x - 3)(x^2 + 3x + 9).$$

Observe that since $x \equiv 3 \pmod 8$, we have

$$x^2 + 3x + 9 \equiv 3 \pmod 8.$$

Therefore, there is a p dividing $x^2 + 3x + 9$ such that $p \equiv \pm 3 \pmod 8$ (same argument as above), and for this p, we have $\left(\frac{2 \cdot 6^2}{p}\right) = 1$, which yields a contradiction in the same manner as the previous case. Thus, all residue classes modulo 8 have been ruled out, and $y^2 = x^3 + 45$ has no solutions in integers.

4.5 Exercises

1. Determine all solutions of the Diophantine equations
   (a) $5x^2 + 2x + 11y + 5 = 0$.
   (b) $3x^2 + 20x + 11y - 3 = 0$.
   (c) $7x^2 + 10x + 13y - 6 = 0$.
2. Show that the Diophantine equation $x^2 - 7xy + y^2 = 3$ has no solutions (look modulo 5).
3. Find all solutions of the following Diophantine equation:
   $$2x^2 + 3xy + 8y^2 + 13z = 0.$$
   Be sure you show that you have all the solutions. *Answer:*
   $$(x,y,z) = (k,k + 13t,-k^2 - 19kt - 104t^2)$$
   or
   $$(x,y,z) = (k,-3k + 13t,-5k^2 + 45kt - 104t^2),$$
   where k, t may be any integers.

5

Arithmetic Functions

5.1 Introduction

In Chapters 2–4, we have studied the most elementary properties of the integers and showed how even the most simple facts about integers allow us to solve certain Diophantine equations. Let us continue the theme of the preceding chapters by studying some of the more subtle properties of the integers, as reflected in the basic facts about arithmetic functions.

To put it simply, an *arithmetic function* is a function which associates to every positive integer n a real number $f(n)$. To get the full spirit of what we have in mind, let us consider a number of examples.

Example 1: All the usual functions of high school algebra are arithmetic functions. For example, $f_1(n) = n$, $f_2(n) = n^2$, $f_3(n) = n^3$, and $f_4(n) = 1/n^2$ are all simple examples of arithmetic functions.

Example 2: Let $\phi(n)$ denote Euler's phi function. That is, $\phi(n)$ equals the number of positive integers less than or equal to n which are relatively prime to n. Then $\phi(n)$ is an arithmetic function.

Example 3: Let $d(n)$ equal the number of positive divisors of the positive integer n. Then $d(1) = 1$, $d(2) = 2$ (the positive divisors of 2 are 1 and 2), $d(6) = 4$ (the positive divisors of 6 are 1, 2, 3, and 6), and if p is a prime, then $d(p) = 2$ (the positive divisors of p are 1 and p).

Example 4: Let $\sigma(n)$ equal the sum of the positive divisors of the positive integer n. Then $\sigma(1) = 1$, $\sigma(2) = 3$, $\sigma(6) = 12$, and if p is a prime, then $\sigma(p) = p + 1$.

Example 5: For a real number r, let $\sigma_r(n)$ equal the sum of the rth powers of the positive divisors of the positive integer n. Then $\sigma_1(n) = \sigma(n)$, where $\sigma(n)$ is as defined in Example 4. Also, $\sigma_0(n) = d(n)$.

We could go on giving further examples of arithmetic functions, but let us refrain from doing so at this moment in favor of the obvious question, why study the properties of arithmetic functions?

First, it has already been necessary for us to define the Euler ϕ function in Chapter 2. It gave us a count on the number of elements in any reduced residue system (Proposition 3.2.25). Also, it was used in the statement of Euler's theorem (Theorem 3.3.2), which, in particular, gave us an explicit formula for an arithmetic inverse. It should be clear, then, that an explicit formula for $\phi(n)$ would be useful. One of the main applications of the theory developed in this chapter is to provide such a formula.

As a second reason we note that arithmetic functions reflect properties of the integers, properties which are often more subtle than we have hitherto met in our study of the integers. For example, consider the function $d(n)$. It measures how far n is away from being a prime, for, as we have seen, if p is a prime, then $d(p) = 2$. Conversely, it is easy to see that if $d(n) = 2$, then n is a prime. Thus, $d(n) = 2$ if and only if n is a prime. Moreover, the larger $d(n)$ gets, the further n is from being prime. Thus, we see that the function $d(n)$ provides some fairly sophisticated information about the integer n.

Another reason for studying arithmetic functions is their intimate connection with Diophantine equations. Let us give an example. Let n be a positive integer. Suppose that we wish to solve the Diophantine equation

$$x^2 + y^2 + z^2 + w^2 = n. \tag{1}$$

In other words, we ask for representations of n as a sum of four perfect squares. Lagrange proved that Eq. (1) always has at least one solution. We shall prove this result in Chapter 6. But how many different solutions (x, y, z, w) does Eq. (1) have? Suppose that we denote the number of solutions by $r_4(n)$. Then $r_4(n)$ is a very interesting arithmetic function. The theorem of Lagrange is equivalent to the assertion that $r_4(n) \geq 1$ for all n. Actually, it is possible to give a quite explicit formula for $r_4(n)$, first discovered by Jacobi, namely: Let $\sigma^*(n)$ denote the sum of the odd, positive divisors of n. Then Jacobi's theorem asserts that

$$r_4(n) = \begin{cases} 8\sigma^*(n) & \text{if } n \text{ is odd,} \\ 24\sigma^*(n) & \text{if } n \text{ is even.} \end{cases}$$

Thus, for example, since the odd divisors of 12 are 1 and 3, we see that $r_4(12) = 24\sigma^*(12) = 24 \cdot (1 + 3) = 96$. Thus, the equation

$$x^2 + y^2 + z^2 + w^2 = 12$$

has 96 different solutions. They are $(\pm 2, \pm 2, \pm 2, 0)$, $(\pm 2, \pm 2, 0, \pm 2)$, $(\pm 2, 0, \pm 2, \pm 2)$, $(0, \pm 2, \pm 2, \pm 2)$, $(\pm 3, \pm 1, \pm 1, \pm 1)$, $(\pm 1, \pm 3, \pm 1, \pm 1)$, $(\pm 1, \pm 1, \pm 3, \pm 1)$, and $(\pm 1, \pm 1, \pm 1, \pm 3)$, where all choices of signs are permitted. (The reader should check this calculation.) Thus, we see that the arithmetic function $\sigma^*(n)$ is the key to the number of solutions of the Diophantine equation (1). This situation is fairly typical of the link between Diophantine equations and arithmetic functions.

Finally, a reason for the study of arithmetic functions lies in their historical and recreational interest. The properties of integers, as expressed through the study of arithmetic functions, have been pursued for many thousands of years by groups of people so diverse as to include both amateur mathematicians and mystics. The properties of certain arithmetic functions were often engraved on good luck charms, called *talismans*, during the Middle Ages. We shall discuss one such example, the perfect and amicable numbers, in Section 5.4. For amateur number theorists, discovering the properties of arithmetic functions has always been a source of interest partially for the ease with which one can compile tables and from these generate empirically derived conjectures. We have included a computer-generated table of $\phi(n)$, $d(n)$, and $\sigma(n)$ (and another, $\mu(n)$, which we shall define in Section 3) for $n \leq 100$ (Table 1).

5.1 Exercises

1. Compute the following without using Table 1 at the end of the book:
   (a) $d(21)$. (b) $d(45)$. (c) $d(305)$.
   (d) $d(180)$. (e) $d(368)$. (f) $d(504)$.
2. Compute the following without using Table 1 at the end of the book:
   (a) $\sigma(38)$. (b) $\sigma(20)$. (c) $\sigma(203)$.
   (d) $\sigma(100)$. (e) $\sigma(128)$. (f) $\sigma(297)$.
3. (a) Compute $d(2^n)$ for $n = 1, 2, \ldots$.
   (b) Compute $d(6^n)$ for $n = 1, 2, \ldots$.
   (c) Show that $d(6^n) = d(2^n)d(3^n)$ for $n = 1, 2, 3, \ldots$.
4. Verify the following relations:
   (a) $\sigma(24) = \sigma(8)\sigma(3)$.
   (b) $\sigma(72) = \sigma(8)\sigma(9)$.
   (c) $\sigma(108) = \sigma(4)\sigma(27)$.
5. Verify the assertion $r_4(n) = 8\sigma^*(n)$ (n odd), $r_4(n) = 24\sigma^*(n)$ (n even) for the special cases $n = 6, 15, 21$.

5.2 Multiplicative Arithmetic Functions

Let us now begin by defining a class of arithmetic functions which contains the most interesting ones for our purposes, namely the *multiplicative arithmetic functions.* We shall show that all the examples discussed in Section 1 are in fact included in this class of arithmetic functions. Therefore, by studying the properties of the multiplicative arithmetic functions in general, we shall, at the same time, be studying the properties of all the special functions introduced in Section 1.

Definition 1: Let $f(n)$ be an arithmetic function. We say that $f(n)$ is *multiplicative* provided that whenever n and m are positive integers such that $\gcd(n, m) = 1$ we have

$$f(nm) = f(n)f(m). \tag{1}$$

Example 2: Let $f(n) = n$ for all positive n. Then

$$f(nm) = nm = f(n)f(m).$$

Thus, we see that Eq. (1) holds even without the hypothesis that $\gcd(n, m) = 1$. Thus, $f(n) = n$ is multiplicative. We call arithmetic functions such that (1) holds for all n, m, *totally multiplicative.* Similarly, if $g(n) = n^r$, where r is any real number (positive, negative, or zero), then $g(n)$ is totally multiplicative.

Before the reader gets the wrong impression, note that not all the arithmetic functions which arise naturally in number theory are totally multiplicative. Indeed, $d(n)$, $\sigma(n)$, and $\phi(n)$ are multiplicative but not totally multiplicative. That $d(n)$, for example, is not totally multiplicative is immediate, since

$$d(12) = d(2 \cdot 6) = 6,$$

whereas

$$d(2) \cdot d(6) = 2 \cdot 4 = 8.$$

It is a very simple matter to prove directly that $d(n)$ is multiplicative. However, one of our main purposes in this chapter is to give a fairly routine procedure for handling arithmetic functions. Thus, we shall use that procedure for $d(n)$ and leave the direct proof as an exercise.

Let us first note our advantage if a function $f(n)$ we are interested in is multiplicative. Write

$$n = p_1^{a_1} p_2^{a_2} \cdots p_t^{a_t}$$

where $p_1, p_2, \ldots, p_t$ are distinct primes and $a_1, a_2, \ldots, a_t$ are positive integers. Then, since $f(n)$ is multiplicative and since $\gcd(p_1^{a_1}, p_2^{a_2} \cdots p_t^{a_t}) = 1$, we have

$$f(n) = f(p_1^{a_1}) f(p_2^{a_2} \cdots p_t^{a_t}).$$

Continuing in this manner we see that

$$f(n) = f(p_1^{a_1}) f(p_2^{a_2}) \cdots f(p_t^{a_t}).$$

Thus, the problem of evaluating $f(n)$ is reduced to the case where n is a power of a prime. That is, usually, a much easier problem. Let us record this last observation.

Proposition 3: Let $f(n)$ be a multiplicative function. Write

$$n = p_1^{a_1} p_2^{a_2} \cdots p_t^{a_t}$$

where $p_1, p_2, \ldots, p_t$ are distinct primes. Then

$$f(n) = f(p_1^{a_1}) f(p_2^{a_2}) \cdots f(p_t^{a_t}).$$

We shall now give the result which is the main cog in our "routine" procedure for deriving formulas for arithmetic functions.

Theorem 4: Let $g(n)$ be a multiplicative arithmetic function. Define another arithmetic function $f(n)$ by

$$f(n) = \sum_{d|n} g(d),$$

where $\sum_{d|n}$ denotes the sum over all positive divisors of n. Then $f(n)$ is a multiplicative arithmetic function.

Note that according to our definition $f(1) = g(1)$, $f(2) = g(1) + g(2)$, $f(3) = g(1) + g(3)$, $f(4) = g(1) + g(2) + g(4)$, $f(5) = g(1) + g(5)$, and $f(12) = g(1) + g(2) + g(3) + g(4) + g(6) + g(12)$. Thus, it is quite surprising that $f(n)$ turns out to be multiplicative.

Proof of Theorem 4: Suppose that $\gcd(m, n) = 1$. Then as d runs over the divisors of m and e runs over the divisors of n, it is easy to see that de runs over the divisors of mn, each divisor occurring exactly once. (*Exercise:* One approach is to use unique factorization.) Let us make use of this simple observation as follows:

$$f(m)f(n) = \Big(\sum_{d|m} g(d)\Big)\Big(\sum_{e|n} g(e)\Big) \qquad \text{(by definition of } f\text{)}$$

$$= \sum_{d|m} \sum_{e|n} g(d)g(e) \qquad \text{(by multiplying out the sums)}$$

$$= \sum_{d|m} \sum_{e|n} g(de) \qquad \text{(since } g \text{ is multiplicative; and since } d|m \text{ and } e|n \text{ and } \gcd(m,n) = 1 \text{ imply that } \gcd(d,e) = 1\text{)}$$

$$= \sum_{c|mn} g(c) \qquad \text{(since } c = de \text{ runs over all divisors of } mn \text{ once and only once as } d \text{ and } e \text{ run over all divisors of } m \text{ and } n\text{, respectively)}$$

$$= f(mn) \qquad \text{(by definition of } f\text{).}$$

Thus, $f(n)$ is multiplicative. ■

Note that the converse of Theorem 4 is also true. See Exercise 12 in Section 5.3.

It would be instructive to write out the steps in the proof for specific values of m and n, say $m = 12$ and $n = 35$.

We shall now write out explicitly our routine procedure for dealing with many arithmetic functions. Let $f(n)$ be an arithmetic function.

Step 1: Express $f(n)$ as a sum over all divisors d of n of a multiplicative function $g(n)$:

$$f(n) = \sum_{d|n} g(d). \tag{2}$$

Conclude from Theorem 4 that $f(n)$ is multiplicative.

Step 2: Let p be a prime and $a \geq 1$ be an integer. Compute $f(p^a)$ from (2). Indeed,

$$f(p^a) = \sum_{i=0}^{a} g(p^i).$$

Step 3: Write $n = p_1^{a_1}p_2^{a_2} \cdots p_t^{a_t}$ and combine Step 2 with Proposition 3, deriving a formula for $f(n)$:

$$f(n) = f(p_1^{a_1})f(p_2^{a_2}) \cdots f(p_t^{a_t}).$$

(Step 1 allows us to apply Proposition 3.)

Let us now give some basic examples of the above procedure. Namely, let us take $f(n) = d(n)$ and $f(n) = \sigma(n)$. In these cases, $f(n)$ is given "naturally" in the form where we can carry out Step 1.

We shall first consider the case of $f(n) = d(n)$. Since $d(n)$ is the number of positive divisors of n, it is the sum over all the positive divisors of n of the constant function 1. That is, if $g(n) = 1$ for all n, then

$$f(n) = \sum_{d|n} g(d);$$

i.e.,

$$d(n) = \sum_{d|n} 1.$$

Since $g(n)$ is clearly multiplicative, we conclude that $f(n) = d(n)$ is multiplicative, and we have Step 1. For Step 2 let p be a prime and $a \geq 1$ be an integer. Then

$$d(p^a) = \sum_{i=0}^{a} 1 = a + 1.$$

That is, it is obvious that p^a has the $a + 1$ divisors $1, p, p^2, \ldots, p^a$. Thus, from Step 3 we obtain

Theorem 5: If $n = p_1^{a_1}p_2^{a_2} \cdots p_t^{a_t}$ where $p_1, p_2, \ldots, p_t$ are distinct primes, then

$$d(n) = (a_1 + 1)(a_2 + 1) \cdots (a_t + 1).$$

Moreover, $d(n)$ is multiplicative.

Next consider the case where $f(n) = \sigma(n)$. Since $\sigma(n)$ is the sum of all the positive divisors of n, it is the sum over the positive divisors of n of the function $g(m) = m$. That is,

$$f(n) = \sum_{d|n} g(d) = \sum_{d|n} d = \sigma(n).$$

Since $g(n) = n$ is clearly multiplicative, we conclude that $f(n) = \sigma(n)$ is multiplicative, and we have Step 1. For Step 2 let p be a prime and $a \geq 1$ be an integer. Then

$$\begin{aligned}\sigma(p^a) &= \sum_{d|p^a} d = \sum_{i=0}^{a} p^i \\ &= 1 + p + p^2 + \cdots + p^a \\ &= \frac{p^{a+1} - 1}{p - 1}.\end{aligned}$$

(The last expression is the sum of the geometric progression.) Thus, from Step 3 we obtain

Theorem 6: If $n = p_1^{a_1} p_2^{a_2} \cdots p_t^{a_t}$ where $p_1, p_2, \ldots, p_t$ are distinct primes, then

$$\begin{aligned}\sigma(n) &= (1 + p_1 + p_1^2 + \cdots + p_1^{a_1})(1 + p_2 + p_2^2 + \cdots + p_2^{a_2}) \cdots \\ &\qquad (1 + p_t + p_t^2 + \cdots + p_t^{a_t}) \\ &= \frac{p_1^{a_1+1} - 1}{p_1 - 1} \frac{p_2^{a_2+1} - 1}{p_2 - 1} \cdots \frac{p_t^{a_t+1} - 1}{p_t - 1}.\end{aligned}$$

Moreover, $\sigma(n)$ is multiplicative.

Example 7: Let $n = 17640 = 2^3 \cdot 3^2 \cdot 5 \cdot 7^2$. Then

$$d(17640) = (3 + 1)(2 + 1)(1 + 1)(2 + 1) = 4 \cdot 3 \cdot 2 \cdot 3 = 72.$$

Also,

$$\begin{aligned}\sigma(17640) &= (1 + 2 + 2^2 + 2^3)(1 + 3 + 3^2)(1 + 5)(1 + 7 + 7^2) \\ &= 15 \cdot 13 \cdot 6 \cdot 57 = 66690\end{aligned}$$

or

$$\begin{aligned}\sigma(17640) &= \frac{2^4 - 1}{2 - 1} \frac{3^3 - 1}{3 - 1} \frac{5^2 - 1}{5 - 1} \frac{7^3 - 1}{7 - 1} \\ &= 15 \cdot 13 \cdot 6 \cdot 57 = 66690.\end{aligned}$$

We shall do one more example.

Example 8: For real numbers r we defined $\sigma_r(n)$ as the sum of the rth powers of the divisors of n. That is,

$$\sigma_r(n) = \sum_{d|n} d^r.$$

Thus, if $g(n) = n^r$, we see that $g(n)$ is multiplicative, and thus $\sigma_r(n)$ is multiplicative. If p is a prime and $a \geq 1$ is an integer, then

$$\sigma_r(p^a) = \sum_{d \mid p^a} d^r = \sum_{i=0}^{a} p^{ir} = \frac{p^{(a+1)r} - 1}{p^r - 1}.$$

(The last expression requires that $r \neq 0$.) Thus, if $n = p_1^{a_1} p_2^{a_2} \cdots p_t^{a_t}$, where $p_1, p_2, \ldots, p_t$ are distinct primes, then

$$\begin{aligned}\sigma_r(n) &= (1 + p_1^r + p_1^{2r} + \cdots + p_1^{a_1 r}) \cdots (1 + p_t^r + p_t^{2r} + \cdots + p_t^{a_t r}) \\ &= \frac{p_1^{(a_1+1)r} - 1}{p_1^r - 1} \cdots \frac{p_t^{(a_t+1)r} - 1}{p_t^r - 1}.\end{aligned}$$

(The latter expression requires that $r \neq 0$.) This example includes $d(n)$ $(r = 0)$ and $\sigma(n)$ $(r = 1)$ as special cases.

5.2 Exercises

1. Write out the proof of Theorem 4, explicitly verifying all the details, for $n = 4$ and $m = 9$.
2. Prove directly that if $n = p_1^{a_1} p_2^{a_2} \cdots p_r^{a_r}$ where $p_1, p_2, \ldots, p_r$ are distinct primes, then

   $$d(n) = (a_1 + 1)(a_2 + 1) \cdots (a_r + 1).$$

   Conclude from this formula that d is multiplicative.
3. Show that $d(n) = 2$ if and only if n is prime.
4. Show that $d(n)$ is odd if and only if n is a perfect square.
5. Show that for every positive integer m there are an infinite number of integers n such that $d(n) = m$. What is the smallest n that works?
6. Show that $\prod_{d \mid n} d = n^{d(n)/2}$. ($\prod_{d \mid n} d$ = product of all the positive divisors of n.)
7. Let $F(n) = \sum_{k \mid n} d(k)$. Derive a formula for $F(n)$. What is $F(n)$ if $n = p_1 \cdots p_r$, p_i distinct primes?
8. Prove that $\sum_{d \mid n} 1/d = \sigma(n)/n$.
9. Let $n = p_1^{a_1} p_2^{a_2} \cdots p_r^{a_r}$, where $p_1, \ldots, p_r$ are distinct primes. Observe that $m \mid n$ if and only if $m = p_1^{v_1} p_2^{v_2} \cdots p_r^{v_r}$ with $0 \leq v_i \leq a_i$ $(1 \leq i \leq r)$ and so

   $$\begin{aligned}\sigma(n) &= \sum_{v_1=0}^{a_1} \sum_{v_2=0}^{a_2} \cdots \sum_{v_r=0}^{a_r} p_1^{v_1} p_2^{v_2} \cdots p_r^{v_r} \\ &= \sum_{v_1=0}^{a_1} \cdots \sum_{v_{r-1}=0}^{a_{r-1}} p_1^{v_1} \cdots p_{r-1}^{v_{r-1}} \sum_{v_r=0}^{a_r} p_r^{v_r}\end{aligned}$$

to derive the formula of Theorem 6 for $\sigma(n)$. Conclude that $\sigma(n)$ is multiplicative from the formula.

10. For any arithmetic function $f(n)$ show that

$$\sum_{d|n} f(d) = \sum_{d|n} f\left(\frac{n}{d}\right).$$

11. If $n = p_1 p_2 \cdots p_k$ ($p_1, p_2, \ldots, p_k$ primes, not necessarily distinct), set

$$\lambda(n) = (-1)^k$$

(so $\lambda(2) = -1$, $\lambda(4) = (-1)^2 = 1$, $\lambda(12) = (-1)^3 = -1$). Set $\lambda(1) = 1$.

(a) Show that $\lambda(n)$ is totally multiplicative.
(b) Show that

$$F(n) = \sum_{d|n} \lambda(d) = \begin{cases} 1 & \text{if } n \text{ is a perfect square,} \\ 0 & \text{otherwise.} \end{cases}$$

12. If $n = p_1 p_2 \cdots p_k$, where $p_1, \ldots, p_k$ are primes, set $\nu(n) = 2^k$ (set $\nu(1) = 1$).

(a) Show that $\nu(n)$ is totally multiplicative.
(b) Derive an expression for $\sum_{d|n} \nu(d)$.

13. If $n = p_1^{a_1} p_2^{a_2} \cdots p_k^{a_k}$ where $p_1, p_2, \ldots, p_k$ are distinct primes and t is an integer, set

$$\omega(n) = t^k.$$

(Set $\omega(1) = 1$.)

(a) Show that $\omega(n)$ is multiplicative.
(b) Show that for n as above

$$\sum_{d|n} \omega(d) = (1 + a_1 t)(1 + a_2 t) \cdots (1 + a_k t).$$

14. Let p be a prime. Define

$$Q(n) = \begin{cases} 0 & \text{if } p \mid n, \\ \left(\dfrac{n}{p}\right) & \text{if } p \nmid n. \end{cases}$$

Let $F(n) = \sum_{d|n} Q(d)$.

(a) Show that Q is multiplicative.
(b) Compute $F(p^k)$ for $k = 1, 2, \ldots$.
(c) Let $q \neq p$ be a prime. Considering the cases where $\left(\frac{q}{p}\right) = 1$ and $\left(\frac{q}{p}\right) = -1$, derive a formula for $F(q^k)$, $k = 1, 2, \ldots$.
(d) Derive an espression for $F(n)$ for $n = 1, 2, \ldots$.
(e) Writing more specifically $F_p(n)$ for the function above, show that if p and q are distinct primes and one of them is congruent to 1 mod 4, then

$$F_p(q^k) = F_q(p^k).$$

Is this result true if $q \equiv p \equiv 3 \pmod 4$?

15. For arithmetic functions f and g, define another arithmetic function denoted $f * g$ (called the *convolution of f and g*) by

$$(f * g)(n) = \sum_{d|n} f(d)g\left(\frac{n}{d}\right).$$

Prove the following results for arithmetic functions f, g, h:

(a) $f * g = g * f$.
(b) $(f * g) * h = f * (g * h)$.
(c) $(f + g) * h = f * h + g * h$. ($(f + g)(n) = f(n) + g(n)$ by definition.)
(d) f and g multiplicative implies that $f * g$ is multiplicative.

Define the special functions $I(n) = n$, $I_k(n) = n^k$, and $\mathbf{1}(n) = 1$.

(e) f is multiplicative implies that $\mathbf{1} * f$ is multiplicative.
(f) $d = \mathbf{1} * \mathbf{1}$.
(g) $\sigma = \mathbf{1} * I$.
(h) $\sigma_k = \mathbf{1} * I_k$.
(i) If f is totally multiplicative, then

$$(f * f)(n) = f(n)d(n).$$

(j) Compute $\sigma * d$.
(k) $I_k * I_\ell = n^\ell \sigma_{k-\ell}(n)$.

16. Show that $d(n) \leq 2\sqrt{n}$. (*Hint:* If $n = k\ell$, then k or ℓ is $\leq \sqrt{n}$.)

17. Show that there are an infinite number of integers n such that $d(n) \geq \log n/\log 2$.

**18. Show that for any integer k there is a constant $C > 0$ such that there are an infinite number of integers n satisfying $d(n) \geq C(\log n)^k$.

19. Show that for all n

$$n \leq \sigma(n) \leq n^2.$$

20. Let $f(n)$ denote the number of distinct solutions modulo n of the congruence $x^2 \equiv a \pmod n$.

(a) Show that $f(n)$ is multiplicative.
(b) Find a formula for $f(n)$.

21. Let $g(x)$ be any polynomial with integer coefficients and let $f_g(n)$ denote the number of distinct solutions (modulo n) of the congruence $g(x) \equiv 0 \pmod n$. Show that $f_g(n)$ is multiplicative.

*22. Let $\lambda(n)$ be the number-theoretic function of Exercise 11. Show that

$$\sum_{n=1}^{x} \lambda(n)\left[\frac{x}{n}\right] = [\sqrt{x}].$$

5.3 The Möbius Inversion Formula

So far we have dealt with arithmetic functions which were essentially defined as a sum, over the divisors of an integer, of a multiplicative function. But this is not always the case. For example, $\phi(n)$ has no obvious expression in this form, and so it is not clear how to apply the procedure outlined above. Applying this procedure to obtain a formula for $\phi(n)$ is one of our primary motivations for this section.

The point is that there is a relatively obvious expression of the required type except that it is "backwards." Namely we shall show in Proposition 7 that

$$n = \sum_{d|n} \phi(d).$$

Thus, the following general question arises: Suppose that

$$f(n) = \sum_{d|n} g(d) \tag{1}$$

for all positive integers n. Can we determine $g(n)$ in terms of $f(n)$? More loosely stated, we ask whether we can "solve" for $g(n)$ in terms of $f(n)$? In this section we shall answer this question in the affirmative. The result is called the *Möbius Inversion Formula*.

Before we go any further, let us convince ourselves that, at least in principle, we can use (1) to compute $g(n)$ from $f(n)$. By (1), for $n = 1$, we see that

$$g(1) = f(1). \tag{2}$$

For $n = 2$, we see that

$$f(2) = g(1) + g(2),$$

so that by (2),

$$g(2) = f(2) - f(1). \tag{3}$$

For $n = 3$, (1) yields

$$f(3) = g(1) + g(3),$$

so that by (2),

$$g(3) = f(3) - f(1).$$

For $n = 4$, (1) yields

$$f(4) = g(1) + g(2) + g(4),$$

so that by (2) and (3)

$$\begin{aligned} g(4) &= f(4) - g(1) - g(2) \\ &= f(4) - f(1) - (f(2) - f(1)) \\ &= f(4) - f(2). \end{aligned}$$

Proceeding in this way, we can compute $g(n)$ in terms of $f(n)$ for arbitrary n. But is there a pattern to the formulas? Indeed, there is, and this pattern was determined by the German mathematician Möbius in 1832. He stated it in terms of what is now called the Möbius function $\mu(n)$. Its definition may look a little strange, but it is very simple, and we should keep in mind that its definition is motivated by the question of solving (1) for $g(n)$ in terms of $f(n)$.

Definition 1: Let n be a positive integer. The *Möbius function* $\mu(n)$ is defined as follows:

$$\mu(1) = 1$$

$$\mu(n) = \begin{cases} 0, & \text{if } n \text{ is divisible by the square of a prime,} \\ (-1)^t, & \text{if } n = p_1 p_2 \cdots p_t, \text{ where } p_1, p_2, \ldots, p_t \text{ are distinct primes.} \end{cases}$$

Thus, for example, $\mu(2) = -1$, $\mu(3) = -1$, $\mu(4) = 0$, $\mu(5) = -1$, and $\mu(6) = 1$. We have included the values of $\mu(n)$ for all $n \leq 100$ in Table 1 at the end of the book. As we shall see shortly, the arithmetic function $\mu(n)$ is an important function. Before we solve our original problem we must derive some properties of $\mu(n)$.

Proposition 2: $\mu(n)$ is multiplicative.

Proof: Let m and n be positive integers such that $\gcd(m, n) = 1$. If $n = 1$, then

$$\mu(mn) = \mu(m) = \mu(m)\mu(1) = \mu(m)\mu(n),$$

since $\mu(1) = 1$. Thus, we may assume that $m > 1$ and $n > 1$. Next suppose that p is a prime and that $p^2 \mid n$. Then, of course, $p^2 \mid mn$, and thus

$$\mu(mn) = 0 = \mu(m)\cdot 0 = \mu(m)\mu(n)$$

since $\mu(n) = 0$. Now we may assume that both m and n are not divisible by the square of any prime. Then we may write $m = p_1 p_2 \cdots p_r$ and $n = q_1 q_2 \cdots q_s$, where $p_1, \ldots, p_r$ are distinct primes and $q_1, \ldots, q_s$ are distinct primes. Then, since $\gcd(m, n) = 1$, no p_i can be a q_j. Therefore, $mn = p_1 \cdots p_r q_1 \cdots q_s$ is a product of primes where all the primes are distinct. Thus,

$$\mu(mn) = (-1)^{r+s} = (-1)^r(-1)^s = \mu(m)\mu(n).$$

■

Proposition 3: Let n be a positive integer. Then

$$\sum_{d \mid n} \mu(d) = \begin{cases} 1 & \textit{if } n = 1, \\ 0 & \textit{if } n > 1. \end{cases}$$

Proof: Define the arithmetic function $f(n)$ by

$$f(n) = \sum_{d \mid n} \mu(d).$$

We compute $f(n)$ by the procedure outlined in Section 2. First, $\mu(n)$ is a multiplicative function. Thus, by Theorem 2.4, $f(n)$ is multiplicative. Now if p is a prime and $a \geq 1$ is an integer, then

$$f(p^a) = \sum_{d \mid p^a} \mu(d) = \mu(1) + \mu(p) + \mu(p^2) + \cdots + \mu(p^a)$$
$$= 1 - 1 = 0$$

since $\mu(1) = 1$ and $\mu(p) = -1$ and $p^2 \mid p^2, p^2 \mid p^3, \ldots, p^2 \mid p^a$, and so the terms $\mu(p^i)$ $(i \geq 2)$ are zero. Finally, if $n = p_1^{a_1} p_2^{a_2} \cdots p_t^{a_t}$ where $p_1, p_2, \ldots, p_t$ are distinct primes, then

$$f(n) = f(p_1^{a_1}) f(p_2^{a_2}) \cdots f(p_t^{a_t}) = 0 \cdot 0 \cdots 0 = 0.$$

As it is clear that $f(1) = \mu(1) = 1$, Proposition 3 is completely proved. ■

Let us now use the Möbius function to answer the question raised at the beginning of this section.

Theorem 4 (Möbius Inversion Formula): Let $g(n)$ be any arithmetic function and let $f(n)$ be given by

$$f(n) = \sum_{d \mid n} g(d).$$

Then

$$g(n) = \sum_{d \mid n} \mu(d) f\left(\frac{n}{d}\right).$$

(Note: It is *not* required that $g(n)$ be multiplicative.)

Proof: We clearly have

$$\sum_{d \mid n} \mu(d) f\left(\frac{n}{d}\right) = \sum_{d \mid n} \mu(d) \Big(\sum_{c \mid (n/d)} g(c) \Big)$$
$$= \sum_{d \mid n} \sum_{c \mid (n/d)} \mu(d) g(c).$$

The last double sum is over all pairs of positive integers (c, d) such that $d \mid n$ and $c \mid (n/d)$. This is the same as the sum over all pairs of positive integers (c, d) such that $cd \mid n$ (exercise). Thus, we compute the double sum in the following way. First, sum over all c dividing n, and for each fixed c, sum over all those d such that $d \mid (n/c)$. Then we see that

$$\sum_{d \mid n} \mu(n) f\left(\frac{n}{d}\right) = \sum_{c \mid n} \sum_{d \mid (n/c)} \mu(d) g(c)$$
$$= \sum_{c \mid n} g(c) \Big(\sum_{d \mid (n/c)} \mu(d) \Big).$$

By Proposition 3, the inner sum equals 0, except when $n/c = 1$, in which case it equals 1. Thus, all terms in the sum over c are 0, except for the term corresponding to $c = n$. Thus,

$$\sum_{d|n} \mu(n) f\left(\frac{n}{d}\right) = g(n) \sum_{d|1} \mu(d) = g(n).$$ ■

Let us give some examples of applications of the Möbius Inversion Formula.

Example 5: Since $d(n) = \sum_{e|n} 1$, we may set $g(n) = 1$ and $f(n) = d(n)$ to deduce that

$$\sum_{e|n} \mu(e) d\left(\frac{n}{e}\right) = 1,$$

a fact which is by no means obvious.

Example 6: Since $\sigma(n) = \sum_{d|n} d$, we may set $g(n) = n$ and $f(n) = \sigma(n)$ to deduce that

$$\sum_{d|n} \mu(d) \sigma\left(\frac{n}{d}\right) = n.$$

Let us now complete our discussion of the Euler phi-function, $\phi(n)$. As noted before, we require the following result:

Proposition 7: Let n be a positive integer. Then

$$n = \sum_{d|n} \phi(d). \tag{4}$$

Before proving Proposition 7, let us use it in conjunction with the Möbius Inversion Formula and our procedure of Section 2 to derive the basic properties of $\phi(n)$.

First, we apply the Möbius Inversion Formula with $g(n) = \phi(n)$ and $f(n) = n$ to Eq. (4) to obtain

$$\begin{aligned} \phi(n) &= \sum_{d|n} \mu(d) \frac{n}{d} \\ &= n \sum_{d|n} \frac{\mu(d)}{d}. \end{aligned} \tag{5}$$

We are now in a position to apply our procedure of Section 2. First, $\mu(n)$ is multiplicative and the function $g(n) = 1/n$ is multiplicative. Thus, the function $h(n) = \mu(n)/n$ is multiplicative. From Eq. (5) and Theorem 2.4 we conclude that $k(n) = \phi(n)/n$ is multiplicative, and thus, finally, $\phi(n) = nk(n)$ is multiplicative. Now if p is a prime and $a \geq 1$ is an integer, then

$$\begin{aligned} \phi(p^a) &= p^a \sum_{d|p^a} \frac{\mu(d)}{d} \\ &= p^a\left(\frac{\mu(1)}{1} + \frac{\mu(p)}{p} + \frac{\mu(p^2)}{p^2} + \cdots + \frac{\mu(p^a)}{p^a}\right) \\ &= p^a\left(1 - \frac{1}{p}\right). \end{aligned}$$

Finally, if $n = p_1^{a_1}p_2^{a_2} \cdots p_t^{a_t}$, where $p_1, p_2, \ldots, p_t$ are distinct primes, then

$$\begin{aligned}\phi(n) &= \phi(p_1^{a_1})\phi(p_2^{a_2}) \cdots \phi(p_t^{a_t}) \\ &= p_1^{a_1}\left(1 - \frac{1}{p_1}\right)p_2^{a_2}\left(1 - \frac{1}{p_2}\right)\cdots p_t^{a_t}\left(1 - \frac{1}{p_t}\right) \\ &= n\left(1 - \frac{1}{p_1}\right)\left(1 - \frac{1}{p_2}\right)\cdots\left(1 - \frac{1}{p_t}\right) \\ &= p_1^{a_1-1}(p_1 - 1)p_2^{a_2-1}(p_2 - 1)\cdots p_t^{a_t-1}(p_t - 1).\end{aligned}$$

We shall record our results in a theorem.

Theorem 8: If $n = p_1^{a_1}p_2^{a_2} \cdots p_t^{a_t}$ where $p_1, p_2, \ldots, p_t$ are distinct primes, then

$$\begin{aligned}\phi(n) &= n\left(1 - \frac{1}{p_1}\right)\left(1 - \frac{1}{p_2}\right)\cdots\left(1 - \frac{1}{p_t}\right) \\ &= p_1^{a_1-1}(p_1 - 1)p_2^{a_2-1}(p_2 - 1)\cdots p_t^{a_t-1}(p_t - 1).\end{aligned}$$

Moreover $\phi(n)$ is multiplicative. Further, we have the formula

$$\phi(n) = n\sum_{d|n}\frac{\mu(d)}{d}.$$

Example 9. $\phi(4320) = \phi(2^5 \cdot 3^3 \cdot 5) = 2^4(2 - 1)3^2(3 - 1)5^0(5 - 1) = 1152.$

All the above reasoning was done on the basis of Proposition 7, which we have yet to prove. Before we can prove Proposition 7, however, we need the following fact:

Lemma 10: Let n be a positive integer and let d be a divisor of n. Then the number of integers k such that $1 \le k \le n$ and $\gcd(k, n) = d$ is equal to $\phi(n/d)$.

Proof: If $\gcd(k, n) = d$, then $d|k$ and $\gcd(k/d, n/d) = 1$. Moreover, if $1 \le k \le n$, then $1 \le k/d \le n/d$. Thus, each integer k satisfying the conditions of the lemma satisfies the conditions $d|k$, $\gcd(k/d, n/d) = 1$ and $1 \le k/d \le n/d$. If k is such an integer, then $k = k'd$ and k' satisfies $\gcd(k', n/d) = 1$ and $1 \le k' \le n/d$. Conversely, given k' such that $\gcd(k', n/d) = 1$ and $1 \le k' \le n/d$, the integer $k = k'd$ satisfies the conditions of the lemma. Therefore, the number of k satisfying the conditions of the lemma equals the number of integers k' such that $1 \le k' \le n/d$ and $\gcd(k', n/d) = 1$. This number is clearly $\phi(n/d)$. ∎

Proof of Proposition 7: First note that every integer k such that $1 \le k \le n$ satisfies $\gcd(k, n) = d$ for one and only one divisor d of n. Therefore, by Lemma 10, we have

$$n = \sum_{d|n} \phi\left(\frac{n}{d}\right) = \sum_{d'|n} \phi(d')$$

since as d runs over all divisors of n, so does $d' = n/d$. ■

5.3 Exercises

1. Let $f(n)$ be an arithmetic function, and let $h(n) = \sum_{d|n} f(d)$. Compute $f(24)$ in terms of h.
2. Prove the Möbius inversion formula by the direct proof given in the text for $n \leq 10$.
3. Calculate $\mu(5)$, $\mu(64)$, $\mu(39)$, and $\mu(30)$.
4. Calculate $\phi(2)$, $\phi(5)$, $\phi(47)$, and $\phi(144)$.
5. Show that $\phi(n)$ is even for $n > 2$.
6. Let $f(n)$ be a multiplicative function.
   (a) Show that $\sum_{d|n} \mu(d)f(d) = \prod_{p|n,\ p \text{ prime}} (1 - f(p))$.
   (b) Show that $\sum_{d|n} \mu(d)/d = \prod_{p|n,\ p \text{ prime}} (1 - (1/p))$.
7. Let $\omega(n)$ denote the number of distinct prime factors of n. Show that
$$\sum_{d|n} |\mu(d)| = 2^{\omega(n)}.$$
8. Show that $\sum_{d|n} \mu(d)\phi(d) = (-1)^{\omega(n)} \prod_{p|n} (p - 2)$, where $\omega(n)$ is defined in Exercise 7.
9. Show that $\sum_{d|n} \mu^2(d)/d = \prod_{p|n} (1 + (1/p))$.
10. Show that the number of reduced fractions a/b such that $0 \leq a/b < 1$, b fixed, is just $\phi(b)$.
11. A famous conjecture due to Mertens states that
$$\left|\sum_{n=1}^{r} \mu(n)\right| < \sqrt{r}$$
for any r. Verify Merten's conjecture for $r \leq 30$. Write a computer program to check Merten's conjecture for r below a given bound. This seemingly innocent conjecture is very deep. In fact, it implies the so-called Riemann hypothesis, one of the most celebrated unsolved problems in mathematics. It seems likely that Merten's conjecture is false, but present-day computers are not fast enough to allow a sufficiently extensive search for a counterexample.
12. Let $f(n)$ be an arithmetic function and define $F(n)$ by
$$F(n) = \sum_{d|n} f(d).$$
Show that if $F(n)$ is multiplicative, then $f(n)$ is multiplicative.

13. Let n be a fixed positive integer. Show that there are only finitely many integers x such that $\phi(x) = n$.

14. Suppose that $f(n)$ is a number-theoretic function and suppose that

$$\sum_{d|n} f(n) = n$$

for all n. Show that $f(n) = \phi(n)$.

15. It is an unsolved problem to show that whenever $\phi(m) | m - 1$, m is prime. Show that if $\phi(m) | m - 1$, then m is square-free.

16. Prove that

$$\sum_{d=1}^{n} \phi(d)\left[\frac{n}{d}\right] = \frac{n(n+1)}{2}.$$

(*Hint:* Use induction on n.)

17. For n, k positive integers, set $\sigma_k^*(n) = \sum_{i=1,\ \gcd(i,n)=1}^{n} i^k$.

    (a) Show that

$$\sum_{d|n} \frac{\sigma_k^*(d)}{d^k} = \frac{1^k + 2^k + \cdots + n^k}{n^k}.$$

    (b) Use part (a) and the Möbius inversion formula to compute $\sigma_1^*(n)$, $\sigma_2^*(n)$, and $\sigma_3^*(n)$.

18. Prove that $\sum_{a=1,\ \gcd(a,n)=1}^{n} a = \frac{1}{2}n\phi(n)$.

19. Let a, b be positive integers and let c denote the product of all primes dividing both a and b.

    (a) Prove that

$$\phi(ab) = \phi(a)\phi(b)\frac{c}{\phi(c)}.$$

    (b) Using part (a), find all pairs of integers a, b for which $\phi(a)\phi(b) = \phi(ab)$.

20. Let $f(n)$ be any arithmetic function and suppose $g(n)$ is defined by

$$g(n) = \sum_{d|n} \mu(d) f\left(\frac{n}{d}\right).$$

Prove that

$$f(n) = \sum_{d|n} g(d).$$

5.4 Perfect and Amicable Numbers

As an application of our theory of arithmetic functions, let us study two classes of numbers which have been known and studied since ancient times, the perfect and amicable numbers.

Definition 1: A *perfect number* is a positive integer n which equals the sum of its proper divisors (i.e., the positive divisors of n, excluding n itself).

Example 2:

(i) The proper divisors of 6 are 1, 2, 3, and $1 + 2 + 3 = 6$, and so 6 is perfect.

(ii) The proper divisors of 28 are 1, 2, 4, 7, 14, and $1 + 2 + 4 + 7 + 14 = 28$, and so 28 is perfect.

(iii) The reader should not have too much difficulty verifying that 496 and 8128 are perfect.

The perfect numbers have occurred throughout history in connection with various forms of numerology. They occurred in the works of the ancient Pythagoreans, who thought that the property which perfect numbers exhibit represented some sort of pinnacle of aesthetics, whence the name "perfect." Much ancient superstition and mystical belief rested on the special properties of integers such as the perfect numbers. And very often in the ancient world, proof by means of numerology substituted for more scientific explanations of natural phenomena. It was in this context that much number theory, including our present discussion of perfect numbers, began.

How can we find perfect numbers? Are there infinitely many of them? These are questions which have been asked by mathematicians for thousands of years and as yet are not completely answered. We shall give same partial answers below.

Since $\sigma(n)$ equals the sum of the divisors of n (including n), we see that n is perfect if and only if $\sigma(n) - n = n$, which is equivalent to

$$\sigma(n) = 2n. \tag{1}$$

There is a class of perfect numbers which was already known to Euclid, namely

Theorem 3: Suppose that $2^a - 1$ is a prime. Then $2^{a-1}(2^a - 1)$ is perfect.

Proof: If $p = 2^a - 1$ is prime, then clearly $a > 1$ and p is odd. Set $n = 2^{a-1}p$. Then $\gcd(2^{a-1}, p) = 1$. Thus, by Theorem 2.6, we have

$$\sigma(n) = \sigma(2^{a-1})\sigma(p) = \frac{2^a - 1}{2 - 1}(p + 1) = 2n. \qquad \blacksquare$$

A prime p of the form $p = 2^a - 1$ is called a *Mersenne prime*. By what we have just proved, the search for perfect numbers is intimately connected with the search for Mersenne primes. The following elementary fact is useful in testing for Mersenne primes:

Proposition 4: If $2^a - 1$ is a prime, then a is a prime.

Proof: Exercise. ■

By Proposition 4, the Mersenne primes are all of the form $2^p - 1$, where p is a prime. The first few are

$$2^2 - 1 = 3$$
$$2^3 - 1 = 7$$
$$2^5 - 1 = 31$$
$$2^7 - 1 = 127$$
$$2^{13} - 1 = 8191$$
$$2^{17} - 1 = 131071$$
$$2^{19} - 1 = 524287.$$

Note, however, that not every integer of the form $2^p - 1$, where p is prime, is itself a prime. We leave it to the reader to show that $2^{11} - 1$ is not prime.

Is it true that the perfect numbers given by Theorem 3 are the only ones? This is not currently known. However, we can prove the following result due to Euler:

Theorem 5: Let n be an even perfect number. Then n is of the form $2^{a-1}(2^a - 1)$, where $2^a - 1$ is a Mersenne prime.

Proof: Write $n = 2^c b$, where $c \geq 1$ and b is odd. Since n is perfect, $\sigma(n) = 2n$, and since $\sigma(n)$ is multiplicative, we have

$$\begin{aligned} 2n = \sigma(n) &= \sigma(2^c b) = \sigma(2^c)\sigma(b) \\ &= (2^{c+1} - 1)\sigma(b). \end{aligned}$$

Therefore,

$$2^{c+1}b = (2^{c+1} - 1)\sigma(b).$$

Note that $\gcd(2^{c+1}, 2^{c+1} - 1) = 1$, and therefore $2^{c+1} \mid \sigma(b)$, say $\sigma(b) = 2^{c+1}d$. Then

$$b = (2^{c+1} - 1)d.$$

Suppose that $d > 1$. Then b has at least the following divisors: b, d, 1. Therefore,

$$\begin{aligned} \sigma(b) \geq b + d + 1 &= (2^{c+1} - 1)d + d + 1 \\ &= 2^{c+1}d + 1, \end{aligned}$$

which contradicts the fact that $\sigma(b) = 2^{c+1}d$. Thus, $d = 1$ and $b = 2^{c+1} - 1$, so that $n = 2^c(2^{c+1} - 1)$. Moreover,

$$\sigma(2^{c+1} - 1) = \sigma(b) = 2^{c+1}d = 2^{c+1}. \tag{2}$$

Now if $2^{c+1} - 1$ is not prime then $\sigma(2^{c+1} - 1) > (2^{c+1} - 1) + 1$ since $2^{c+1} - 1$ has a divisor other than $2^{c+1} - 1$ and 1. But this is a contradiction of Eq. (2), so that $2^{c+1} - 1$ is a prime. Thus, if we set $a = c + 1$, we see that $n = 2^{a-1}(2^a - 1)$, where $2^a - 1$ is a Mersenne prime. ■

There are two basic questions about perfect numbers which mathematicians cannot answer at the present time. First, it is not known if there exists an odd perfect number. Second, it is not known whether there are infinitely many perfect numbers. One way of settling the latter would be to prove that there are infinitely many Mersenne primes. But this is not known to be true or false. Using high-speed computers, some recent work has been done by a number of people, including D. H. and E. Lehmer, J. Selfridge, and J. Brillhart. A recent result of B. Tuckerman is that $2^{19937} - 1$ is a Mersenne prime. It has 6002 digits and is the largest known prime. Moreover, it is only the twenty-fourth known Mersenne prime. Computers have also been used to factor completely the numbers $2^p - 1$ for primes p. A recent result of D. H. and E. Lehmer and J. Selfridge, for example, asserts that

$$2^{157} - 1 = 852133201 \cdot 60726444167 \cdot 1654058017289 \cdot 2134387368610417.$$

Another sort of number which came to be considered by number theorists by way of the mystics and astrologers is the so-called amicable numbers.

Definition 6: A pair of numbers m, n is said to be *amicable* if the sum of the proper divisors of m equals n and the sum of the proper divisors of n equals m.

By reasoning as we did with perfect numbers, we see that for m, n to be amicable, it is both necessary and sufficient that

$$\sigma(m) = m + n = \sigma(n).$$

Amicable numbers were used in preparing horoscopes and preparing talismans (good luck charms) and were supposed to have the power of creating ties between individuals. An example of an amicable pair is $m = 220$, $n = 284$, since

$$\sigma(n) = \sigma(m) = 504 = 220 + 284.$$

Another is given by $m = 17296$ and $n = 18416$, an example due to Fermat. Several hundred amicable pairs are known, but it is not known whether there are an infinite number.

5.4 Exercises

1. Show that if $2n$ is a perfect number, then

$$\sum_{d|n} \frac{1}{d} = 2.$$

2. Find all even perfect numbers less than 1000.
3. Show that an odd perfect number cannot be a prime or a product of two primes.
4. Show that if m, n are amicable numbers, then $\sigma(n) = \sigma(m) = n + m$.

6

A Few Diophantine Equations

6.1 Introduction

In this chapter, we shall study in detail the theory behind five particular Diophantine equations. All but one of these Diophantine equations are *quadratic* in the following sense. Let us suppose that integers $a_1, a_2, \ldots, a_k, n$ are given. Let us define the polynomial $f(x_1, \ldots, x_k)$ in the variables $x_1, \ldots, x_k$ by

$$f(x_1, \ldots, x_k) = a_1x_1^2 + \cdots + a_kx_k^2.$$

Then we may consider the quadratic Diophantine equation

$$f(x_1, \ldots, x_k) = n. \tag{1}$$

All but one of the equations we shall consider in this chapter will be special cases of this general Diophantine equation. For example, in Section 2, we shall consider the Pythagorean equation

$$x^2 + y^2 = z^2, \tag{2}$$

which, in this context, can be written $x_1^2 + x_2^2 - x_3^2 = 0$, so that (2) is a special case of (1) with $k = 3$, $a_1 = a_2 = 1$, $a_3 = -1$, $n = 0$. In Section 4, we shall consider the problem of representing a positive integer n as a sum of two squares, corresponding to the Diophantine equation

$$x^2 + y^2 = n. \tag{3}$$

This is a special case of (1) in which $k = 2$, $a_1 = a_2 = 1$. In Section 5, we shall study the problem of representing a positive integer n as a sum of four squares, corresponding to the Diophantine equation

$$x^2 + y^2 + z^2 + w^2 = n. \tag{4}$$

This is a special case of (1) in which $k = 4$, $a_1 = a_2 = a_3 = a_4 = 1$. Finally, in Section 6, we shall consider Pell's equation

$$x^2 - dy^2 = 1, \tag{5}$$

which is a special case of (1), where $k = 2$, $a_1 = 1$, $a_2 = -d$, $n = 1$. We shall also consider, in Section 3, Fermat's Last Theorem in the case of exponent 4:

$$x^4 + y^4 = z^4.$$

Although this equation is not a special case of (1), it nevertheless relies on the theory of the Pythagorean equation (2), and thus is closely connected with Eq. (1) in any case.

Before proceeding with a study of our five Diophantine equations, let us consider the general equation (1) and let us get some idea of the sort of questions we can ask. The first and most obvious question is, does the Diophantine equation (1) have any solutions? If the answer to this question is no (for particular choice of $a_1, \ldots, a_k, n$), then we are faced with the problem of proving that the answer is no. In this chapter, we shall meet a novel technique for doing just this: Fermat's method of infinite descent.

If the Diophantine equation (1) has solutions (for particular $a_1, \ldots, a_k$, n), then the obvious problem confronting us is to find them all. This is usually a very difficult problem, but occasionally it can be solved. For example, we shall show how to find all solutions of the Pythagorean and Pell equations. We shall show that in each case there are an infinite number of solutions and we shall determine them completely.

A final problem which is suggested by Eqs. (3) and (4) is to find all n for which (1) has a solution (for fixed $a_1, \ldots, a_k$). In the case of Eq. (3), this amounts to describing all those integers which can be written as a sum of two squares. In case of Eq. (4), this amounts to describing all those integers which can be written as a sum of four squares.

It is possible to generalize the Diophantine equation (1) by using a more general quadratic polynomial in $x_1, \ldots, x_k$. Indeed, suppose that a_{ij} $(1 \leq i, j \leq k)$ are given integers. Let

$$g(x_1, \ldots, x_k) = a_{11}x_1^2 + a_{12}x_1x_2 + a_{13}x_1x_3 + \cdots + a_{kk}x_k^2.$$

Then $g(x_1, \ldots, x_k)$ is the most general quadratic polynomial in k variables with integer coefficients. One can study the Diophantine equation

$$g(x_1, \ldots, x_k) = n$$

for a given integer n. This leads to a whole branch of number theory in which research is still going on. The reader will get some taste for this theory in the second half of this book, where we shall consider the special case $k = 2$.

That is, we shall develop the theory of Diophantine equations of the form

$$a_{11}x_1^2 + a_{12}x_1x_2 + a_{22}x_2^2 = n. \tag{6}$$

The questions we shall ask will be the same questions as we asked above: Are there any solutions (for given $a_{11}, a_{12}, a_{22}, n$)? If there are solutions, describe them explicitly. Suppose that a_{11}, a_{12}, a_{22} are given. For which n, does (6) have solutions? These questions will lead us very rapidly into the theory of quadratic fields, first studied by Gauss.

In Section 5, we shall prove the remarkable result that Eq. (4) is solvable for *every* positive integer n. In other words, every positive integer n is the sum of four perfect squares. This leads to a natural question: Can every positive integer n be represented as a sum of a fixed number of perfect cubes? That is, we ask if there is a positive integer k such that the equation

$$x_1^3 + \cdots + x_k^3 = n$$

is solvable for every positive integer n. We can ask a similar question about fourth powers, fifth powers, etc. In 1775, the English mathematician Waring asserted that every positive integer n could be written as a sum of 4 squares, 9 cubes, 19 fourth powers, and so forth. Waring had no proof for his assertion, and the statement remained unproved until 1909, when David Hilbert finally proved it. Hilbert showed that for every $k \geq 2$ there is an integer N_k such that every integer is the sum of N_k kth powers. Let $g(k)$ be the smallest possible value of N_k. Then from our results of Section 5, we shall show that $g(2) = 4$. It is also known that $g(3) = 9$. However, the values of $g(4)$ and $g(5)$ are not known. The values of $g(k)$ for $k \geq 6$ are all known, however.

6.1 Exercises

1. Find all solutions to the following Diophantine equations:
   (a) $x^2 + y^2 = 8$.
   (b) $x^2 + y^2 = 51$.
   (c) $x^2 + y^2 + z^2 = 10$.
   (d) $x^2 + y^2 + z^2 + w^2 = 18$.
   (e) $x^2 + 2xy + 2y^2 = 17$.
2. Let (x_0, y_0) be a solution of the equation $x^2 - 6y^2 = 1$.
   (a) Show that $(5x_0 + 12y_0, 5y_0 + 2x_0)$ is a solution.
   (b) Use part (a) to compute at least five different solutions to $x^2 - 6y^2 = 1$.
   (c) Show that $x^2 - 6y^2 = 1$ has infinitely many solutions.
3. Show that not all positive integers are sums of two squares. Of three squares.
4. Write all integers ≤ 20 as sums of four squares.

5. For each integer $x < 64$, determine the smallest number of perfect cubes whose sum is x.
6. Determine which primes $p \leq 30$ are sums of two squares. Can you make a general conjecture?
7. Determine which primes $p \leq 30$ are sums of three squares. Can you make a general conjecture?
8. Let a, b be integers and set $x = a^2 - b^2$, $y = 2ab$, and $z = a^2 + b^2$. Show that $x^2 + y^2 = z^2$. Conclude that there is an infinite number of distinct Pythogorean triples.
9. Assume that $x^n + y^n = z^n$ has no solution for x, y, z nonzero whenever $n = 4$ or $n =$ a prime $p > 2$. Show that $x^n + y^n = z^n$ has no solution for x, y, z nonzero for all $n \geq 3$.
10. Let $f(x, y)$ be any polynomial with integer coefficients. Let a, b, c, d be integers such that $ad - bc = \pm 1$. Let

$$g(x, y) = f(ax + by, cx + dy).$$

Show that $f(x, y) = n$ is solvable in integers x, y if and only if $g(x, y) = n$ is solvable in integers. This gives a general method for changing one equation into another without changing the solutions. You should do some special cases with, say, the equation of Exercise 2.

6.2 The Equation $x^2 + y^2 = z^2$

One of the oldest of all Diophantine problems is the determination of all right triangles whose sides are integers. If x, y, z are the lengths of the three sides with $z =$ the length of the hypotenuse, then the Pythagorean theorem asserts that

$$x^2 + y^2 = z^2. \tag{1}$$

Thus, it suffices to solve the Diophantine equation (1). Although Eq. (1) is usually associated with the Pythagorean school (around 570 B.C.), it appears that the ancient Babylonians more than 1000 years before knew more about its solution than the Pythagoreans did. The Babylonians, for philosophical and computational reasons, dealt only with rational numbers whose denominators they could handle easily in their base 60 number system. They used solutions to (1) to prepare a rudimentary table of trigonometric functions for angles differing by about 1 degree with the values of x, y, z chosen so that the entries in the table had finite expansions in their number system. That required extremely sophisticated knowledge of Eq. (1) and of mathematics in general.

Let us now determine all integer solutions to (1). First, we observe that if (x, y, z) is a solution, then so is $(\pm x, \pm y, \pm z)$ for any choice of signs. Thus, we may assume that $x > 0, y > 0, z > 0$. Next, note that if x, y, and z have a common factor d, then if we set $x_0 = x/d$, $y_0 = y/d$, $z_0 = z/d$, we have

$$x_0^2 + y_0^2 = z_0^2, \tag{**}$$

so that (x_0, y_0, z_0) is also a solution of (1). We may clearly choose d so that x_0, y_0, and z_0 have no common factors, and $x_0 > 0$, $y_0 > 0$, $z_0 > 0$. Assume that such a choice of d has been made. We claim that

$$\gcd(x_0, y_0) = \gcd(x_0, z_0) = \gcd(y_0, z_0) = 1. \tag{2}$$

Indeed, suppose that $e \mid x_0$ and $e \mid y_0$. Then $e^2 \mid x_0^2 + y_0^2$, so that $e^2 \mid z_0^2$. Thus, $e \mid z_0$, and e is a common factor of x_0, y_0, and z_0. Therefore, by assumption, $e = \pm 1$ and $\gcd(x_0, y_0) = 1$. We may prove similarly that $\gcd(x_0, z_0) = 1$ and $\gcd(y_0, z_0) = 1$.

Next let us observe that either x_0 or y_0 must be even, for if x_0 and y_0 were both odd, then $x_0^2 \equiv y_0^2 \equiv 1 \pmod 4$. Thus,

$$z_0^2 = x_0^2 + y_0^2 \equiv 1 + 1 = 2 \pmod 4.$$

As we observed many times before, a perfect square cannot be congruent to $2 \pmod 4$. Thus, either x_0 or y_0 must be even. Note that by (2), both cannot be even since $\gcd(x_0, y_0) = 1$. Thus, let us assume that x_0 is even. Then y_0 is odd, and also, since $z_0^2 = x_0^2 + y_0^2$, we see that z_0 is odd. Thus, $y_0 - z_0$ and $y_0 + z_0$ are both even, and we see that

$$\left(\frac{x_0}{2}\right)^2 = \frac{z_0 - y_0}{2} \frac{z_0 + y_0}{2}. \tag{3}$$

Next, note that

$$\gcd\left(\frac{z_0 - y_0}{2}, \frac{z_0 + y_0}{2}\right) = 1, \tag{4}$$

since any common factor of $(z_0 - y_0)/2$ and $(z_0 + y_0)/2$ divides

$$\frac{z_0 + y_0}{2} + \frac{z_0 - y_0}{2} = z_0$$

and

$$\frac{z_0 + y_0}{2} - \frac{z_0 - y_0}{2} = y_0,$$

and we have proved that $\gcd(y_0, z_0) = 1$.

By (3), the product of $(z_0 - y_0)/2$ and $(z_0 + y_0)/2$ is a perfect square. Moreover, since $z_0 > 0$ and $y_0 > 0$, we have $(z_0 + y_0)/2 > 0$. And from (3), we see that $(z_0 - y_0)/2 > 0$, so that, by (4), $(z_0 - y_0)/2$ and $(z_0 + y_0)/2$ are perfect squares. Thus, there are integers a and b such that

$$\frac{z_0 - y_0}{2} = b^2$$

$$\frac{z_0 + y_0}{2} = a^2.$$

Solving these equations for y_0 and z_0 gives

$$y_0 = a^2 - b^2, \qquad z_0 = a^2 + b^2.$$

We then obtain from (**) that $x_0 = 2ab$. Thus, we see that

$$x = 2abd, \qquad y = (a^2 - b^2)d, \qquad z = (a^2 + b^2)d. \tag{5}$$

It is a simple matter to check that (5) yields a solution to (1) for any choice of a, b, d. All the above reasoning is for x_0 even. If y_0 were even, then we would get the same formulas as (5), except that x and y would be interchanged. Thus, we have proved the following result:

Theorem 1: Every solution of the Diophantine equation $x^2 + y^2 = z^2$ is of the form

$$x = \pm 2abd, \qquad y = \pm(a^2 - b^2)d, \qquad z = \pm(a^2 + b^2)d \tag{*}$$

or of a similar form with x and y interchanged. Conversely, if a, b, d are any integers, then (*), for any choice of signs, is a solution to the Diophantine equation.

From Theorem 1, we can deduce a result which will be of use to us in the next section. Let us first, however, make a definition.

Definition 2: Suppose that (x, y, z) is a solution of $x^2 + y^2 = z^2$ such that x, y, and z have no common factors greater than 1. Then (x, y, z) is said to be a *primitive solution*.

Suppose that (x, y, z) is a primitive solution of $x^2 + y^2 = z^2$. Let us continue the notation of Theorem 1. It is clear that d is a common factor of x, y, z, and thus $d = 1$. Moreover, if e is a common factor of a and b, then e is a common factor of x, y, z, and so $e = 1$ also. That is, $\gcd(a, b) = 1$. Furthermore, one of a or b must be even. For if a and b are both odd, then 2 is a common factor of y and z, which means that $2 \mid z^2 - y^2 = x^2$, so that x is even, which contradicts the primitivity of the solution (x, y, z). Suppose, furthermore, that $x > 0$, $y > 0$, $z > 0$. Then we may clearly choose a and b positive, $a > b$. Thus, we have proved the following result.

Theorem 3: Let (x, y, z), with $x > 0$, $y > 0$, $z > 0$, be a primitive solution of the Pythagorean equation. Then there exist positive integers a, b with $\gcd(a, b) = 1$, one of a, b even, $a > b$, such that either

$$x = 2ab, \qquad y = a^2 - b^2, \qquad z = a^2 + b^2,$$

or the corresponding formulas hold with x and y interchanged.

6.2 Exercises

1. Write down five different primitive Pythogorean triples.
2. Find all solutions of the Diophantine equation $x^2 + 4y^2 = z^2$.
3. Find all angles θ such that $\sin\theta$ and $\cos\theta$ are rational numbers.

4. Find all solutions of the Diophantine equation $5x^2 + 10xy + 10y^2 = z^2 + 2z + 1$.
5. Find all solutions of $x^2 + y^2 = z^4$.
6. Show by induction on n that for all $n \geq 1$, $x^2 + y^2 = z^{2^n}$ has an infinite number of solutions for which $xyz \neq 0$.
7. Show that $x^2 + y^4 = z^2$ has an infinite number of solutions for which $xyz \neq 0$ and $\gcd(x, y) = 1$.
8. Let $n \geq 3$ be given. Show that there is a Pythogorean triple (x, y, z) such that one of x, y, z is n.
9. Determine all right triangles with sides of integral length such that one leg differs by 2 or 3 from the hypotenuse.
10. Find all solutions of
    (a) $x^2 + 2y^2 = z^2$.
    (b) $x^2 + 5y^2 = z^2$.
11. Find all solutions of $x^2 + py^2 = z^2$, where p is a prime.
12. Show that the Diophantine equation $x^2 - y^2 = m^3$ is always solvable for x, y given m.
13. Show that the Diophantine equation $x^2 - y^2 = m^k$ is solvable for x, y for any given m, $k \geq 3$.
14. When is $x^2 - y^2 = m$ solvable for x, y?
15. For which m is $x^2 - y^2 = m^2$ solvable for x, y?

6.3 The Equation $x^4 + y^4 = z^2$

We showed in Section 2 that the Diophantine equation $x^2 + y^2 = z^2$ has infinitely many solutions. This leads us naturally to inquire about the solutions of

$$x^n + y^n = z^n \qquad (n \geq 3). \tag{1}$$

As we mentioned in Chapter 1, Fermat conjectured that (1) has no solutions in nonzero x, y, z.

Lemma 1: It suffices to prove Fermat's conjecture for $n = p$ an odd prime and $n = 4$.

Proof: This is Exercise 9 of Section 1. ■

In this section, we shall prove Fermat's conjecture for $n = 4$. In fact, since it is no more difficult, we shall prove that the equation

$$x^4 + y^4 = z^2 \tag{2}$$

has no nonzero solutions in integers. It is clear that this implies the truth of the Fermat conjecture for $n = 4$.

The proof of the theorem makes essential use of the result concerning Pythagorean triples of Section 2. Of course, if $x^4 + y^4 = z^2$, then (x^2, y^2, z) is a Pythagorean triple. Another way of stating the result then is that there can be no right triangle whose sides all have integral length with the lengths of both legs being perfect squares.

More important than the result itself is the general method employed in its proof. It has many applications in number theory. It is called the method of *infinite descent*, and goes as follows. Suppose that (x_1, y_1, z_1) is an arbitrary solution of (2) with $z_1 > 0$. Suppose that from this solution we may derive another solution (x_2, y_2, z_2) of (2) with $0 < z_2 < z_1$. Then we may of course derive another solution (x_3, y_3, z_3) of (2) with $0 < z_3 < z_2 < z_1$. Continuing in this way, we arrive at a contradiction because we cannot have an arbitrarily long string of integers between 0 and z_1. Thus, there could not have been a solution in the first place.

We can state the method more precisely. Suppose that (x_1, y_1, z_1) is a solution of (2) with $z_1 > 0$ least. (If there are nonzero solutions at all, then there must be one with $z_1 > 0$ least by the well-ordering principle.) Then from this solution we derive (x_2, y_2, z_2) as a solution to (2) with $0 < z_2 < z_1$, and we immediately have a contradiction.

The proof of the insolubility of (2) is due to Fermat. He wrote out the proof using his method of infinite descent in another of his famous marginal notes in his copy of Diophantus' works. Fermat himself was very pleased with his method and states, "The proof of this theorem I have reached only after elaborate and ardent study. I reproduce the proof here, since this kind of demonstration will make possible wonderful progress in number theory." It is a good thing that he did decide to write out his proof. Fermat often announced his results without giving proofs, thus leaving the proofs to be supplied by later mathematicians; often their attempts failed and success only followed arduous efforts.

Theorem 2: The Diophantine equation $x^4 + y^4 = z^2$ has no solutions in integers x, y, z unless $x = 0$ or $y = 0$.

Proof: The proof is long, but all the steps are simple. We suppose that there is a solution (x, y, z) with $x \neq 0$, $y \neq 0$, $z > 0$, and assume that this solution is such that z is least. We shall apply Theorem 2.3 twice to derive a solution with a smaller z.

The statement that z is least immediately implies that $\gcd(x, y) = 1$, for if $d = \gcd(x, y)$, then $d^4 \mid x^4 + y^4 = z^2$, so that $d^2 \mid z$, which implies that

$$\left(\frac{x}{d}\right)^4 + \left(\frac{y}{d}\right)^4 = \left(\frac{z}{d^2}\right)^2.$$

If $d > 1$, the last equation would give a solution of $x^4 + y^4 = z^2$ with a smaller value of z. Thus, $d = 1$.

We then see that x^2, y^2, z can have no common factor; that is, they form a primitive solution of $X^2 + Y^2 = Z^2$. Thus, from Theorem 2.3, there are integers a, b, $0 < b < a$, with $\gcd(a, b) = 1$ and one of a, b even such that

$$x^2 = 2ab, \qquad y^2 = a^2 - b^2, \qquad z = a^2 + b^2, \tag{3}$$

or the corresponding equations hold with x and y interchanged. Since our original equation is symmetric in x and y, we may assume that (3) holds.

Notice that a is odd, for if a is even, then b is odd, since x, y, z have no common factor > 1, and so

$$y^2 \equiv -b^2 \equiv -1 \pmod 4,$$

which is impossible. Thus, b is even, and

$$b^2 + y^2 = a^2,$$

where b, y, z have no common factors (since $\gcd(a, b) = 1$), and so form another primitive solution of $X^2 + Y^2 = Z^2$. Again from Theorem 2.3 there are integers u, v such that $\gcd(u, v) = 1$ and

$$b = 2uv, \qquad y = u^2 - v^2, \qquad a = u^2 + v^2.$$

Now

$$x^2 = 2ab = 4uv(u^2 + v^2). \tag{4}$$

Since $\gcd(u, v) = 1$, we also have that $\gcd(u, u^2 + v^2) = \gcd(v, u^2 + v^2) = 1$. Thus, all of $u, v, u^2 + v^2$ must be perfect squares; i.e.,

$$u = r^2, \qquad v = s^2, \qquad u^2 + v^2 = t^2.$$

Thus,

$$s^4 + r^4 = t^2, \tag{5}$$

and we have another solution to (2).

We need only check that this solution satisfies $0 < t < z$ with r and s nonzero. But $r = 0$ implies that $u = 0$, which implies by (4) that $x = 0$, which we have assumed is not the case. Similarly, $s \neq 0$. Moreover, since $b \neq 0$ (because $x^2 = 2ab$, $x \neq 0$), we have

$$t^2 = u^2 + v^2 = a \leq a^2 < a^2 + b^2 = z,$$

and so $t < z$. Thus, z was not least after all, and Theorem 1 is proved. ■

6.3 Exercises

1. Show that $(1/x^4) + (1/y^4) = 1/z^4$ has no solutions in integers.
2. Show that $(x^2 + 1)^4 + (y^2 + 2)^4 = (z + 4)^2$ has no integer solutions.
3. Determine all solutions of the Diophantine equation

$$(x^4 + 1)^4 + y^{12} = (z^2 + 1)^4.$$

4. Determine all solutions of the Diophantine equation
$$(x^2 + y^2 - 2)^4 + 15 = z^2.$$
5. Show that $x^4 + 16 = (y^2 + z^2 - 2)^2$ has no solutions in integers.
6. Show that $x^4 - y^4 = z^2$ has no solutions in integers such that $xyz \neq 0$.
7. Show that the following Diophantine equations have no solutions in integers $xyz \neq 0$.
   (a) $x^4 + 2y^4 = z^2$
   (b) $x^4 + 5y^4 = z^2$

6.4 The Equation $x^2 + y^2 = n$

In this section, we shall answer the following question: Which integers can be written as a sum of two perfect squares? (*Note:* 0 is a perfect square.) This is the same as asking for those positive integers n for which the Diophantine equation
$$x^2 + y^2 = n$$
is solvable. We shall give a complete solution to this problem. The results of this section were first given as another of those famous marginal notes of Fermat in his copy of Diophantus' works.

Before we begin, let us make a few remarks. First of all, note that 3 cannot be written as a sum of two squares, and so the above equation is certainly not solvable for all n. Second, the reader should observe that here, for the first time, we are using the theory of congruences in a nontrivial way to show that a given Diophantine equation actually has a solution.

First, let us show how to reduce considerations to the case of $n = p$ a prime. We start with the following identity:
$$(x_1^2 + y_1^2)(x_2^2 + y_2^2) = (x_1x_2 + y_1y_2)^2 + (x_1y_2 - y_1x_2)^2. \qquad (*)$$
This identity does not look mysterious if you are familiar with complex numbers. It is simply the statement that the absolute value of the product of two complex numbers equals the product of the absolute values of the two complex numbers.* In any case, even if you are not familiar with complex numbers, you may verify the identity directly by multiplying out both sides.

What does this identity do for us? It says, in particular, the following:

Lemma 1: If n and m can both be written as a sum of two squares, then so can nm. .

Example 2: Since $13 = 2^2 + 3^2$ and $29 = 2^2 + 5^2$ we know that $13 \cdot 29 = 377$ is a sum of two squares. Indeed, here for $n = 13$, $m = 29$ we have $x_1 = 2$,

---

*The two complex numbers we have in mind are $x_1 - \sqrt{-1}y_1$ and $x_2 + \sqrt{-1}y_2$.

$y_1 = 3$, $x_2 = 2$, $y_2 = 5$, and so

$$377 = (2 \cdot 2 + 3 \cdot 5)^2 + (2 \cdot 5 - 2 \cdot 3)^2$$

or

$$377 = 19^2 + 4^2.$$

It should be clear from Lemma 1 that we should attempt to determine which primes can be written as a sum of two squares. We first use congruences in a familiar way to find a negative result.

Lemma 3: Let p be a prime. Suppose that $p \equiv 3 \pmod 4$. Then p cannot be written as a sum of two squares.

Proof: Let us suppose that, on the contrary, we have integers x and y such that

$$x^2 + y^2 = p. \tag{1}$$

Then $x^2 + y^2 \equiv 3 \pmod 4$, which is clearly impossible since a square is congruent to 0 or $1 \pmod 4$. ■

We shall now prove the converse of Lemma 3. This is the main and the most difficult step in determining which integers can be written as a sum of two squares.

Lemma 4: Let p be a prime. Suppose that $p = 2$ or $p \equiv 1 \pmod 4$. Then p can be written as a sum of two squares.

Proof: If $p = 2$, then $p = 2 = 1^2 + 1^2$, and so this case is settled.

Suppose that $p \equiv 1 \pmod 4$. Then by Theorem 3.3.5, we can solve the congruence

$$x^2 \equiv -1 \pmod p.$$

Since the integers $0, \pm 1, \pm 2, \ldots, \pm(p-1)/2$ form a complete residue system modulo p, we may, of course, assume that the integer x is one of these integers. Thus, we have integers x and t, where $|x| \leq (p-1)/2 < p/2$, such that

$$x^2 + 1 = tp.$$

Then $t > 0$, and

$$t = \frac{x^2 + 1}{p} < \frac{(p/2)^2 + 1}{p} = \frac{p}{4} + \frac{1}{p} < p.$$

Since $1 = 1^2$ we have shown the following: There are integers x, y, t ($y = 1$) such that

$$x^2 + y^2 = tp \quad \text{and} \quad 1 \leq t < p. \tag{2}$$

This is not quite what we need. We need the same statement with $t = 1$. We shall show that if $t > 1$, then t may be reduced in value (by using a different x and y). In this way, we obtain (2) with $t = 1$. (Note the similarity with the method of infinite descent.)

Let $k \geq 1$ be the least integer such that kp is a sum of two squares. Thus, we have integers x_1 and y_1 such that

$$x_1^2 + y_1^2 = kp. \tag{3}$$

We assume that $k > 1$ and derive a smaller value of k, and thus we have a contradiction. We know from (2) that $1 < k < p$.

Choose integers x_2 and y_2 such that

$$x_2 \equiv x_1 (\text{mod } k) \quad \text{and} \quad y_2 \equiv y_1 (\text{mod } k). \tag{4}$$

Using the complete residue system $0, \pm 1, \pm 2, \ldots$ we may assume that

$$|x_2| \leq \frac{k}{2} \quad \text{and} \quad |y_2| \leq \frac{k}{2}. \tag{5}$$

Moreover, we cannot have $x_2 = y_2 = 0$ since then $x_1 \equiv y_1 \equiv 0(\text{mod } k)$ and thus $x_1^2 \equiv y_1^2 \equiv 0(\text{mod } k^2)$. But then, from (3), we have

$$kp = x_1^2 + y_1^2 \equiv 0(\text{mod } k^2),$$

which implies that $k|p$, violating the assumption that $1 < k < p$. Now

$$x_2^2 + y_2^2 \equiv x_1^2 + y_1^2 \equiv 0(\text{mod } k)$$

using (3) and (4). Thus, there is an m such that

$$x_2^2 + y_2^2 = km. \tag{6}$$

Then

$$m = \frac{x_2^2 + y_2^2}{k} \geq 1,$$

since one of x_2, y_2 is nonzero and $k > 0$. Moreover, by using (5), we have

$$m \leq \frac{(k/2)^2 + (k/2)^2}{k} = \frac{1}{2}k < k.$$

That is, $1 \leq m < k$. Combining (3) and (6) with our basic identity (∗), we have

$$k^2 mp = (x_1x_2 + y_1y_2)^2 + (x_1y_2 - x_2y_1)^2.$$

Using (4) again, we have

$$x_1x_2 + y_1y_2 \equiv x_1^2 + y_1^2 \equiv 0(\text{mod } k)$$

and

$$x_1y_2 - x_2y_1 \equiv x_1y_1 - x_1y_1 = 0(\text{mod } k).$$

Thus,

$$mp = \left(\frac{x_1x_2 + y_1y_2}{k}\right)^2 + \left(\frac{x_1y_2 - x_2y_1}{k}\right)^2,$$

where $(x_1x_2 + y_1y_2)/k$ and $(x_1y_2 - x_2y_1)/k$ are both integers. We have now reached the desired contradiction. Namely, mp is a sum of two squares and $1 \leq m < k$. ■

Let us now combine Lemmas 1, 3, and 4 to determine precisely which integers may be written as a sum of two squares. First, any perfect square s^2 can trivially be so represented:

$$s^2 + 0^2 = s^2 \qquad (x = s, y = 0).$$

Thus, if n is a positive integer, write $n = s^2 n_0$, where n_0 has no square factors. Write $n_0 = p_1 p_2 \cdots p_t$, where $p_1, \ldots, p_t$ are distinct primes. Applying Lemmas 1 and 4 we see that if for each i, $p_i = 2$ or $p_i \equiv 1 \pmod 4$, then $n = s^2 p_1 p_2 \cdots p_t$ can be written as a sum of two squares.

Conversely, suppose that n can be written as a sum of two squares. Let us prove that if we write $n = s^2 p_1 \cdots p_t$ with $p_1, \ldots, p_t$ distinct primes, then either $p_i = 2$ or $p_i \equiv 1 \pmod 4$ for $1 \leq i \leq t$. Let us reason by contradiction. Suppose that some p_i is congruent to $3 \pmod 4$. Without loss of generality, suppose that $p_1 \equiv 3 \pmod 4$. Then, if $x^2 + y^2 = n$, we have $x^2 + y^2 \equiv 0 \pmod{p_1}$. Assume that $p_1 \nmid y$. Then there exists an arithmetic inverse y^* of y mod p_1. Then we have $(xy^*)^2 \equiv -1 \pmod{p_1}$, so that -1 is a quadratic residue mod p_1, which contradicts the fact that $p_1 \equiv 3 \pmod 4$. Thus, $p_1 \nmid y$ leads to a contradiction. Similarly, $p_1 \nmid x$ leads to a contradiction. Thus, we conclude that $p_1 \mid x$ and $p_1 \mid y$. Thus,

$$p_1^2 \mid x^2 + y^2 = n = s^2 p_1 \cdots p_t.$$

Since $p_1, \ldots, p_t$ are different primes, we must have $p_1 \mid s^2$, so that $p_1 \mid s$. Thus,

$$\left(\frac{x}{p_1}\right)^2 + \left(\frac{y}{p_1}\right)^2 = \left(\frac{s}{p_1}\right)^2 p_1 \cdots p_t.$$

Replacing n by n/p_1^2, we see that n/p_1^2 is a sum of two squares and $n/p_1^2 = w^2 p_1 \cdots p_t$. By repeating the same argument with n replaced by n/p_1^2, we may remove another factor of p_1 from w. By repeating this process, we eventually arrive at an integer n_1 such that n_1 is a sum of two squares, say $n_1 = a^2 + b^2$, $n_1 = v^2 p_1 \cdots p_t$, with $p_1 \nmid v$. But then, applying the same argument one more time, we get $p_1 \mid v$, which is a contradiction. Thus, $p_1 \not\equiv 3 \pmod 4$. (Note that the argument we have just used is another application of infinite descent.)

We have now completely proved the following result:

Theorem 5: Let n be a positive integer. Write $n = s^2 n_0$, where n_0 has no square factors. Then n can be written as a sum of two squares if and only if the only prime factors of n_0 are among the primes 2 and the primes $p \equiv 1 \pmod 4$.

Example 6: $888 = 2^3 \cdot 3 \cdot 37$. Thus, $888 = 2^2(2 \cdot 3 \cdot 37)$. Since $3 \equiv 3 \pmod 4$, we see that 888 cannot be written as a sum of two squares.

Example 7: $332514 = 2 \cdot 3^2 \cdot 7^2 \cdot 13 \cdot 29 = (3 \cdot 7)^2(2 \cdot 13 \cdot 29)$. Since $13 \equiv 29 \equiv 1 \pmod 4$, we see that 332514 can be written as a sum of two squares. Let us use the identity (*) to actually carry out the computations to get some idea concerning its efficiency. In Example 2 we already used (*) to show that

$$13 \cdot 29 = 19^2 + 4^2.$$

Combining this with $2 = 1^2 + 1^2$, we obtain

$$2 \cdot 13 \cdot 29 = (19 \cdot 1 + 4 \cdot 1)^2 + (19 \cdot 1 - 4 \cdot 1)^2 = 23^2 + 15^2.$$

Then finally

$$\begin{aligned} 332514 &= (3 \cdot 7)^2(23^2 + 15^2) \\ &= (3 \cdot 7 \cdot 23)^2 + (3 \cdot 7 \cdot 15)^2 \\ &= 483^2 + 315^2. \end{aligned}$$

6.4 Exercises

1. Suppose that $\gcd(a, b) = 1$. Show that if a is not a sum of two squares, then ab is not a sum of two squares.

2. Prove that there exist an infinite number of Pythagorean triples using the results of this section.

3. Determine whether the following integers can be written as sums of two squares. In each case determine all possible representations as a sum of two squares. (Don't forget to include squares of negative integers.)
   (a) $n = 3$.
   (b) $n = 5$.
   (c) $n = 49$.
   (d) $n = 60$.
   (e) $n = 85$.
   (f) $n = 29$.

4. Using the results of Exercise 3, explicitly represent $85 \cdot 29 = 2465$ as a sum of two squares.

5. For n a positive integer, let $r_2(n)$ denote the number of representations of n in the form $x^2 + y^2$. Let us agree to count $3^2 + 5^2$ and $5^2 + 3^2$, for example, as different representations.
   (a) Using Exercise 3, compute $r_2(3)$, $r_2(5)$, $r_2(49)$, $r_2(60)$, $r_2(85)$, and $r_2(29)$.
   (b) Using the identity of the text, show that $r_2(n)$ is multiplicative.
   (c) Show that for p a prime,

$$r_2(p) = \begin{cases} 8, & p \text{ odd and } \equiv 1 \pmod 4, \\ 4, & p = 2; \\ 0, & \text{otherwise.} \end{cases}$$

   (*Hint:* It suffices to consider the first case. Moreover, it suffices to show that there is only one solution of $x^2 + y^2 = p$ with $0 < x <$

$y < p$. Let (x, y) and (x_1, y_1) be two such solutions. Show that $(xy^*)^2 \equiv (x_1y_1^*)^2 \equiv -1 (\text{mod } p)$ and then that $x = x_1$, $y = y_1$.)

6. Show that the Diophantine equation $5x^2 + 14xy + 10y^2 = n$ is solvable in x, y if and only if n is representable as a sum of two squares.

*7. Let $n > 0$ be given. Determine a necessary and sufficient condition that $x^2 + 2y^2 = n$ be solvable in integers x, y.

6.5 The Equation $x^2 + y^2 + z^2 + w^2 = n$

In the last section we completely settled the question of which integers could be written as a sum of two perfect squares. In particular we observed that not all integers could be written in this way. Then, can we write every integer as a sum of three squares? For example, 3 is not the sum of two squares but $3 = 1^2 + 1^2 + 1^2$, and 43 is not the sum of two squares but $43 = 3^2 + 3^2 + 5^2$. But, alas, 7 is not the sum of three squares, as is readily checked. Now $7 = 2^2 + 1^2 + 1^2 + 1^2$ is the sum of four squares. Bachet in 1621 conjectured that every integer is a sum of four squares.

Now we come to yet another of the famous marginal notes of Fermat. He stated that he had indeed proved the result by his method of infinite descent. Unfortunately he gave no further indication of how the proof was accomplished. Later he challenged the other mathematicians of his day to establish the validity of the result to determine, as he stated, "whether I value my discovery higher than it deserves." There is little doubt that Fermat could prove the result, but it was not for more than 100 years after Fermat's death that a proof was finally given in 1770 by the French mathematician J. L. Lagrange after the great Euler had tried and failed. The result is, in fact, usually credited to Lagrange, not Fermat.

From our present point of view the proof can be made to parallel quite closely the proof given concerning the sum of two squares. We first write down an identity (*) (discovered by Euler) which says that if n and m can be written as sums of two squares, then so can nm. This identity is not an obvious application of complex numbers as was the one in the last section, and its discovery was a substantial achievement.* Its verification, however, is trivial as one need only multiply out both sides. It illustrates J. E. Littlewood's famous comment that any identity is trivial if written down by somebody else.

With the identity, it suffices to show that any prime p can be written as a sum of four squares. To accomplish this we first show that a certain congruence can be solved (Lemma 3) to show that some multiple of p is a

*It does, however, come from a more recondite system of numbers called the *quaternions*. The quaternions, discovered by Hamilton in the mid-nineteenth century, were not known at the time of Euler.

sum of four squares. We then show, in a proof that closely parallels the similar proof concerning two squares, that the smallest multiple of p that can be written as a sum of four squares is $1 \cdot p = p$ and the proof is complete.

We first give the identity:

$$(x_1^2 + y_1^2 + z_1^2 + w_1^2)(x_2^2 + y_2^2 + z_2^2 + w_2^2) = (x_1x_2 + y_1y_2 + z_1z_2 + w_1w_2)^2 + (x_1y_2 - y_1x_2 + z_1w_2 - w_1z_2)^2 + (x_1z_2 - z_1x_2 + w_1y_2 - y_1w_2)^2 + (x_1w_2 - w_1x_2 + y_1z_2 - z_1y_2)^2. \quad (*)$$

As we noted above, the identity may be verified directly by multiplying out both sides. Thus, we have

Lemma 1: If n and m can be written as sums of four squares, then so can nm. In particular, if we show that every prime can be written as a sum of four squares, then we have established that every integer can be written as a sum of four squares.

Example 2: Since $30 = 1^2 + 2^2 + 3^2 + 4^2$ and $29 = 2^2 + 5^2 = 2^2 + 5^2 + 0^2 + 0^2$, we know that $30 \cdot 29 = 870$ is a sum of four squares. Indeed, here for $n = 30$, $m = 29$, we have $x_1 = 1$, $y_1 = 2$, $z_1 = 3$, $w_1 = 4$ and $x_2 = 2$, $y_2 = 5$, $z_2 = 0$, $w_2 = 0$, and so

$$870 = (2 + 10)^2 + (5 - 4)^2 + (-6 + 20)^2 + (-8 - 15)^2$$

or

$$870 = 12^2 + 1^2 + 14^2 + 23^2.$$

We shall now give the lemma on congruences. Its proof is an interesting and simple exercise in quadratic residues.

Lemma 3: Let p be a prime. Then there are integers x and y such that

$$x^2 + y^2 \equiv -1 \pmod{p}.$$

Proof: If $p = 2$, let $x = 1$ and $y = 0$. If $p \equiv 1 \pmod 4$, then we know that $\left(\frac{-1}{p}\right) = 1$, and we may take $y = 0$ and x as a solution of $x^2 \equiv -1 \pmod{p}$.

Finally, the case where $p \equiv 3 \pmod 4$ remains. In this case $\left(\frac{-1}{p}\right) = -1$. We want to find integers x and y such that

$$x^2 \equiv -(y^2 + 1) \pmod{p}.$$

That is, we want to find an integer y such that

$$\left(\frac{-(y^2 + 1)}{p}\right) = 1.$$

Now

$$\left(\frac{-(y^2 + 1)}{p}\right) = \left(\frac{-1}{p}\right)\left(\frac{y^2 + 1}{p}\right) = -\left(\frac{y^2 + 1}{p}\right).$$

Thus, we want an integer y such that

$$\left(\frac{y^2+1}{p}\right) = -1.$$

Since the integers y^2 are just the quadratic residues mod p, our problem is now the following: Find a quadratic residue a mod p such that $a + 1$ is a quadratic nonresidue. That is, find a such that $\left(\frac{a}{p}\right) = 1$ and $\left(\frac{a+1}{p}\right) = -1$. It is obvious that such an a must exist. Indeed, if a did not exist, then $\left(\frac{1}{p}\right) = 1$ implies that $\left(\frac{1+1}{p}\right) = \left(\frac{2}{p}\right) = 1$ implies that $\left(\frac{2+1}{p}\right) = \left(\frac{3}{p}\right) = 1$, and so forth. Then, every integer would be a quadratic residue mod p. We know that this is not so. ■

We may now prove the following theorem:

Theorem 4: Let n be a positive integer. Then n can be written as a sum of four squares.

Proof: By Lemma 1 we may assume that $n = p$, a prime. If $p = 2$, then $p = 1^2 + 1^2 + 0^2 + 0^2$, and so we may assume that $p > 2$ and so is odd.

From Lemma 3, we may solve the congruence

$$x^2 + y^2 \equiv -1 (\text{mod } p).$$

Since the collection of integers $0, \pm 1, \ldots, \pm(p-1)/2$ is a complete residue system modulo p, we may, of course, assume that the integers x, y are among these integers. So we have integers x, y, t, where $|x|, |y| \leq (p-1)/2 < p/2$ such that

$$x^2 + y^2 + 1 = tp.$$

Then $t > 0$ and

$$t = \frac{x^2 + y^2 + 1}{p} < \frac{(p/2)^2 + (p/2)^2 + 1}{p} = \frac{p}{2} + \frac{1}{p} < p.$$

Since $1 = 1^2$, $0 = 0^2$, we have shown the following: There are integers x, y, z, w, t ($z = 1$ and $w = 0$) such that

$$x^2 + y^2 + z^2 + w^2 = tp \quad \text{and} \quad 1 \leq t < p. \tag{1}$$

We need the statement (1) with $t = 1$. We again show that if $t > 1$, then t may be reduced in value. Or, what amounts to the same thing, we show that the least value of t that works in (1) is $t = 1$.

Thus, let $k \geq 1$ be the least integer such that kp is a sum of four squares. Then we have integers x_1, y_1, z_1, w_1 such that

$$x_1^2 + y_1^2 + z_1^2 + w_1^2 = kp. \tag{2}$$

We assume that $k > 1$. We then have from (1) that $1 < k < p$.

Choose integers x_2, y_2, z_2, w_2 from the complete residue system $0, \pm 1,$

$\pm 2, \ldots$ mod k such that

$$x_2 \equiv x_1 (\text{mod } k), \quad y_2 \equiv y_1 (\text{mod } k), \quad z_2 \equiv z_1 (\text{mod } k), \quad \text{and} \quad w_2 \equiv w_1 (\text{mod } k). \tag{3}$$

Then

$$|x_2| \leq \frac{k}{2}, \quad |y_2| \leq \frac{k}{2}, \quad |z_2| \leq \frac{k}{2}, \quad \text{and} \quad |w_2| \leq \frac{k}{2}. \tag{4}$$

Moreover, we cannot have $x_2 = y_2 = z_2 = w_2 = 0$ since then $x_1 \equiv y_1 \equiv z_1 \equiv w_1 \equiv 0 (\text{mod } k)$, and so from (2)

$$kp = x_1^2 + y_1^2 + z_1^2 + w_1^2 \equiv 0 (\text{mod } k^2),$$

which implies that $k | p$, violating the assumption that $1 < k < p$.

Again it is not possible that $|x_2| = |y_2| = |z_2| = |w_2| = k/2$ since then k is even, and so $k/2 \equiv -(k/2)(\text{mod } k)$, and thus $x_1 \equiv y_1 \equiv z_1 \equiv w_1 \equiv (k/2)(\text{mod } k)$, and so $x_1^2 \equiv y_1^2 \equiv z_1^2 \equiv w_1^2 \equiv (k^2/4)(\text{mod } k^2)$, which again gives by (2)

$$kp = x_1^2 + y_1^2 + z_1^2 + w_1^2 \equiv \frac{k^2}{4} + \frac{k^2}{4} + \frac{k^2}{4} + \frac{k^2}{4} = k^2 \equiv 0 (\text{mod } k^2),$$

which is again a contradiction.

Now

$$x_2^2 + y_2^2 + z_2^2 + w_2^2 \equiv x_1^2 + y_1^2 + z_1^2 + w_1^2 \equiv 0 (\text{mod } k)$$

using (2) and (3). Thus, there is an m such that

$$x_2^2 + y_2^2 + z_2^2 + w_2^2 = km. \tag{5}$$

Further,

$$m = \frac{x_2^2 + y_2^2 + z_2^2 + w_2^2}{k} \geq 1,$$

since one of x_2, y_2, z_2, w_2 is nonzero and $k > 0$. Moreover, by (4) and the fact that not all of $|x_2|, |y_2|, |z_2|, |w_2|$ can be $k/2$, we have

$$m < \frac{(k/2)^2 + (k/2)^2 + (k/2)^2 + (k/2)^2}{k} = k.$$

That is, $1 \leq m < k$. Combining (2) and (5) with our basic identity (*), we have

$$k^2 mp = A^2 + B^2 + C^2 + D^2,$$

where A, B, C, D are the four terms on the right-hand side of (*). Using (3) again,

$$A = x_1 x_2 + y_1 y_2 + z_1 z_2 + w_1 w_2 \equiv x_1^2 + y_1^2 + z_1^2 + w_1^2 \equiv 0 (\text{mod } k)$$

and

$$B = x_1 y_2 - y_1 x_2 + z_1 w_2 - w_1 z_2 \equiv x_1 y_1 - y_1 x_1 + z_1 w_1 - w_1 z_1 \equiv 0 \quad (\text{mod } k)$$

and similarly $C \equiv 0 \pmod{k}$ and $D \equiv 0 \pmod{k}$. Thus,

$$mp = \left(\frac{A}{k}\right)^2 + \left(\frac{B}{k}\right)^2 + \left(\frac{C}{k}\right)^2 + \left(\frac{D}{k}\right)^2,$$

where A/k, B/k, C/k, D/k are all integers. We have now reached the desired contradiction. Namely, mp is a sum of four squares and $1 \leq m < k$. ■

6.5 Exercises

1. (a) Represent 5, 7, and 11 as sums of four squares.
   (b) Use the identity of the text to represent $5 \cdot 7$ and $7 \cdot 11$ as sums of four squares.
2. Let $r_4(n)$ denote the number of representations of n as a sum of four squares. Let us agree to count the order of the squares in such a representation. Thus, $r_4(1) = 8$, corresponding to the eight representations $1 = (\pm 1)^2 + 0^2 + 0^2 + 0^2 = 0^2 + (\pm 1)^2 + 0^2 + 0^2 = 0^2 + 0^2 + (\pm 1)^2 + 0^2 = 0^2 + 0^2 + 0^2 + (\pm 1)^2$.
   (a) Compute $r_4(5)$, $r_4(7)$, and $r_4(10)$.
   (b) Show that $r_4(n)$ is multiplicative.
   (c) Show that the formula

   $$r_4(n) = 8 \sum_{\substack{d \mid n \\ 4 \nmid d}} d$$

   holds for $n = 5, 7, 10$. (This formula is due to Jacobi. For a proof, see Hardy and Wright, *The Theory of Numbers*, Oxford University Press, Inc., New York, p. 314.)
3. Show that if $n \equiv 7 \pmod{8}$, then n is not the sum of three squares.
4. Complete the following outline to give another proof of Lemma 3: There are $(p + 1)/2$ distinct residue classes of numbers x^2 and $(p + 1)/2$ distinct residue classes of numbers $-(y^2 + 1)$ so there must be an x, y such that $x^2 \equiv -(y^2 + 1) \pmod{p}$.

6.6 Pell's Equation: $x^2 - dy^2 = 1$

Let us now discuss the equation $x^2 - dy^2 = 1$, which is known in the literature as *Pell's equation*. Here, we assume that d is a given *positive* integer. There is one case of Pell's equation which is trivial. Namely, suppose that $d = a^2$ for some integer a. Then

$$1 = x^2 - dy^2 = x^2 - a^2y^2 = (x - ay)(x + ay),$$

which holds if and only if $x - ay = \pm 1$, $x + ay = \pm 1$, which is equivalent to $x = \pm 1, y = 0$. *Thus, henceforth, let us always consider Pell's equation for* $d > 0$ *not equal to a perfect square*. We shall prove that if d is not equal to a perfect square, then Pell's equation has an infinite number of solutions. We shall prove this fact, as well as give a fairly explicit description of all the solutions. We shall see that the most difficult part of obtaining such a description is determining *one* solution of Pell's equation other than $(\pm 1, 0)$. From one solution other than $(\pm 1, 0)$ we shall show how to compute infinitely many others. To give the reader a feel for this procedure, let us begin with a numerical example.

Example 1: Let us find an infinite number of solutions of $x^2 - 2y^2 = 1$. It is clear that $x = 3$, $y = 2$ is one solution. Thus, solutions other than $x = \pm 1$, $y = 0$ exist. Let x_0, y_0 be any solution. Then, we assert that $x = 3x_0 + 4y_0$, $y = 2x_0 + 3y_0$ is also a solution. Indeed,

$$\begin{aligned} x^2 - 2y^2 &= (3x_0 + 4y_0)^2 - 2(2x_0 + 3y_0)^2 \\ &= x_0^2 - 2y_0^2 = 1. \end{aligned}$$

For example, if $x_0 = 3$, $y_0 = 2$, then we obtain the solution $x = 17$, $y = 12$. Then, if we set $x_0 = 17$, $y_0 = 12$, we find $x = 99$, $y = 70$. Continuing in this way, we obtain an infinite number of solutions of $x^2 - 2y^2 = 1$. The solutions are all different, since the value of y keeps increasing. We shall see later in this section that there is nothing accidental about the above technique for generating solutions.

The equation $x^2 - dy^2 = 1$ was labeled Pell's equation by Euler. However, this is one instance (among many in mathematics) where a name has been attached in error to a result and the name was so commonly accepted in the literature that it stuck. Actually, Pell had little to do with Pell's equation. The first mention of the equation seems to date back to Archimedes, although it is not known what he knew about it. A method for solving it was given by the English mathematician Lord Brouncker in 1657 using what are called continued fractions. Wallis and again Fermat claimed to have proved that there is always a solution other than $x = \pm 1$, $y = 0$, and Fermat first noted that there are always infinitely many solutions. As usual, Fermat did not publish his proof. The first published proof was given by Lagrange in 1766.

The proof that $x^2 - dy^2 = 1$ can be solved for all nonsquare d parallels, to a certain extent, the proofs given in the preceding sections of the two-square and four-square theorems. First there is an identity. Then we show that $x^2 - dy^2 = t$ can be solved. Next we show, using congruences and the identity, that we may take $t = 1$. However, in the case of Pell's equation, there are more complications than arose in the preceding sections. For example, it is not enough to show that we can solve $x^2 - dy^2 = t$ for some t. We need a t which gives an infinite number of solutions. Moreover, we need a new idea to find t. (We solved congruences to get the t in the two- and four-square theorems.) Both in obtaining the t and in many other analyses concerning Pell's equation, we must come to grips with the *irrational number* $\sqrt{d}$. That the arithmetic of irrational numbers should be able to tell us anything about the integers is far from obvious and the present section as well as Appendix B (where the existence of the t will be shown) will provide some motivation for the extensive study of irrational numbers of certain types which we shall undertake in the second half of this book.

Let us begin our study of Pell's equation with the identity we promised:

$$(x_1^2 - dy_1^2)(x_2^2 - dy_2^2) = (x_1x_2 - dy_1y_2)^2 - d(x_1y_2 - y_1x_2)^2. \qquad (*)$$

To verify this identity, merely multiply out both sides. The identity allows us to draw a similar conclusion to the ones we drew in the two- and four-square problems (Lemmas 4.1 and 5.1). Namely, if (x_1, y_1) and (x_2, y_2) are both solutions of $x^2 - dy^2 = 1$, then we may obtain another solution (x_3, y_3) by the formulas

$$x_3 = x_1x_2 - dy_1y_2, \qquad y_3 = x_1y_2 - y_1x_2.$$

For example, if $d = 2$ and $x_1 = 3$, $y_1 = 2$, $x_2 = 99$, and $y_2 = 70$, then we see that $x_3 = 17$, $y_3 = 12$ is also a solution. Thus, $(*)$ allows us to generate new solutions out of given ones. Let us now turn to the problem of determining at least one solution (other than $x = \pm 1$, $y = 0$). As a first step in this direction, let us state a result which corresponds to the results of the previous sections which asserted that certain congruences are solvable. (Namely, the assertion of Section 4 that $p \equiv 1 \pmod 4$ implies that $x^2 \equiv -1 \pmod p$ may be solved, and the assertion of Section 5 that $x^2 + y^2 \equiv -1 \pmod p$ may be solved (Lemma 5.3).)

Lemma 2: Set $B = 2\sqrt{d} + 1$. Then there are infinitely many distinct pairs of integers (x, y) such that

$$|x^2 - dy^2| \leq B.$$

For example, if $d = 2$, then $B = 2\sqrt{2} + 1 \leq 4$, so that Lemma 2 asserts that there exist infinitely many pairs of integers (x, y) such that $|x^2 - 2y^2| \leq 4$. Of course, the lemma has little interest in itself, since we shall prove that $x^2 - dy^2 = 1$ has an infinite number of solutions, which clearly implies the lemma. However, the lemma is a technical device to get to this point. To prove the lemma, some quite new ideas are needed, mainly concerning the approximation of irrational numbers by rational numbers. To put these ideas in the proper context, let us not interrupt our study of Pell's equation to discuss them at this point. Rather, let us postpone the proof of Lemma 2 until Appendix B and let us immediately show how Lemma 2 implies that Pell's equation has an infinite number of solutions.

Theorem 3: Let $d > 0$ be a nonsquare. Then the Diophantine equation

$$x^2 - dy^2 = 1 \tag{1}$$

has an infinite number of distinct solutions in integers x, y.

Proof: By Lemma 2, there are infinitely many pairs of integers (x, y) such that $|x^2 - dy^2| \leq B$, where $B = 2\sqrt{d} + 1$. Since there are only a finite number of integers k such that $|k| \leq B$ and since each of the numbers $x^2 - dy^2$ are integers, there is an integer k such that there are infinitely many pairs of integers (x, y) with

$$x^2 - dy^2 = k. \tag{2}$$

If $k = 0$, then $d = (x/y)^2$, contradicting the fact that d is not a perfect square. (Why is x/y an integer?) Therefore, $k \neq 0$.

Let us now look at the solutions (x, y) to (2) modulo $|k|$. For any integers x, y there are integers a, b such that $0 \leq a, b < |k|$ and such that $x \equiv a(\text{mod } |k|)$, $y \equiv b(\text{mod } |k|)$. There are only k^2 possibilities for the pair (a, b). Thus, since (2) has an infinite number of solutions, we see that we can find a, b such that infinitely many solutions (x, y) of (2) satisfy $x \equiv a(\text{mod } |k|)$, $y \equiv b(\text{mod } |k|)$. For any pair (x_1, y_1), (x_2, y_2) of these solutions, we have

$$x_1^2 - dy_1^2 = k, \qquad x_2^2 - dy_2^2 = k$$

and

$$x_1 \equiv a \equiv x_2(\text{mod } |k|), \qquad y_1 \equiv b \equiv y_2(\text{mod } |k|). \tag{3}$$

Using our basic identity (*), we obtain

$$k^2 = (x_1x_2 - dy_1y_2)^2 - d(x_1y_2 - y_1x_2)^2.$$

However,

$$x_1x_2 - dy_1y_2 \equiv x_1^2 - dy_1^2 \equiv 0(\text{mod } |k|)$$

$$x_1y_2 - y_1x_2 \equiv x_1y_1 - y_1x_1 \equiv 0(\text{mod } |k|).$$

Therefore,

$$1 = \left(\frac{x_1x_2 - dy_1y_2}{k}\right)^2 - d\left(\frac{x_1y_2 - y_1x_2}{k}\right)^2,$$

so that we get a solution (x, y) to (1), with

$$x = \frac{x_1x_2 - dy_1y_2}{k}, \qquad y = \frac{x_1y_2 - y_1x_2}{k}.$$

Let us fix one solution (x_1, y_1) of (2) and (3). Since $k \neq 0$, we clearly see that one of x_1 or y_1 is nonzero. Let (x_2, y_2) run over an infinite set of solutions to (2) and (3). Then either x_2 or y_2 assumes infinitely many values, so that either x or y assumes an infinite number of values. (Why?) Thus, (x, y) runs over infinitely many solutions to (1). ∎

Let us now give a technique for generating infinitely many solutions of (1). For this purpose, we require the following two identities:

$$(x_1 + \sqrt{d}\,y_1)(x_2 + \sqrt{d}\,y_2) = (x_1x_2 + dy_1y_2) + \sqrt{d}\,(x_1y_2 + y_1x_2)$$
$$(x_1 - \sqrt{d}\,y_1)(x_2 - \sqrt{d}\,y_2) = (x_1x_2 + dy_1y_2) - \sqrt{d}\,(x_1y_2 + y_1x_2). \qquad (**)$$

The import of these identities is that the product of two numbers of the form $x + \sqrt{d}\,y$ (respectively, $x - \sqrt{d}\,y$) is again a number of the same form.

Suppose that x_1 and y_1 are integers. Then, by using (**) repeatedly, we see that

$$(x_1 + \sqrt{d}\,y_1)^n = x_n + \sqrt{d}\,y_n \qquad (4)$$

for integers x_n and y_n. In fact, since for $n \geq 2$ we have

$$\begin{aligned}(x_1 + \sqrt{d}\,y_1)^n &= (x_1 + \sqrt{d}\,y_1)(x_1 + \sqrt{d}\,y_1)^{n-1}\\ &= (x_1 + \sqrt{d}\,y_1)(x_{n-1} + \sqrt{d}\,y_{n-1}),\end{aligned}$$

(**) implies that

$$x_n = x_1x_{n-1} + dy_1y_{n-1}, \qquad y_n = x_1y_{n-1} + y_1x_{n-1}. \qquad (5)$$

Similarly,

$$(x_1 - \sqrt{d}\,y_1)^n = x_n - \sqrt{d}\,y_n,$$

where x_n and y_n may also be computed from x_1 and y_1 using Eq. (5).

Lemma 4: Suppose that $n > 0$ and that (x_1, y_1) is a solution to Pell's equation $x^2 - dy^2 = 1$. Let x_n and y_n be defined by the condition

$$(x_1 + \sqrt{d}\,y_1)^n = x_n + \sqrt{d}\,y_n.$$

Then (x_n, y_n) is a solution of $x^2 - dy^2 = 1$.

Proof: We have

$$(x_1 - \sqrt{d}\,y_1)^n = x_n - \sqrt{d}\,y_n,$$

so that

$$\begin{aligned}x_n^2 - dy_n^2 &= (x_n + \sqrt{d}\,y_n)(x_n - \sqrt{d}\,y_n)\\ &= (x_1 + \sqrt{d}\,y_1)^n(x_1 - \sqrt{d}\,y_1)^n\\ &= ((x_1 + \sqrt{d}\,y_1)(x_1 - \sqrt{d}\,y_1))^n\\ &= (x_1^2 - dy_1^2)^n = 1^n = 1.\end{aligned}$$

■

Example 5: Again consider $x^2 - 2y^2 = 1$. Set $x_1 = 3$, $y_1 = 2$. Then we define x_n and y_n by

$$x_n + \sqrt{2}\,y_n = (3 + 2\sqrt{2})^n,$$

so that

$$\begin{aligned}(3 + 2\sqrt{2})^2 &= 17 + 12\sqrt{2}\\ (3 + 2\sqrt{2})^3 &= (17 + 12\sqrt{2})(3 + 2\sqrt{2}) = 99 + 70\sqrt{2},\end{aligned}$$

which yield the solutions (17, 12), (99, 70) given in Example 1. Note that the formulas of Example 1 are precisely the relations (5), namely for $n \geq 2$,

$$x_n = 3x_{n-1} + 4y_{n-1}, \qquad y_n = 2x_{n-1} + 3y_{n-1}.$$

For $n = 2$, we see that

$$x_2 = 3x_1 + 4y_1, \qquad y_2 = 2x_1 + 3y_1$$

is a solution of (1) for any solution (x_1, y_1) of (1).

Now that we have a machine for generating solutions from a single solution, let us find all solutions. First, note that if (x, y) is a solution, then so is $(\pm x, \pm y)$ for any choice of signs. Therefore, it suffices to determine all those solutions for which $x \geq 0$, $y \geq 0$. Moreover, it is trivial to see that the only solutions for which either x or y is zero are $(\pm 1, 0)$. Thus, it suffices to determine all those solutions for which $x > 0$, $y > 0$. Such solutions are called *positive solutions*. Let us find a simple way of determining whether or not a solution is positive.

Lemma 6: Suppose that (x, y) is any solution of $x^2 - dy^2 = 1$. Then (x, y) is a positive solution if and only if

$$x + \sqrt{d}\,y > 1. \tag{6}$$

Proof: If $x > 0$ and $y > 0$, then $x \geq 1$ and $y \geq 1$, so that

$$x + \sqrt{d}\,y \geq 1 + \sqrt{d} \geq 2 > 1.$$

Conversely, suppose that (6) holds. If one of x or y is zero, then $(x, y) = (\pm 1, 0)$ for which (6) does not hold. Thus, assume that $x \neq 0$, $y \neq 0$. Let us consider four cases for x and y.

Case 1: $x < 0$ and $y < 0$. In this case $x + \sqrt{d}\,y < 0$ and (6) does not hold.

Case 2: $x > 0$ and $y < 0$. In this case, $x - \sqrt{d}\,y \geq 1 + \sqrt{d} > 1$, so that by (6) and (1), we have

$$1 = x^2 - dy^2 = (x - \sqrt{d}\,y)(x + \sqrt{d}\,y) > 1 \cdot 1 = 1,$$

which is absurd.

Case 3: $x < 0$ and $y > 0$. Then $-x + \sqrt{d}\,y > 1$, and again by (6) and (1), we have

$$-1 = -x^2 + dy^2 = (-x + \sqrt{d}\,y)(x + \sqrt{d}\,y) > 1 \cdot 1 = 1,$$

which is also absurd.

Thus, the only possible case is $x > 0$ and $y > 0$. ■

Let us now tell the complete story about the solutions to Pell's equation We know that the equation $x^2 - dy^2 = 1$ has at least one solution (x_0, y_0) in integers with $x_0 \neq 0$, $y_0 \neq 0$ by Theorem 3. Moreover, since $(\pm x_0, \pm y_0)$

is also a solution for any choice of signs, we see that there is at least one positive solution, say (x_0', y_0'). Set $M = x_0' + \sqrt{d}\,y_0'$. If (x_1, y_1) is any positive solution of $x_1^2 - dy_1^2 = 1$, then the condition

$$x_1 + \sqrt{d}\,y_1 \leq M \tag{7}$$

implies that $x_1 \leq M$ and $y_1 \leq M$. Thus, in particular, there are only finitely many choices for x_1 and y_1. Let us choose the positive solution (x_1, y_1) for which $x_1 + \sqrt{d}\,y_1$ is least. This is possible since (1) has only finitely many positive solutions. Call (x_1, y_1) the *fundamental solution* of Pell's equation. We may now state our main result on Pell's equation.

Theorem 7: Let (x_1, y_1) be the positive solution to Pell's equation for which $x_1 + \sqrt{d}\,y_1$ is least. For each positive integer n, define x_n and y_n by

$$x_n + \sqrt{d}\,y_n = (x_1 + \sqrt{d}\,y_1)^n.$$

Then all the solutions of Pell's equation $x^2 - dy^2 = 1$ are given by

$$(x, y) = (\pm 1, 0) \quad \text{and} \quad (x, y) = (\pm x_n, \pm y_n), \tag{8}$$

where all choices of signs are allowed. Moreover, all these solutions are different.

Proof: We know from Lemma 4 that all the (x_n, y_n) of (4) are, in fact, solutions of $x^2 - dy^2 = 1$. Let us next prove that the (x, y) in (8) are all different. From Eq. (5), we see, since $x_1 > 0$, $y_1 > 0$, that $x_n > 0$, $y_n > 0$ for all n. (Use induction.) Thus, no one of $(\pm x_n, \pm y_n)$ can be equal to $(\pm 1, 0)$. Thus, let us prove that all of $(\pm x_n, \pm y_n)$ are different. It clearly suffices to show that all of (x_n, y_n) are different since $x_n > 0$, $y_n > 0$. However, since (x_1, y_1) is a positive solution, we have $x_1 + \sqrt{d}\,y_1 > 1$ by Lemma 6. However, if $(x_n, y_n) = (x_m, y_m)$ with say $n < m$, then we must have

$$(x_1 + \sqrt{d}\,y_1)^n = x_n + \sqrt{d}\,y_n = x_m + \sqrt{d}\,y_m = (x_1 + \sqrt{d}\,y_1)^m,$$

so that

$$(x_1 + \sqrt{d}\,y_1)^{m-n} = 1,$$

which is impossible since $x_1 + \sqrt{d}\,y_1 > 1$. Thus, all the (x, y) of (5) are different.

Let (u, v) be any solution of $x^2 - dy^2 = 1$. Since $(\pm 1, 0)$ are the only solutions with one of x or y equal to zero, and since if (x, y) is a solution, then so is $(\pm x, \pm y)$, we may assume, without loss of generality, that $u > 0$, $v > 0$. Let us show that $(u, v) = (x_n, y_n)$ for some $n \geq 1$.

Since (x_1, y_1) was chosen as the positive solution of $x^2 - dy^2 = 1$ for which $x_1 + \sqrt{d}\,y_1$ is least, we see that

$$x_1 + \sqrt{d}\,y_1 \leq u + \sqrt{d}\,v. \tag{9}$$

We assert that there is a positive integer n such that

$$(x_1 + \sqrt{d}\,y_1)^n \leq u + \sqrt{d}\,v < (x_1 + \sqrt{d}\,y_1)^{n+1}. \tag{10}$$

Indeed, we have observed that $x_1 + \sqrt{d}y_1 > 1$, so that the powers of $x_1 + \sqrt{d}y_1$ become arbitrarily large. Thus, there is a largest value of n for which $u + \sqrt{d}v \geq (x_1 + \sqrt{d}y_1)^n$. By (9), this largest value of n is at least 1. Moreover, it is clear that this largest value of n forces (10) to hold.

Let us multiply (10) by $(x_1 - \sqrt{d}y_1)^n$, which is positive since $x_1 + \sqrt{d}y_1 > 0$ and since $1 = x_1^2 - dy_1^2 = (x_1 + \sqrt{d}y_1)(x_1 - \sqrt{d}y_1)$. Then we see that

$$1 \leq (u + \sqrt{d}v)(x_1 - \sqrt{d}y_1)^n < x_1 + \sqrt{d}y_1. \tag{11}$$

Recall that $(x_1 - \sqrt{d}y_1)^n = x_n - \sqrt{d}y_n$. Set $u_1 = ux_n - dvy_n$, $v_1 = vx_n - y_nu$. Then $(u + \sqrt{d}v)(x_1 - \sqrt{d}y_1)^n = u_1 + \sqrt{d}v_1$. Moreover, a simple calculation shows that

$$u_1^2 - dv_1^2 = (u^2 - dv^2)(x_n^2 - dy_n^2) = 1.$$

Therefore, (u_1, v_1) is a solution of $x^2 - dy^2 = 1$. Moreover, (11) asserts that

$$1 \leq u_1 + \sqrt{d}v_1 < x_1 + \sqrt{d}y_1. \tag{12}$$

If $u_1 + \sqrt{d}v_1 > 1$, then by Lemma 6, (u_1, v_1) is a positive solution of $x^2 - dy^2 = 1$. However, if this is the case, then (12) contradicts the way in which (x_1, y_1) was chosen. Therefore, we must have $u_1 + \sqrt{d}v_1 \leq 1$. However, (12) implies that $u_1 + \sqrt{d}v_1 = 1$. Thus,

$$(u + \sqrt{d}v)(x_1 - \sqrt{d}y_1)^n = 1.$$

Multiplying both sides of this equation by $(x_1 + \sqrt{d}y_1)^n$, we see that

$$u + \sqrt{d}v = (x_1 + \sqrt{d}y_1)^n = x_n + \sqrt{d}y_n. \tag{13}$$

Thus, $u - x_n = \sqrt{d}(y_n - v)$. If $y_n - v \neq 0$, we have

$$\sqrt{d} = \frac{u - x_n}{y_n - v},$$

so that

$$d = \left(\frac{u - x_n}{y_n - v}\right)^2,$$

which contradicts the fact that d is not a perfect square. (Why is $(u - x_n)/(y_n - v)$ an integer?) Thus, a contradiction is reached and $y_n - v = 0$. Then, by (13), we have $u = x_n$, so we have proved that $(u, v) = (x_n, y_n)$. ■

Example 8: We again consider $x^2 - 2y^2 = 1$. We have observed before that $x = 3$, $y = 2$ is a solution. Thus, the least solution (x_1, y_1) must satisfy

$$x_1 + \sqrt{2}y_1 \leq 3 + \sqrt{2} \cdot 2 < 6,$$

and so $x_1 \leq 5$ and $y_1 \leq 5$. It is easily seen that the only pair x, y solving $x^2 - 2y^2 = 1$ satisfying $0 < x \leq 5$ and $0 < y \leq 5$ is $x = 3$, $y = 2$ (try all the cases). Thus, $x_1 = 3$ and $y_1 = 2$, and so all solutions of $x^2 - 2y^2 = 1$ are given by $x = \pm x_n$, $y = \pm y_n$ (and, of course $x = \pm 1$, $y = 0$), where

$$x_n + \sqrt{2}y_n = (3 + \sqrt{2} \cdot 2)^n.$$

We observe that we may obtain x_n, y_n recursively from x_{n-1}, y_{n-1} more easily by (5):

$$x_n = 3x_{n-1} + 4y_{n-1}, \qquad y_n = 2x_{n-1} + 3y_{n-1}.$$

Compare this result to Example 1.

It should be pointed out that Theorem 7 allows us to determine all solutions to (1) once we know the least solution. Also from the discussion above, if we know any solution to (1) with $y \neq 0$, it is a simple task to determine the least solution. However, in practice, determining a solution to (1) can be an arduous task as even for small d's the least x_1, y_1 may be very large. For example, for $d = 46$, $x_1 = 24335$ and $y_1 = 3588$. We have given no method for determining a solution. One method does exist, and it is called the method of continued fractions. The interested reader should see *An Introduction to the Theory of Numbers* by G. H. Hardy and E. M. Wright, Oxford U. Press, 1960, pp. 129–153.

There is a very simple result that we can obtain now concerning the more general equation

$$x^2 - dy^2 = k \tag{14}$$

for fixed integer k. As we have observed many times before, (14) may have no solutions at all. As a general example, if $d = p$ is a prime, $p \nmid k$ and $\left(\frac{k}{p}\right) = -1$, then (14) can have no solutions, as a solution to (14) would give a solution to

$$x^2 \equiv k (\text{mod } p),$$

violating the assumption that $\left(\frac{k}{p}\right) = -1$. In the second half of this book we shall deal with the question of deciding for which k (14) can be solved. For now we shall settle for the following result:

Theorem 9: If $x^2 - dy^2 = k$ has one solution, then it has infinitely many.

Proof: If x_1, y_1 is a solution to (14) and x_2, y_2 is a solution to (1) then $x = x_1x_2 + dy_1y_2$, $y = x_1y_2 + y_1x_2$ is a solution of (14). As x_2, y_2 ranges through the solutions of (1) (there are infinitely many with $x_2 > 0$ and $y_2 > 0$), we easily see that we obtain an infinite number of solutions of (14). ■

Example 10: Since $5^2 - 2\cdot 3^2 = 7$, define x_n, y_n ($n \geq 0$) by

$$x_n + \sqrt{2}\,y_n = (3 + \sqrt{2}\cdot 2)^n(5 + \sqrt{2}\cdot 3).$$

Then for all $n \geq 0$, $x_n^2 - 2y_n^2 = 7$. For example, if $n = 1$, then

$$x_1 = 3\cdot 5 + 2\cdot 2\cdot 3 = 27$$
$$y_1 = 3\cdot 3 + 2\cdot 5 = 19,$$

and $27^2 - 2\cdot 19^2 = 7$.

6.6 Exercises

1. Determine all solutions of $x^2 - a^2y^2 = n$ for fixed integers a, n.

2. Find infinitely many solutions of the Diophantine equation $x^2 - 3y^2 = 6$.

3. Compute fundamental solutions of the following Pell equations:

   (a) $x^2 - 8y^2 = 1$.
   (b) $x^2 - 7y^2 = 1$.
   (c) $x^2 - 3y^2 = 1$.
   (d) $x^2 - 15y^2 = 1$.
   (e) $x^2 - 17y^2 = 1$.

4. Compute all solutions of the Diophantine equations (a)–(e) of Exercise 3.

5. Let (x_1, y_1) be the positive solution to Pell's equation for which $x_1 + \sqrt{d}\,y_1$ is least. For each integer n (positive, negative, or zero), define x_n and y_n by

$$x_n + \sqrt{d}\,y_n = (x_1 + \sqrt{d}\,y_1)^n.$$

   Then show that all solutions of $x^2 - dy^2 = 1$ are given by $(x, y) = (x_n, y_n)$ and $(x, y) = (-x_n, -y_n)$. Moreover, all these solutions are different.

6. Let d be a nonsquare integer, and assume that the Diophantine equation $x^2 - dy^2 = -1$ has a solution (x_0, y_0).

   (a) Show how to determine all solutions of this equation in terms of the least solution (x, y) for which $x + y\sqrt{d} > 1$ (a so-called fundamental solution).
   (b) Show that a fundamental solution always exists.
   (c) Show that, if (x_1, y_1) is a fundamental solution of $x^2 - dy^2 = -1$, then upon setting $(x_1 + y_1\sqrt{d})^2 = x + y\sqrt{d}$, we have that (x, y) is a fundamental solution for the Pell equation $x^2 - dy^2 = 1$.

7. Show that there exists an infinite number of solutions (x, y) of Pell's equation for which $y \equiv 0 (\text{mod } N)$, N a given integer.

8. Let Γ be a collection of positive real numbers satisfying the following three properties: (i) If γ_1, γ_2 belong to Γ, so does $\gamma_1\gamma_2$; (ii) if γ belongs to Γ, so does γ^{-1}; (iii) there exists an interval containing 1 which contains only finitely many elements of Γ. Show that Γ is precisely the set of all powers (positive, negative, and zero) of some element of Γ.

9. We shall derive a formula for the Fibonacci numbers defined as follows: $f_0 = 0, f_1 = 1$ and inductively for $n \geq 2, f_n = f_{n-1} + f_{n-2}$.

(a) Given any integers a, b and a positive integer n, show there are integers c, d such that

$$\left(a + \frac{1+\sqrt{5}}{2}b\right)^n = c + \frac{1+\sqrt{5}}{2}d.$$

(b) Defining integers c_n, d_n by

$$\left(\frac{1+\sqrt{5}}{2}\right)^n = c_n + \frac{1+\sqrt{5}}{2}d_n,$$

show that $d_n = f_n$ for $n \geq 0$.

(c) Show that

$$\left(\frac{1-\sqrt{5}}{2}\right)^n = c_n + \frac{1-\sqrt{5}}{2}d_n$$

(same c_n, d_n as in part (b)).

(d) Show that

$$f_n = \frac{1}{\sqrt{5}}\left(\frac{1+\sqrt{5}}{2}\right)^n - \frac{1}{\sqrt{5}}\left(\frac{1-\sqrt{5}}{2}\right)^n.$$

10. (a) Let $d > 0$ be a square-free integer such that $d \equiv 1 \pmod 4$. Let $f(x, y) = x^2 + xy + ((1 - d)/4)y^2$. Show that $f(x, y) = 1$ has an infinite number of solutions. (*Hint:* Observe that $x + y\sqrt{d} = (x - y) + ((1 + \sqrt{d})/2)(2y)$ and that $f(x, y) = (x + ((1 + \sqrt{d})/2)y)(x + ((1 - \sqrt{d})/2)y)$.)

(b) Show that if $f(x_0, y_0) = 1$ and x_n, y_n are defined by

$$x_n + \frac{1+\sqrt{d}}{2}y_n = \left(x_0 + \frac{1+\sqrt{d}}{2}y_0\right)^n$$

(n any integer—positive or negative), then x_n, y_n are integers and $f(x_n, y_n) = 1$.

(c) Show that $f(x, y) = 1$ has a solution x_0, y_0 such that $x_0 > 0$, $y_0 > 0$ and that $x_0 + ((1 + \sqrt{d})/2)y_0$ is least.

(d) Show that with (x_0, y_0) determined in part (c) and then (x_n, y_n) determined in part (b), all the solutions of $f(x, y) = 1$ are given by $(x, y) = (x_n, y_n)$ or $(x, y) = (-x_n, -y_n)$ $(n = 0, \pm 1, \pm 2, \ldots)$.

Appendix B

Diophantine Approximations

Let us begin by considering the number π, which was defined by the ancient Greeks as the ratio of the circumference of a circle to its diameter. As we all learned in secondary school, π "is" $\frac{22}{7}$. In fact, one state legislature went so far as to legislate the value of π to be $\frac{22}{7}$ (after all, "there is no need making math any harder than necessary"). Actually, there is a great deal to be said for the approximation $\frac{22}{7}$ for π which was first given by Archimedes in 212 B.C. The number

$$\pi = 3.1415926535\ldots$$

is an infinite decimal and so is awkward to use in hand computations. Certainly one feels more comfortable using $\frac{22}{7}$, because we feel more comfortable dealing with integers (and their ratios), especially with small integers. Indeed, $\frac{22}{7}$ is the ratio of two small integers (22 and 7) and

$$\frac{22}{7} = 3.142857\ldots$$

is a fairly good approximation of π (the difference is 0.00126...). It can, in fact, be shown that the difference $|7\pi - 22|$ is smaller than the difference $|q\pi - p|$ for all fractions p/q with $1 \leq q \leq 105$. π, in fact, has another extraordinary approximation, namely 355/113. This fraction has the property that the difference $|113\pi - 355|$ is smaller than the difference $|q\pi - p|$ for all fractions p/q with $1 \leq q \leq 33{,}101$. It is accurate to six decimal places (error is less than 0.000000268) as

$$\frac{355}{113} = 3.1415929203\ldots.$$

These approximations for π are extremely good, in fact better than the approximations enjoyed by most irrational numbers. How well can we expect to be able to approximate a real number by rational numbers with small denominators? Explicitly, let α be a real number and let $N > 1$ be a fixed integer. Among all fractions p/q with $1 \leq q \leq N$, how small can we make $|\alpha - p/q|$? This will be answered in Theorem 1, but first let us study another example.

Let us consider the problem of setting up a calendar. The time it takes the earth to complete its orbit about the sun has been calculated to be 365 days, 5 hours, 48 minutes, and 46 seconds. Since we want our year to have an integer number of days, we must vary the length of the years. What is an efficient method for doing this? Since there are 86,400 seconds in a day, and 5 hours 48′46″ equals 20,926 seconds, the actual time for the earth to orbit the sun exceeds the calendar year of 365 days by 20,926/86,400 of a day. Thus, the calendar could be corrected by adding 20,926 days at the end of 864 centuries. This obviously is not appropriate. What we need is a rational approximation of the number 20,926/86,400 which is easier to use to correct the calendar. It would be convenient to have an approximation p/q with a small q. Many approximations have been used. In 45 B.C. Julius Ceasar used the approximation of $\frac{1}{4}$ by letting 1 year in every 4 have 366 days. This is a fair approximation and is incorrect by 1 day in 128 years. The fraction $\frac{8}{33}$ was proposed by Omar Alkhayami in 1079 A.D. and is called the Persian intercalation. It is more accurate than the Julian intercalation. The calendar we use works according to the following scheme. There is a leap year in every year divisible by 4 except those divisible by 100 and not divisible by 400 (e.g., 1900 was not a leap year but 2000 will be). This corresponds to approximating 20,926/86,400 by 97/400. This calendar is off by 26 seconds every year. The advantage of 97/400, over, say, $\frac{8}{33}$ which is closer than 97/400, is that the corrections can be made in a manner easier to remember. It has, in fact, been proposed that the leap year in the year 4000 be omitted and in all succeeding years which are multiples of 4000. This would make the calendar inaccurate by only 1 day in 200 centuries.

Before we begin the more formal material of this appendix let us consider the question of actually determining these good rational approximations. In the sequel we shall show how well we can approximate but shall give no method for finding the approximations. There is, in fact, a constructive method for finding the approximations called *continued fractions*. This is a topic commonly included in elementary number theory books (for example, I. Niven and H. Zuckerman, *An Introduction to the Theory of Numbers*, 3rd ed. Wiley, 1973.), and we urge the readers to pursue this fascinating subject.

We shall now show how small we can make $|\alpha - p/q|$. The result is usually called Dirichlet's theorem, although it was certainly known to others before him.

Theorem 1: Let α be a real number. Let $N > 1$ be an integer. Then there is an integer q such that $1 \leq q \leq N$ and

$$\left|\alpha - \frac{p}{q}\right| < \frac{1}{qN}. \tag{1}$$

The proof of Theorem 1 relies on a very useful and simpleminded statement which goes under various names such as the *pigeon hole principle* or the *(Dirichlet) box principle:*

If we put $N + 1$ objects into N boxes, then at least one of these boxes must contain more than one of the objects.

Some of the results of Chapter 6 could have been proved in this way; for example, we suggested an alternative proof of Lemma 6.5.3 in Exercise 4 in Section 6.5.

It is convenient to rewrite the conclusion (1) of Theorem 1 in the following way:

There is an integer q such that $1 \leq q \leq N$ and

$$|q\alpha - p| \leq \frac{1}{N}.$$

Proof of Theorem 1: Consider the numbers $0 = 0 \cdot \alpha$, $\alpha - [\alpha]$, $2\alpha - [2\alpha]$, $\ldots$, $N\alpha - [N\alpha]$. That is, consider the numbers $k\alpha - [k\alpha]$ for $0 \leq k \leq N$. These $N + 1$ numbers will be our objects. They all lie between 0 and 1.

Divide up the interval from 0 to 1 into N equal subintervals (See Fig. B-1):

$$0=\frac{0}{N}\text{ to }\frac{1}{N},\ \frac{1}{N}\text{ to }\frac{2}{N},\ \frac{2}{N}\text{ to }\frac{3}{N},\ \ldots,\ \frac{N-1}{N}\text{ to }\frac{N}{N}=1.$$

0 1/6 2/6 3/6 4/6 5/6 1

Figure B-1. $N = 6$.

These N subintervals are the boxes. There are N of them, and so two of the numbers $k\alpha-[k\alpha]$ $(0\le k\le N)$ must lie in the same subinterval; that is, they can be no farther apart than $1/N$. Say these two numbers are $n\alpha-[n\alpha]$ and $m\alpha-[m\alpha]$, where $0\le n<m\le N$. Set $q=m-n$ and $p=[m\alpha]-[n\alpha]$. Then

$$|q\alpha-p|=|(n\alpha-[n\alpha])-(m\alpha-[m\alpha])|\le\frac{1}{N}\cdot$$

Moreover, $q=m-n\le m\le N$, and $q\ge 1$, since $m-n>0$. ■

Corollary 2: There are an infinite number of integers $q\ge 1$ and p such that

$$\left|\alpha-\frac{p}{q}\right|\le\frac{1}{q^2}\cdot \tag{2}$$

Proof: It is an easy exercise to prove the result if α is rational. Thus, we assume that α is irrational.

Setting $N_1=1$ in Theorem 1 we obtain p_1 and q_1 $(=1)$ such that

$$\left|\alpha-\frac{p_1}{q_1}\right|\le\frac{1}{q_1N_1}=\frac{1}{q_1^2}\cdot$$

Since α is irrational, $|q_1\alpha-p_1|\ne 0$, and so we may choose N_2 such that

$$\frac{1}{N_2}<|q_1\alpha-p_1|. \tag{3}$$

Then, by Theorem 1, there are integers p_2, q_2 such that $1\le q_2\le N_2$ and

$$\left|\alpha-\frac{p_2}{q_2}\right|\le\frac{1}{q_2N_2}\le\frac{1}{q_2^2}\cdot$$

Thus, by (3) $|q_2\alpha-p_2|\le 1/N_2<|q_1\alpha-p_1|$, and so (q_2,p_2) is different from (q_1,p_1). Again α is irrational, and so $|q_2\alpha-p_2|\ne 0$, and we may choose N_3 so that

$$\frac{1}{N_3}<|q_2\alpha-p_2|.$$

By Theorem 1 we can find integers p_3, q_3 such that $1\le q_3\le N_3$ and

$$\left|\alpha-\frac{p_3}{q_3}\right|\le\frac{1}{q_3N_3}\le\frac{1}{q_3^2}\cdot$$

Also, $|q_3\alpha-p_3|\le 1/N_3<|q_2\alpha-p_2|<|q_1\alpha-p_1|$, and so (q_3,p_3) is different from (q_2, p_2) and (q_1, p_1). Continuing in this way (induction) we construct an infinite number of solutions to (2). ■

Example 3: We observed before that

$$\left|\pi - \frac{22}{7}\right| = 0.00126\ldots < \frac{1}{7^2}$$

and that

$$\left|\pi - \frac{355}{113}\right| \leq 0.000000268 < \frac{1}{113^2}.$$

Moreover, by assuming the statement that $|7\pi - 22|$ is smaller than $|q\pi - p|$ for any other fraction p/q with $1 \leq q \leq 105$, we see from Theorem 1 with $N = 105$ that we must have

$$\left|\pi - \frac{22}{7}\right| \leq \frac{1}{7 \cdot 105} = 0.0013605\ldots,$$

which is close to the true order of magnitude of the difference.

The next question we pose is whether we can do better than (2). For example, can we guarantee that there are an infinite number of integers q, p such that $|\alpha - p/q| < 1/q^3$ or some other function of q smaller than $1/q^2$. The answer is no in the sense that there are irrational numbers α such that

$$\left|\alpha - \frac{p}{q}\right| \geq \frac{1}{3q^2} \tag{4}$$

for all q, p. However one can show that $|\alpha - p/q| \leq 1/\sqrt{5}\,q^2$ always has an infinite number of solutions; this is called the Hurwitz theorem and cannot be proved here. We shall give an illustration of (4) in the next example, which will lead into the main topic of this appendix, namely the proof of Lemma 6.6.2.

Example 4: Let $\alpha = \sqrt{2}$. Given any integers $q \geq 1$ and p, we have $p - q\alpha \neq 0$ and $p + q\alpha \neq 0$. Thus,

$$(p - q\alpha)(p + q\alpha) = p^2 - 2q^2 \neq 0.$$

(This, of course, was observed in Section 6.6.) Observe that if n is a nonzero integer, then $|n| \geq 1$. Thus,

$$1 \leq |p^2 - 2q^2| = |q\alpha - p||q\alpha + p|.$$

If $|\alpha - p/q| \leq 1/q^2$, then $|q\alpha - p| \leq 1/q$. Therefore, since $q \geq 1$ and $\alpha > 0$, we would have $p > 0$ and $p \leq q\alpha + (1/q)$. Thus,

$$1 \leq |q\alpha - p||q\alpha + p| \leq |q\alpha - p|\left(q\alpha + q\alpha + \frac{1}{q}\right) \leq |q\alpha - p| \cdot (3q)$$

for $q \geq 3$. Thus for $q \geq 3$, we have

$$|q\alpha - p| \geq \frac{1}{3q},$$

so that

$$\left|\alpha - \frac{p}{q}\right| \geq \frac{1}{3q^2}.$$

Exploiting the idea of Example 4, we shall prove Lemma 6.6.2.

Lemma 5: Let $B = 2\sqrt{d} + 1$, where d is a positive integer. Then there exist an infinite number of integers x and y such that

$$|x^2 - dy^2| \leq B.$$

Proof: From Corollary 2 we have an infinite number of integers x and y such that $y \geq 1$ and

$$|y\sqrt{d} - x| \leq \frac{1}{y}.$$

Thus, $|\sqrt{d} - x/y| \leq 1/y^2$ and $|x/y| \leq \sqrt{d} + 1/y^2 \leq \sqrt{d} + 1$. Finally,

$$|x^2 - dy^2| = |y\sqrt{d} - x||y\sqrt{d} + x| \leq \frac{|y\sqrt{d} + x|}{y}$$

$$= \left|\sqrt{d} + \frac{x}{y}\right| \leq \sqrt{d} + \left|\frac{x}{y}\right| \leq \sqrt{d} + \sqrt{d} + 1 = B. \quad \blacksquare$$

Exercises

1. Let $a_1, a_2, a_3, a_4, a_5, a_6$ be integers, not all zero, such that $|a_i| \leq A$ for $1 \leq i \leq 6$. Show by the pigeon hole principle that the linear equations

$$a_1x + a_2y + a_3z = 0$$
$$a_4x + a_5y + a_6z = 0$$

have a solution x, y, z with at least one of x, y, z nonzero such that

$$|x|, |y|, |z| \leq 72A^2.$$

2. Generalize Exercise 1 to a system of linear equations in more variables as follows: Let

$$a_{11}x_1 + \cdots + a_{1n}x_n = 0$$
$$\vdots$$
$$a_{r1}x_1 + \cdots + a_{rn}x_n = 0$$

be r equations in n unknowns with $n > r$ and with a_{ij} integers and $|a_{ij}| \leq A$ for all i, j. Then there are integers $x_1, \ldots, x_n$, not all zero, such that

$$|x_i| \leq 2(2nA)^{r/(n-r)} \qquad (1 \leq i \leq n).$$

3. Let α be a rational number. Show that $|q\alpha - p| < 1/q$ has an infinite number of solutions in integers q, p.

4. Let $d > 0$ be a nonsquare integer. Show that there is a constant C (depending on d) such that

$$|q\sqrt{d} - p| > \frac{C}{q}$$

for all integers $q \geq 1$ and p.

5. Use the pigeon hole principle to prove the following result of Dirichlet. Let $\alpha_1, \ldots, \alpha_n$ be real numbers. Show that there is a constant $C > 0$ such that for all integers $B \geq 1$ there are integers $q, p_1, p_2, \ldots, p_n$ such that

$$|q\alpha_i - p_i| \leq \frac{C}{B^{1/n}} \qquad (1 \leq i \leq n,\ 1 \leq q \leq B).$$

6. Deduce from Exercise 5 that the inequalities $|q\alpha_i - p_i| < C/q^{1/n}$ $(1 \leq i \leq n)$ have an infinite number of solutions in integers $q \geq 1$, $p_1, \ldots, p_n$, for some constant $C > 0$.

7. A famous result of Thue says that there are only a finite number of integers q, p such that

$$|q\sqrt[3]{2} - p| < \frac{1}{q^{3/2}}.$$

Use this result to show that the Diophantine equation

$$x^3 - 2y^3 = 1$$

has only a finite number of solutions.

Introduction to Chapters 7–11

In Chapters 1–6, we developed the most elementary facts about the integers and have used those facts to study a variety of Diophantine equations. In particular, in Chapter 6, our study included the equations

$$x^2 + y^2 = n \tag{1}$$

$$x^2 - dy^2 = 1, \tag{2}$$

associated with the names of Fermat and Pell, respectively. The reader may have noted in the first 6 Chapters that the methods used to solve Diophantine equations vary from equation to equation. Therefore, it seems reasonable to ask if there is some uniform method for solving all Diophantine equations. It has recently been proved that no such method exists (using the methods of logic).* Thus, this problem has a definitive, but very unsatisfactory, solution. Suppose that we restrict the problem to a smaller class of Diophantine equations. Can a uniform method then be obtained? In some cases, certainly. Take, for instance, all linear equations of the form $ax + by = n$, which we solved in Chapter 2. Our method there was a uniform one, based on the Euclidean algorithm. Are there other classes of equations which can be

*The problem of finding such a method was proposed by David Hilbert in 1900 and has recently been settled by the combined efforts of Martin Davis, Hillary Putnam, and Julia Robinson, with the final step being taken by the Russian mathematician Matiyashevich in 1970.

handled? For example, let us consider the class of all quadratic Diophantine equations of the form

$$ax^2 + bxy + cy^2 = n. \tag{3}$$

This class contains Eqs. (1) and (2) as special cases. As we have seen, the solutions of (1) and (2) are quite different. For example, (1) has only a finite number of solutions, whereas (2) has an infinite number. Therefore, any uniform method for solving (3) must be quite sophisticated in order to detect the difference between Eqs. (1) and (2). It is our plan to take up the study of the family of equations (3) in the rest of this book. As we shall see, the theory of these equations is much more complicated than that for linear equations, and there are still simple questions about equations of the form (3) which defy solution. However, we shall go a long way in the study of quadratic Diophantine equations in two variables. For example, we shall give an algorithm for determining all solutions of (3) when a, b, c, and n are given.

Although scattered results about quadratic Diophantine equations (e.g., Eqs. (1) and (2)) had been given in the seventeenth and eighteenth centuries by Fermat, Euler, Lagrange, and others, the first systematic attempt at tackling the general equation (3) was undertaken by Gauss in 1799 in his *Disquisitiones Arithmeticae*. His results are some of the greatest achievements in the history of the theory of numbers. Gauss not only organized centuries of number-theoretic thought into a coherent subject; he charted the course which has guided number-theoretic research up to the present time. In Chapters 7–11, we shall undertake an exposition, in suitably modern language, of Gauss' results on quadratic Diophantine equations.

The revolution in number theory which Gauss began centered on a systematic use of irrational numbers, such as $\sqrt{2}$, $\sqrt{5}$, $\sqrt{-5}$, and $\sqrt{-1}$. At first, it may seem philosophically troubling that one can get new and interesting results about the integers by using numbers which are not integers and which seem to have little if any connection with the integers. Nevertheless, there is a deep connection. The basic idea is to introduce certain "generalized integers" and to study the properties of these integers. In turn, these properties yield information about Diophantine equations.

A second independent thread in nineteenth-century number theory was the work of Ernst Kummer on Fermat's Last Theorem. Kummer thought that he could prove Fermat's infamous conjecture by using properties of certain generalized integers formed from roots of unity. It turned out that he was in error about the properties of the generalized integers, and so his alleged proof was false. However, his later attempts to resurrect his proof led him to study the generalized integers in a very deep way and led him to the invention of *ideals* or *ideal numbers.*

Finally, it was Dirichlet and Dedekind who fused Gauss' work on quadratic Diophantine equations and Kummer's work on Fermat's last theorem into the field now known as *algebraic number theory*. It is our purpose in Chapters 7–11 to give an introduction to algebraic number theory for what are called *quadratic number fields*. This should be enough to give the reader a taste of what this beautiful branch of number theory is about and give him the background to read further if he is so inclined.

We shall organize the rest of this book as follows: Chapter 7 will be devoted to the study of one system of generalized integers, the Gaussian integers. We shall see that we are naturally led to consider the Gaussian integers from the Diophantine equation $x^2 + y^2 = n$. In Chapter 8, we shall build on Chapter 7 and study the general theory of quadratic numbers and give the algorithm for solving Eq. (3) mentioned above. In Chapter 9 we shall complete our theory of quadratic integers, deriving a certain type of unique factorization theory. In Chapter 10 we shall apply this theory to certain Diophantine equations other than (3) and outline how the theory generalizes to cubic and higher-degree numbers. Finally in Chapter 11 we shall give a systematic account of the theory of determining the integers n for which (3) can be solved (for a fixed a, b, c).

Throughout the rest of this book, lowercase italic Roman letters will always denote rational numbers.

Starting at Chapter 8 we shall assume that the reader is familiar with the material given in an undergraduate course in abstract algebra (see, for example, L. J. Goldstein, *Abstract Algebra, A First Course*, Prentice-Hall, Englewood Cliffs, N. J., 1973.) and with the most elementary facts concerning vector spaces over the rational numbers.

7

The Gaussian Integers

7.1 Introduction

In Chapter 6, we studied the problem of representing an integer as a sum of two squares, that is, the problem of determining all solutions of the Diophantine equation $x^2 + y^2 = n$ (for fixed n). Our main result (Theorem 6.4.5) was a characterization of all positive integers n which are sums of two squares. Our methods probably seemed very ad hoc and somewhat unenlightening, in that there was little behind them except for a number of fortuitous accidents and a great deal of algebraic manipulation. In this chapter, we shall reconsider the two-square problem in a much more natural setting—the theory of Gaussian integers. By studying the Gaussian integers and their properties, we shall derive a completely independent proof of the two-square theorem.

Suppose that n is a positive integer and that x and y are integers such that $x^2 + y^2 = n$. Then, if i denotes the complex number $\sqrt{-1}$, we may factor $x^2 + y^2$ into $(x + iy)(x - iy)$. Thus, we have

$$(x + iy)(x - iy) = n. \tag{1}$$

This formula suggests that we look at the set of all numbers of the form

$$x + iy, \qquad x, y \text{ integers}.$$

This collection of numbers is called the *Gaussian integers*. Equation (1) suggests a fundamental connection between the Gaussian integers and the problem of representing a positive integer as a sum of two squares.

In this chapter, we shall explore the properties of the Gaussian integers, with a view toward developing for them an arithmetic similar to that developed for the ordinary integers in Chapter 2. At the end of the chapter, we shall return to the two-square problem and shall show how our theory of arithmetic in the Gaussian integers can be used to give a very natural proof of Theorem 6.4.5.

Let us begin with a formal definition:

Definition 1: A *Gaussian integer* is a complex number of the form $x + iy$, where x, y are integers and $i = \sqrt{-1}$. The set of all Gaussian integers will be denoted $\mathbf{Z}[i]$.

We shall use Greek letters $\alpha, \beta, \gamma, \ldots$ for Gaussian integers.

For example, $1 = 1 + 0i$, $2 + i$, and $5 - 3i$ are all examples of Gaussian integers. If n is an ordinary integer, then n is also a Gaussian integer, since $n = n + 0 \cdot i$. We shall see in this chapter that the Gaussian integers possess most of the arithmetic properties of the ordinary integers. For example, we shall prove a division algorithm and a unique factorization theorem.

Proposition 2: Let α, β be Gaussian integers. Then $\alpha + \beta$, $\alpha - \beta$, and $\alpha \cdot \beta$ are Gaussian integers.

Proof: Suppose that $\alpha = x + iy$, $\beta = w + iz$. Then

$$\alpha \pm \beta = (x \pm w) + i(y \pm z)$$

$$\alpha \cdot \beta = (xw - yz) + i(yw + xz).$$

The point, of course, is that $x \pm w$, $y \pm z$, $xw - yz$, and $yw + xz$ are integers. ■

Note, however, that the quotient of two Gaussian integers is not usually a Gaussian integer. For example,

$$\frac{1}{1+i} = \frac{1}{1+i} \cdot \frac{1-i}{1-i} = \frac{1-i}{2} = \frac{1}{2} - \frac{1}{2}i.$$

On the other hand, for certain Gaussian integers α, $1/\alpha$ is a Gaussian integer. For example,

$$\frac{1}{i} = -i.$$

If x is an ordinary integer with the property that $1/x$ is an integer, then $x = \pm 1$. However, we have just seen that there are other Gaussian integers with this property.

Definition 3: Let α be a nonzero Gaussian integer. If $1/\alpha$ is a Gaussian integer, then α is said to be a *unit*.

From what we have said above, we see that $\pm 1, \pm i$ are units of $\mathbf{Z}[i]$. We shall come to the arithmetic significance of units later in the chapter.

Let us recall a few facts about complex numbers. If $\alpha = x + iy$, x, y real, is a complex number, then the complex number $\alpha' = x - iy$ is called the *conjugate* of α. We shall denote the conjugate of α by α' rather than the more customary $\bar{\alpha}$ in order to make our notation easily adaptable to the setting of more general quadratic fields, which we shall study in Chapter 8.

If $\alpha = x + iy$, x, y real, is a complex number, then the *absolute value* of α, denoted $|\alpha|$, is given by

$$\begin{aligned} |\alpha| &= \sqrt{x^2 + y^2} \\ &= \sqrt{\alpha\alpha'}. \end{aligned}$$

For our purposes, it is more convenient to use the square of $|\alpha|$ rather than $|\alpha|$ itself. Let us define the *norm* of α, denoted $N(\alpha)$, by

$$\begin{aligned} N(\alpha) &= |\alpha|^2 \\ &= x^2 + y^2 \\ &= \alpha\alpha'. \end{aligned}$$

A few elementary properties of the norm are summarized in the following result:

Proposition 4: Let α, β be complex numbers.

(i) $N(\alpha)$ is a nonnegative real number.

(ii) $N(\alpha) = 0$ if and only if $\alpha = 0$.

(iii) $N(\alpha\beta) = N(\alpha)N(\beta)$.

(iv) If α is a Gaussian integer, then $N(\alpha)$ is an ordinary integer.

Proof:

(i) Let $\alpha = x + iy$, x, y real. Then $N(\alpha) = x^2 + y^2$ is a nonnegative real number. In case α is a Gaussian integer, $N(\alpha)$ is an ordinary integer, whence (iv).

(ii) $N(\alpha) = 0$ if and only if $x = y = 0$.

(iii) $N(\alpha\beta) = (\alpha\beta)(\alpha\beta)' = (\alpha\beta)(\alpha'\beta') = (\alpha\alpha')(\beta\beta') = N(\alpha)N(\beta)$.

(Here we have made use of the elementary property of complex conjugation: $(\alpha\beta)' = \alpha'\beta'$.) ■

The student should view the norm of α as measuring the "size" of α, in much the same way that the usual absolute value for real numbers measures the size of a real number. Let us apply the notion of a norm to find all units in $\mathbf{Z}[i]$.

Proposition 5: Let α be a Gaussian integer. Then α is a unit if and only if $N(\alpha) = 1$.

Proof: First note that α is a unit if and only if $1/\alpha$ is a Gaussian integer. Thus, if α is a unit, both $N(\alpha)$ and $N(1/\alpha)$ are ordinary integers by Proposition 4, part (iv). But then by Proposition 4, part (iii), we have

$$N(\alpha)N\left(\frac{1}{\alpha}\right) = N\left(\alpha \cdot \frac{1}{\alpha}\right) = N(1) = 1,$$

so that $N(\alpha) = \pm 1$. But since $N(\alpha) \geq 0$ by Proposition 4, part (i), we have $N(\alpha) = 1$. Conversely, suppose that $N(\alpha) = 1$. Then $\alpha\alpha' = 1$ and $\alpha' = 1/\alpha$. But if $\alpha = x + iy$, x, y integers, we then have $\alpha' = 1/\alpha = x - iy$, and so $1/\alpha$ is a Gaussian integer. ■

Theorem 6: The units of the Gaussian integers are $\pm 1, \pm i$.

Proof: Let $\alpha = x + iy$ be a unit of the Gaussian integers. Then by Proposition 5 we have

$$N(\alpha) = x^2 + y^2 = 1.$$

But since x and y are ordinary integers, the only solutions to the last equation are $(x, y) = (\pm 1, 0), (0, \pm 1)$, corresponding to $\alpha = \pm 1, \pm i$. ■

7.1 Exercises

1. Let d be an integer which is not a perfect square and let S_d denote the set of all numbers of the form $a + b\sqrt{d}$, where a, b are integers.
   (a) Show that the sum, difference, and product of elements of S_d belong to S_d.
   (b) A *unit* of S_d is an element η of S_d such that $1/\eta$ belongs to S_d. Show that $\eta = a + b\sqrt{d}$ is a unit of S_d if and only if $a^2 - b^2 d = \pm 1$.
   (c) Let $d < 0$. Show that S_d contains only a finite number of units.
   (d) Let $d > 0$. Show that S_d contains infinitely many units.
2. Refer to Exercise 1.
   (a) Determine all units of S_{-3}.
   (b) Determine all units of S_{-2}.
3. Prove that the units of S_2 are precisely the numbers $\pm(1 + \sqrt{2})^n$, n any integer. (*Hint:* Refer to Theorem 6.6.7).

7.2 The Fundamental Theorem of Arithmetic in the Gaussian Integers

Let us now try to build up a theory of divisibility and unique factorization for the Gaussian integers. With only small changes in detail, we shall parallel our discussion after the corresponding discussion for the ordinary integers. Let us start with the notion of divisibility of Gaussian integers.

Definition 1: Let α and β be Gaussian integers. We say that α *divides* β (or α is a *divisor* of β), denoted $\alpha \mid \beta$, if there exists a Gaussian integer γ such that $\beta = \alpha\gamma$. If α does not divide β, we write $\alpha \nmid \beta$.

Note that since every ordinary integer is a Gaussian integer, if a and b are ordinary integers and if a divides b (in the sense of divisibility defined in Chapter 2), then a divides b in the sense of Definition 1. Thus, our two uses of "divisibility" are consistent and cannot lead to confusion.

Example 2:

(i) $1 \mid \alpha$ for every Gaussian integer α.

(ii) Since $(2 + i)(3 - i) = 7 + i$, we have $2 + i \mid 7 + i$.

In looking over our development of the results in Chapter 2, we can hardly fail to notice the central rôle played by the division algorithm. Let us now prove an analogue for the Gaussian integers. The division algorithm for the ordinary integers asserts that if a and b are integers, $b \neq 0$, then we can divide a by b to get a quotient q and a remainder r such that the remainder is small compared to $|b|$. Specifically, $a = qb + r$, where $0 \leq r < |b|$. It is clear what we mean by quotient and remainder in the Gaussian integers. But what do we mean when we require that the remainder be "small compared to $|b|$"? Recall that we measure the size of Gaussian integers by their norm. Thus, we can phrase the division algorithm for the Gaussian integers in the following form:

Theorem 3 (division algorithm): Let α, β be Gaussian integers, $\beta \neq 0$. Then there exist Gaussian integers γ, δ such that

$$\alpha = \beta\gamma + \delta$$

and $0 \leq N(\delta) < N(\beta)$.

Proof: Let $\alpha = a + bi$, $\beta = c + di$ where a, b, c, d are ordinary integers. Then

$$\frac{\alpha}{\beta} = \frac{a + bi}{c + di} \cdot \frac{c - di}{c - di} = \frac{ac + bd}{c^2 + d^2} + \frac{bc - ad}{c^2 + d^2} i = e + fi,$$

where e and f are the rational numbers

$$e = \frac{ac + bd}{c^2 + d^2}, \qquad f = \frac{bc - ad}{c^2 + d^2}.$$

There exist ordinary integers g, h such that

$$|g - e| \leq \tfrac{1}{2}, \qquad |h - f| \leq \tfrac{1}{2}.$$

Set $\gamma = g + hi$. Then

$$\frac{\alpha}{\beta} = \gamma + (e - g) + (f - h)i,$$

so that

$$\alpha = \beta\gamma + \{(e - g) + (f - h)i\}\beta.$$

Set $\delta = \{(e - g) + (f - h)i\}\beta$. Then $\alpha = \beta\gamma + \delta$, and since γ is clearly a Gaussian integer, we see that $\delta = \alpha - \beta\gamma$ is a Gaussian integer. Moreover,

$$\begin{aligned} N(\delta) &= N((e - g) + (f - h)i)N(\beta) \\ &= N(\beta)\cdot\{(e - g)^2 + (f - h)^2\} \\ &\leq N(\beta)\{\tfrac{1}{4} + \tfrac{1}{4}\} \\ &= \tfrac{1}{2}N(\beta) \\ &< N(\beta), \end{aligned}$$

since $N(\beta) \neq 0$ (because $\beta \neq 0$). ∎

One of the fundamental consequences of the division algorithm for ordinary integers was the existence of gcd's. Let us make a similar use of the division algorithm for Gaussian integers. First, however, we must define the notion of greatest common divisor in the Gaussian integers.

Definition 4: Let α and β be Gaussian integers. A *greatest common divisor* (gcd) *of* α *and* β is a Gaussian integer γ such that

(i) $\gamma \mid \alpha$ and $\gamma \mid \beta$.

(ii) If δ is any Gaussian integer such that $\delta \mid \alpha$ and $\delta \mid \beta$, then $\delta \mid \gamma$.

Note that this definition of gcd is almost the same as the one given for the ordinary integers. The only difference is that for the case of the ordinary integers we required that the gcd be positive. However, that requirement must be omitted since there is no way to define what it means for a Gaussian integer such as $3 + i$ to be positive. We make a sacrifice for this dropped requirement: Whereas the gcd is unique in the ordinary integers, it is not in the Gaussian integers.

Example 5: Let α and β be Gaussian integers, γ a gcd of α and β. Then $\pm\gamma$, $\pm i\gamma$ are gcd's of α and β (exercise).

Note from Example 5 that we may multiply a gcd by an arbitrary unit and get another gcd. The converse of this statement is also true. (See Proposition 7.)

Definition 6: Let α and β be Gaussian integers. We say that α and β are *associates* if there is a unit ϵ such that $\alpha = \epsilon\beta$. In other words, α and β are associates if α is one of $\beta, -\beta, i\beta, -i\beta$.

Proposition 7: Let α and β be Gaussian integers not both 0. Then any two gcd's of α and β are associates of one another.

Proof: Assume that $\alpha \neq 0$. Let γ_1 and γ_2 be two gcd's of α and β. Then, using the definition of a gcd, we see that $\gamma_1 | \alpha$, $\gamma_1 | \beta$, $\gamma_2 | \alpha$, $\gamma_2 | \beta$, so that $\gamma_1 | \gamma_2$ and $\gamma_2 | \gamma_1$. But since α is nonzero, we see that $\gamma_1 \neq 0$. Moreover, $\gamma_1 | \gamma_2$ and $\gamma_2 | \gamma_1$ imply that

$$\gamma_2 = \eta\gamma_1, \qquad \gamma_1 = \lambda\gamma_2,$$

for Gaussian integers η, λ. Therefore, $\gamma_1 = \eta\lambda\gamma_1$, and $\eta\lambda = 1$ (since $\gamma_1 \neq 0$). Thus, $\lambda = 1/\eta$ is a Gaussian integer, and so η is a unit. Thus, since $\gamma_2 = \eta\gamma_1$, we see that γ_1 and γ_2 are associates.* ■

Let us now prove the existence of greatest common divisors.

Theorem 8: Let α, β be Gaussian integers, not both 0. Then α and β have a greatest common divisor.

Proof: Let S denote the set of all Gaussian integers of the form

$$\alpha\lambda + \beta\eta,$$

where λ, η are Gaussian integers. Since 1, 0 are Gaussian integers, we see that $\alpha = \alpha \cdot 1 + \beta \cdot 0$ and $\beta = \alpha \cdot 0 + \beta \cdot 1$ belong to S. In particular, S contains nonzero numbers. Choose γ in S such that $N(\gamma)$ is a positive integer which is as small as possible. (We can find γ by applying the well-ordering principle to the set of norms of nonzero elements of S.) We assert that γ is a gcd of α and β. Since γ is in S, we have

$$\gamma = \alpha\lambda_0 + \beta\eta_0$$

for some Gaussian integers λ_0, η_0. Therefore, if $\delta | \alpha$ and $\delta | \beta$, we have $\alpha = \delta\theta$ and $\beta = \delta\zeta$, so that $\gamma = \delta(\theta\lambda_0 + \zeta\eta_0)$ and $\delta | \gamma$. Thus, property (ii) of the definition of a gcd is true. To prove property (i), we show that every element of S is a multiple of γ. Then, since α and β belong to S, this would imply that $\gamma | \alpha$ and $\gamma | \beta$, which is property (i). We first observe that if ϵ and ρ belong to S and θ is any Gaussian integer, then $\epsilon - \theta\rho$ belongs to S. Indeed, if $\epsilon = \alpha\lambda_1 + \beta\eta_1$ and $\rho = \alpha\lambda_2 + \beta\eta_2$, then

$$\epsilon - \theta\rho = \alpha(\lambda_1 - \theta\lambda_2) + \beta(\eta_1 - \theta\eta_2)$$

belongs to S. Now let ω be any element of S. By the division algorithm, we may write $\omega = \gamma\zeta + \rho$ for Gaussian integers ζ, ρ, $0 \leq N(\rho) < N(\gamma)$. By the above, ω and γ are in S, so that $\rho = \omega - \gamma\zeta$ is in S. But by the choice of γ, $N(\rho) = 0$, and so $\rho = 0$ and $\omega = \gamma\zeta$. Thus, every element of S is a multiple of γ. ■

---

*Note that the proof of Proposition 7 yields half of the following useful result: Let α, β be nonzero Gaussian integers. Then $\alpha | \beta$ and $\beta | \alpha$ if and only if α and β are associates.

Corollary 9: Let α, β be Gaussian integers, not both 0, and γ a gcd of α and β. Then there exist Gaussian integers η, λ such that $\gamma = \alpha\eta + \beta\lambda$.

The reader should note the parallel between the proof of Theorem 8 and the proof of the corresponding result for the ordinary integers. Let us now define the notion of primes in the Gaussian integers. Note that every Gaussian integer γ has the following divisors: $\pm\gamma$, $\pm i\gamma$, ± 1, $\pm i$ (exercise).

Definition 10: A *Gaussian prime* is a Gaussian integer π, which is not a unit and whose only divisors are associates of π and units. That is, the only divisors of π are $\pm\pi$, $\pm i\pi$, ± 1, and $\pm i$.

The simplest way to locate Gaussian primes is to use the following easy lemma:

Lemma 11: Let π be a Gaussian integer such that $N(\pi) = p$, an ordinary prime. Then π is a Gaussian prime.

Proof: Suppose that $\delta \mid \pi$. Then $\pi = \delta\gamma$ for some γ, so that $N(\pi) = N(\delta)N(\gamma)$. But since $N(\pi) = p$, a prime, and since $N(\delta)$, $N(\gamma)$ are positive (ordinary) integers, we must have either $N(\delta) = 1$ or $N(\gamma) = 1$. Thus, either γ or δ must be a unit. Thus, δ must be one of $\pm\pi$, $\pm i\pi$ (γ a unit) or ± 1, $\pm i$ (δ a unit). ■

Example 12:

(i) $2 + i$ is a Gaussian prime, since $N(2 + i) = 5$.

(ii) Note that not every Gaussian prime π has norm equal to an ordinary prime. For example, 3 is a Gaussian prime (Exercise), but $N(3) = 9$.

In Chapter 2, we showed that a prime p satisfies Euclid's lemma. Namely, if $p \mid ab$, a, b integers, then either $p \mid a$ or $p \mid b$. A similar statement holds for Gaussian primes.

Theorem 13 (Euclid's lemma for the Gaussian integers)**:** Let π be a Gaussian prime, α, β Gaussian integers. If $\pi \mid \alpha\beta$, then $\pi \mid \alpha$ or $\pi \mid \beta$.

Proof: Assume that $\pi \nmid \beta$, and let us show that $\pi \mid \alpha$. Since the only divisors of π are ± 1, $\pm i$, $\pm\pi$, and $\pm i\pi$ and since $\pi \nmid \beta$, we see that the only possibility for a gcd of π and β is a unit. Thus, 1 is a gcd of β and π by Example 5. Thus, by Corollary 9, we may write

$$1 = \pi\eta + \beta\lambda,$$

so that

$$\alpha = \pi(\eta\alpha) + (\alpha\beta)\lambda.$$

But since $\pi(\eta\alpha)$ and $(\alpha\beta)\lambda$ are multiples of π, so is α. Thus, $\pi \mid \alpha$. ■

Let us now try to factor a Gaussian integer into a product of Gaussian primes. Just as we do not try to factor 0, ± 1 in the ordinary integers, we do not try to factor 0, ± 1, $\pm i$ in the Gaussian integers.

Proposition 14: Let γ be a Gaussian integer which is neither a unit nor zero. Then γ can be expressed as a product of Gaussian primes.

Proof: Let us proceed by induction on $N(\gamma)$. Since $\gamma \neq 0, \pm 1, \pm i$, we see that $N(\gamma) \geq 2$. If $N(\gamma) = 2$, then γ is a Gaussian prime by Lemma 11. Thus, in this case γ is a product of Gaussian primes. Thus, assume that $N(\gamma) > 2$ and that every Gaussian integer of norm less than $N(\gamma)$ can be written as a product of Gaussian primes. If γ is a Gaussian prime, then γ can certainly be written as a product of Gaussian primes. Thus, assume that γ is not a Gaussian prime. Then we can write $\gamma = \alpha\beta$, where neither of α, β is a unit. Then, $1 < N(\alpha), N(\beta) < N(\gamma)$. By the induction hypothesis, we may write

$$\alpha = \pi_1 \cdots \pi_s$$

$$\beta = \eta_1 \cdots \eta_t,$$

where $\pi_1, \ldots, \pi_s, \eta_1, \ldots, \eta_t$ are Gaussian primes. Thus,

$$\gamma = \alpha\beta = \pi_1 \cdots \pi_s \eta_1 \cdots \eta_t$$

is a product of Gaussian primes. This completes the induction. ∎

Let us now inquire as to whether the factorization into Gaussian primes is unique. It certainly cannot be if we interpret uniqueness too strictly. For example, if $\gamma = \pi_1\pi_2\pi_3$ for Gaussian primes π_1, π_2, π_3, then we can also write

$$\gamma = (-\pi_1)(-\pi_2)\pi_3 = (-\pi_1)\pi_2(-\pi_3) = (i\pi_1)(-\pi_2)(i\pi_3) = (i\pi_1)(\pi_2)(-i\pi_3),$$

etc. Thus, we can get many factorizations from a given one just by inserting units, more or less at will. However, note that in all the above factorizations the first (respectively, second, third) primes are always associates of one another. This is the brand of uniqueness which we would expect—uniqueness up to taking associates. More precisely, we can prove the following analogue of the fundamental theorem of arithmetic:

Theorem 15: Let γ be a Gaussian integer which is neither a unit nor zero. Then γ can be written as a product of Gaussian primes. Moreover, if

$$\gamma = \pi_1 \cdots \pi_s = \eta_1 \cdots \eta_t$$

are two expressions of γ as a product of Gaussian primes, then $s = t$, and, after possible renumbering of $\eta_1, \ldots, \eta_s$, we have that π_1 is an associate of η_1, π_2 is an associate of η_2, and so forth.

Proof: By Proposition 14, we need prove only the second statement. Let us proceed by induction on $N(\gamma)$. Since γ is not zero and not a unit, we see that $N(\gamma) \geq 2$. If $N(\gamma) = 2$, γ is a prime, and the result is clear. Assume that

$N(\gamma) > 2$ and that the result is true for all Gaussian integers of norm less than $N(\gamma)$. Suppose that

$$\gamma = \pi_1 \cdots \pi_s = \eta_1 \cdots \eta_t,$$

where $\pi_1, \ldots, \pi_s, \eta_1, \ldots, \eta_t$ are Gaussian primes. Without loss of generality, suppose that $s > 1$. Then $\pi_1 | \pi_1 \cdots \pi_s$, so that $\pi_1 | \eta_1 \cdots \eta_t$. By using Theorem 13 repeatedly, we see that $\pi_1 | \eta_j$ for some j. Let us renumber so that $\pi_1 | \eta_1$. But since η_1 is a Gaussian prime, this implies that π_1 and η_1 are associates, say $\eta_1 = \pi_1 \epsilon$, ϵ a unit. Then, we have

$$\pi_1 \pi_2 \cdots \pi_s = \pi_1(\epsilon\eta_2) \cdots \eta_t,$$

so that

$$\pi_2 \pi_3 \cdots \pi_s = (\epsilon\eta_2)\eta_3 \cdots \eta_t. \tag{*}$$

Since $N(\pi_1) \geq 2$, we see that (since $s > 1$)

$$1 < N(\pi_2\pi_3 \cdots \pi_s) < N(\pi_1\pi_2 \cdots \pi_s) = N(\gamma).$$

Therefore, we may apply the induction to the factorization (*) to get that $s - 1 = t - 1$ and, upon reordering, that π_2 is an associate of η_2, π_3 is an associate of η_3, and so forth. Thus, the induction is completed. ■

7.2 Exercises

1. Determine which of the following divisibility relations hold:
   (a) $1 + i | 2$.
   (b) $2 + 3i | 5 - i$.
   (c) $3 | (2 - i)(3 + i)$.
   (d) $3 - 2i | 26$.
2. Use the division algorithm for the Gaussian integers to determine the quotient and remainder when β is divided by α, where
   (a) $\alpha = 5 - 2i$, $\beta = 6 + i$.
   (b) $\alpha = 3 + 15i$, $\beta = 187 + 46i$.
3. Show that every Gaussian integer γ has the divisors $\pm\gamma$, $\pm i\gamma$, ± 1, $\pm i$.
4. Let α and β be Gaussian integers. Show that if $\alpha | \beta$, then $N(\alpha) | N(\beta)$. Does the converse hold?
5. Make a table of all Gaussian primes π such that $N(\pi) \leq 10$. (*Hint:* Use Exercise 4.)
6. Let α be a nonzero Gaussian integer which is not a unit or a prime. Show that there exists $\beta | \alpha$ such that $1 < N(\beta) \leq N(\alpha)^{1/2}$.
7. Using Exercises 5 and 6, factor all Gaussian integers of norm < 10 into Gaussian prime factors.
8. Prove that there exist infinitely many Gaussian primes.

9. By imitating the Euclidean algorithm, devise a computational technique for computing gcd's of Gaussian integers.

10. Find $\gcd(5 + i, 2 - i)$, $\gcd(2 + 4i, 6 - 2i)$.

*11. Prove the analogues of Theorems 3 and 15 for the set of numbers $a + b\sqrt{-2}$, a, b integers.

12. Let α, β, γ be Gaussian integers, $\gamma \neq 0$. We say that α *is congruent to* β *modulo* γ (denoted $\alpha \equiv \beta(\text{mod } \gamma)$) if $\gamma | \alpha - \beta$. Prove the following properties of congruences:

    (a) If $\alpha \equiv \beta(\text{mod } \gamma)$, then $\beta \equiv \alpha(\text{mod } \gamma)$.
    (b) $\alpha \equiv \alpha(\text{mod } \gamma)$ for all α.
    (c) If $\alpha \equiv \beta(\text{mod } \gamma)$ and $\beta \equiv \delta(\text{mod } \gamma)$, then $\alpha \equiv \delta(\text{mod } \gamma)$.
    (d) If $\alpha_1 \equiv \beta_1(\text{mod } \gamma)$ and $\alpha_2 \equiv \beta_2(\text{mod } \gamma)$, then $\alpha_1 \pm \alpha_2 \equiv \beta_1 \pm \beta_2(\text{mod } \gamma)$, and $\alpha_1\alpha_2 \equiv \beta_1\beta_2(\text{mod } \gamma)$.

13. Let γ be a nonzero Gaussian integer. Define a *complete residue system* modulo γ to be a set S of Gaussian integers with the property that every Gaussian integer is congruent modulo γ to exactly one element of S.

    (a) Find complete residue systems modulo 3, modulo $1 + i$.
    *(b) Show that a complete residue system modulo γ contains $N(\gamma)$ elements.

14. Let γ be a nonzero Gaussian integer. A *reduced residue system* modulo γ is a collection T of Gaussian integers such that every Gaussian integer α for which $\gcd(\alpha, \gamma) = 1$ is congruent to precisely one element of T. Find reduced residue systems modulo 4, $1 - i$, $2 + i$.

15. Let π be a Gaussian prime. Show that a reduced residue system modulo π contains $N(\pi) - 1$ elements. (Note: Exercise 13(b).)

16. Let π be a Gaussian prime, and a be a positive integer. Show that a reduced residue system modulo π^a contains $N(\pi)^{a-1}(N(\pi) - 1)$ elements.

17. Let $\phi(\gamma)$ denote the number of elements in a reduced residue system modulo γ.

    (a) Show that if $\gcd(\gamma, \alpha) = 1$, then $\alpha^{\phi(\gamma)} \equiv 1(\text{mod } \gamma)$.
    (b) Show that if π is a Gaussian prime, $\pi \nmid \alpha$, then $\alpha^{N(\pi)-1} \equiv 1(\text{mod } \pi)$.

18. Let $T = \{\alpha_1, \ldots, \alpha_t\}$ be a reduced residue system modulo the Gaussian prime π. Show that

$$\alpha_1 \cdots \alpha_t \equiv -1(\text{mod } \pi).$$

    (This is the analogue of Wilson's theorem for Gaussian integers.)

19. Let γ be a nonzero Gaussian integer, and α be a Gaussian integer such that $\gcd(\alpha, \gamma) = 1$. Show that there exists a Gaussian integer α^* such that $\alpha\alpha^* \equiv 1(\text{mod } \gamma)$.

20. Find necessary and sufficient conditions that the linear congruence $\alpha x \equiv \beta (\text{mod } \gamma)$ be solvable for Gaussian integers x.

21. Solve the linear congruences
    (a) $2x \equiv 1 - 2i (\text{mod } 3)$.
    (b) $3x \equiv 1 (\text{mod } 1 - i)$.

22. Solve the Diophantine equation $2x + (2 + i)y = 11 - 3i$ in the Gaussian integers.

23. Find a necessary and sufficient condition that the Diophantine equation $\alpha x + \beta y = \gamma$ be solvable in the Gaussian integers.

24. State and prove an analogue of the Chinese remainder theorem for the Gaussian integers.

*25. Use Exercises 16 and 24 to prove that the number of elements $\phi(\gamma)$ in a reduced residue system modulo the nonzero Gaussian integer γ is just $N(\pi_1)^{a_1-1}(N(\pi_1) - 1) \cdots N(\pi_t)^{a_t-1}(N(\pi_t) - 1)$, where $\gamma = \pi_1^{a_1} \cdots \pi_t^{a_t}$ is a factorization of γ into a product of powers of distinct Gaussian primes $\pi_1, \ldots, \pi_t$.

7.3. The Two-Square Problem Revisited

In this section, we shall give a new proof for the two-square theorem of Chapter 6. That is, we shall find all positive integers n such that the Diophantine equation

$$x^2 + y^2 = n$$

is solvable. This is equivalent to the problem of determining all Gaussian integers of norm n, since

$$x^2 + y^2 = N(x + iy).$$

To solve the latter problem, let us explicitly describe all the Gaussian primes. Since an associate of a Gaussian prime is a Gaussian prime, it suffices to determine all Gaussian primes up to taking associates.

Lemma 1: Let π be a Gaussian prime. Then there is one and only one ordinary prime p such that $\pi | p$.

Proof: Note that $N(\pi)$ is an ordinary positive integer. Therefore, we may write $N(\pi) = p_1 \cdots p_t$, where $p_1, \ldots, p_t$ are ordinary primes. Thus, since $N(\pi) = \pi\pi'$, we see that $\pi | p_1 \cdots p_t$, and by Euclid's lemma (Theorem 2.13), we see that $\pi | p_i$, for some i. Thus, π divides some ordinary prime. Now we show that π cannot divide more than one ordinary prime, for if $\pi | p$ and $\pi | q$ for distinct ordinary primes p, q, then we can write $1 = px + qy$ for ordinary integers x, y since p, q are relatively prime (as ordinary integers). Thus, since p and q are both multiples of π, 1 is a multiple of π, say $1 = \pi\eta$. But then $\eta = 1/\pi$ is a Gaussian integer, and so π is a unit, which is a contradiction. ∎

From Lemma 1, we see that we can find all Gaussian primes (at least up to associates) by factoring all ordinary primes p. The case $p = 2$ is easiest:

$$2 = -i(1 + i)^2,$$

and $1 + i$ is a Gaussian prime since $N(1 + i) = 2$. Thus, the only Gaussian primes π such that $\pi \mid 2$ are associates of $1 + i$.

Henceforth let us assume that p is odd and assume that $\pi = x + iy$ divides p. Then $\pi\eta = p$ for some Gaussian integer η. But then

$$p^2 = N(p) = N(\pi)N(\eta),$$

so that $N(\pi) = x^2 + y^2$ equals either p or p^2. Since x and y are ordinary integers, x^2 and y^2 can be congruent only to 0 or 1(mod 4). Thus,

$$x^2 + y^2 \equiv 0, 1, \text{ or } 2 (\text{mod } 4).$$

Thus, if $x^2 + y^2 = p$, then p cannot be congruent to 3(mod 4). Consequently, if $p \equiv 3(\text{mod } 4)$, then $x^2 + y^2 = p^2$. Therefore,

$$p^2 = N(p) = N(\pi)N(\eta) = p^2N(\eta),$$

so that $N(\eta) = 1$ and η is a unit. Thus, π is an associate of p if $p \equiv 3(\text{mod } 4)$. In this case, p does not factor any further in the Gaussian integers; that is, p itself is a Gaussian prime.

Finally, assume that $p \equiv 1(\text{mod } 4)$. By Theorem 3.3.5 we then know that the congruence $z^2 \equiv -1(\text{mod } p)$ is solvable. Let z be a solution. Then $p \mid z^2 + 1$. Therefore, since $\pi \mid p$, we see that $\pi \mid z^2 + 1$. However, $z^2 + 1 = (z - i)(z + i)$, so that

$$\pi \mid (z - i)(z + i).$$

But then, by Euclid's lemma,

$$\pi \mid z - i \quad \text{or} \quad \pi \mid z + i. \tag{$*$}$$

Note that $p \nmid z - i$ and $p \nmid z + i$ since

$$\frac{1}{p}z \pm \frac{1}{p}i$$

are not Gaussian integers. Thus, by ($*$), π and p are not associates. In particular, $N(\pi) \neq N(p) = p^2$ (since η is not a unit.) And since we showed above that $N(\pi)$ is either p or p^2, we conclude that if $p \equiv 1(\text{mod } 4)$, then $N(\pi) = p$. Thus, $\pi\pi' = p$ and p is divisible by π and π'.

To complete the determination of all Gaussian primes, we must decide whether π and π' are associates. They are not, as a simple case-by-case analysis will show. Suppose that $\pi = x + iy$, so that $N(\pi) = x^2 + y^2 = p$. Suppose that π and π' are associates, say $\pi = \epsilon\pi'$, where $\epsilon = 1, -1, i, -i$. Recall that $\pi' = x - iy$. Thus, $\epsilon = 1$ implies that $x + iy = x - iy$, so that $y = 0$ and $x^2 = p$, which contradicts the fact that p is a prime. Similarly, $\epsilon = -1$ implies that $x = 0$ and $y^2 = p$, which is again absurd. Also $\epsilon = i$

implies that $x + iy = i(x - iy) = y + ix$, so that $x = y$ and $p = x^2 + y^2 = 2x^2$, which contradicts the fact that p is odd. Finally, $\epsilon = -i$ implies that $x = -y$, and so again $2x^2 = p$. Thus, in all cases, $\pi = \epsilon\pi'$, for a unit ϵ, leads to a contradiction.

Let us summarize the investigations.

Theorem 2: Let p be an ordinary prime. Then p factors in the Gaussian integers in the following manner:

Case 1: $p = 2$: $p = -i\pi^2$, where $\pi = 1 + i$ is a Gaussian prime, $N(\pi) = 2$.

Case 2: $p \equiv 3 \pmod 4$: $p = \pi$ is a Gaussian prime, $N(\pi) = p^2$.

Case 3: $p \equiv 1 \pmod 4$: $p = \pi\pi'$, where π and π' are nonassociated Gaussian primes and $N(\pi) = N(\pi') = p$.

Example 3:

(i) $7 \equiv 3 \pmod 4$, and so 7 is a Gaussian prime.

(ii) $5 = (2 + i)(2 - i)$, with $2 \pm i$ Gaussian primes.

Let us now turn to the problem of determining what positive integers are norms of Gaussian integers. Let α be a nonzero nonunit. Then, we may write

$$\alpha = \pi_1 \cdots \pi_s,$$

where $\pi_1, \ldots, \pi_s$ are Gaussian primes. But then

$$N(\alpha) = N(\pi_1) \cdots N(\pi_s).$$

Suppose that $\pi_1 | p_1, \ldots, \pi_s | p_s$, where p_i is the ordinary prime which π_i divides. (The p_i may be repeated.) Then

$$N(\alpha) = p_1^{a_1} p_2^{a_2} \cdots p_s^{a_s},$$

where $a_i = 2$ if $p_i \equiv 3 \pmod 4$ and $a_i = 1$ if $p_i = 2$ or $p_i \equiv 1 \pmod 4$. Thus, we see that

$$N(\alpha) = m^2 q_1 \cdots q_t, \qquad (**)$$

where m is an ordinary integer and $q_1, \ldots, q_t$ are distinct ordinary primes equal to 2 or $\equiv 1 \pmod 4$. Conversely, reading the same argument backwards shows that every integer of the form (**) with $q_1, \ldots, q_t$ distinct ordinary primes equal to 2 or $\equiv 1 \pmod 4$ is the norm of a Gaussian integer. Thus, we have finally rederived our two-square theorem (Theorem 6.4.5).

7.3 Exercises

1. Factor the following rational primes in the Gaussian integers:

   (a) 3. (b) 5. (c) 7. (d) 11. (e) 71.

2. Construct a table of Gaussian primes of norm ≤ 50 using the factorizations of rational primes.

3. Show that if $p \equiv 1 \pmod 4$, p prime, then the Diophantine equation $x^2 + y^2 = p$ has exactly four solutions.
4. Let $p_1, \ldots, p_t$ be distinct primes, $p_i \equiv 1 \pmod 4$. Show that the number of solutions of the Diophantine equation $x^2 + y^2 = p_1 \cdots p_t$ is 4^t.
5. Compute all solutions to the Diophantine equations $x^2 + y^2 = 30$, $x^2 + y^2 = 65$, and $x^2 + y^2 = 32045$.
6. Determine the number of solutions of the Diophantine equation $x^2 + y^2 = n$, n a given ordinary integer. (*Hint:* Write $n = m^2 p_1 \cdots p_t$, where $p_1, \ldots, p_t$ are distinct primes.)
7. Use the arithmetic of the Gaussian integers to determine all solutions to the Diophantine equation $x^2 + y^2 = z^2$.
8. Use the arithmetic of the Gaussian integers to determine all solutions to the Diophantine equation $x^2 + y^2 = z^3$. Note that this Diophantine equation can be written in the form of the Bachet equation $y^2 = x^3 + k$.
9. Use the arithmetic of the Gaussian integers to determine all solutions to the Diophantine equation $x^2 + y^2 = z^n$ $(n \geq 2)$.

8

Arithmetic in Quadratic Fields

8.1 Introduction

In Chapter 7, we studied the Diophantine equation $x^2 + y^2 = t$, for given t. Our approach was to factor $x^2 + y^2$ into $(x - \sqrt{-1}y)(x + \sqrt{-1}y)$. In this way, the problem of determining all solutions of the Diophantine equation is transformed into the problem of determining all Gaussian integers $x + \sqrt{-1}y$ having norm t. We then solved the latter problem by studying the arithmetic of the Gaussian integers. More specifically, we derived a theory of unique factorization in the Gaussian integers, which closely parallels the theory in the integers.

Let d be any integer and let us consider the Pell-type Diophantine equation $x^2 - dy^2 = t$ for given t. As was the case in Chapter 7, we can factor the left-hand side:

$$x^2 - dy^2 = (x + \sqrt{d}\,y)(x - \sqrt{d}\,y).$$

We are immediately led to consider the set of complex numbers of the form $x + y\sqrt{d}$, where x and y are integers. In the case $d = -1$, this set of numbers is just the Gaussian integers. For general d, they form an interesting setting in which to investigate arithmetic. We now ask to what extent we can develop factorization theory for this set of numbers. In the back of our minds, we hope that we can use such a factorization theory to solve the Diophantine equation $x^2 - dy^2 = t$ in much the same manner as we proceeded in Chapter

7. Unfortunately, the theory for arbitrary d is a good deal more complicated than the theory for $d = -1$. It will take the rest of the book to give an extensive but incomplete theory. However, in the present chapter, we shall begin to study the theory of quadratic numbers, that is, numbers of the form $s + t\sqrt{d}$, where s and t are rational numbers. The material of this chapter is necessary background in order to develop the factorization theory of Chapter 9. However, we do enough in this chapter to reap some direct information about quadratic Diophantine equations. Specifically, we shall give an algorithm for finding all solutions of a Diophantine equation of the form

$$ax^2 + bxy + cy^2 = m,$$

where a, b, c, m are given integers.

8.2 Quadratic Fields

Let D be a rational number and let $\mathbf{Q}(\sqrt{D})$ denote the set of all complex numbers of the form

$$s + t\sqrt{D}, \qquad s, t \text{ rational}.$$

If D is the square of a rational number, then $\mathbf{Q}(\sqrt{D})$ consists of just the rational numbers. Therefore, let us assume throughout this section that *D is not the square of a rational number*. Then $\mathbf{Q}(\sqrt{D})$ is called the *quadratic field* belonging to D. Note that $\mathbf{Q}(\sqrt{D})$ contains all rational numbers (set $t = 0$). In this section, we shall record some of the most elementary facts about quadratic fields.

First, let us find a better description of the elements of $\mathbf{Q}(\sqrt{D})$. Write $D = x/y$ with x, y integers and $y > 0$, $\gcd(x, y) = 1$. It is clear that x and y are uniquely determined by D. Then $\sqrt{D} = (1/y)\sqrt{xy}$. Moreover, if we write $xy = u^2d$, where u, d are integers and d is not divisible by the square of an integer, then* $\sqrt{D} = (u/y)\sqrt{d}$. Moreover, since we have assumed that D is not the square of any rational number, we see that $d \neq 1, 0$. Now,

$$s + t\sqrt{D} = s + \frac{ut}{y}\sqrt{d},$$

and as t runs over all rational numbers, so does ut/y. Thus, we have

$$\mathbf{Q}(\sqrt{D}) = \mathbf{Q}(\sqrt{d}),$$

and we see that it suffices to study the quadratic fields $\mathbf{Q}(\sqrt{d})$, where d is a nonzero integer, not divisible by a perfect square, other than 1. Such d's are called *square-free integers*. Thus, all the quadratic fields are those corre-

*Here $\sqrt{d}$ is chosen so that the equation is true; note that this is always possible.

sponding to

$$d = -1, \pm 2, \pm 3, \pm 5, \pm 6, \pm 7, \pm 10, \pm 11, \pm 13, \pm 14,$$
$$\pm 15, \pm 17, \pm 19, \pm 21, \ldots.$$

Henceforth, we shall always restrict our attention to quadratic fields written in the form $\mathbf{Q}(\sqrt{d})$ for d a square-free integer.

We shall always denote the elements of $\mathbf{Q}(\sqrt{d})$ *by lower case Greek letters, such as* $\alpha, \beta, \gamma, \delta, \ldots$.

Lemma 1: Let α belong to $\mathbf{Q}(\sqrt{d})$. Then α can be written in one and only one way in the form $\alpha = a + b\sqrt{d}$ with a, b rational. In particular, if $a + b\sqrt{d} = a^* + b^*\sqrt{d}$ with a, b, a^*, b^* rational, then $a = a^*$, $b = b^*$.

Proof: Suppose that $a + b\sqrt{d} = a^* + b^*\sqrt{d}$. If $b \neq b^*$, then

$$d = \left(\frac{a - a^*}{b^* - b}\right)^2.$$

Thus, $(a - a^*)/(b^* - b)$ is an integer and $d \neq 1$ is a perfect square, contradicting the fact that d is square-free. Moreover $b = b^*$ implies that $a = a^*$. ■

Next, note that if $d > 0$, then $s + t\sqrt{d}$ is a real number, so that $\mathbf{Q}(\sqrt{d})$ is contained in the set of real numbers. On the other hand, if $d < 0$, then $\sqrt{d}$ is a complex number but not a real number, so that $\mathbf{Q}(\sqrt{d})$ is contained in the complex numbers but not in the real numbers.

Since, in any case, $\mathbf{Q}(\sqrt{d})$ consists of complex numbers, we may perform the usual operations of arithmetic on the elements of $\mathbf{Q}(\sqrt{d})$.

Proposition 2: Let α, β be contained in $\mathbf{Q}(\sqrt{d})$. Then $\alpha \pm \beta$ and $\alpha\beta$ are contained in $\mathbf{Q}(\sqrt{d})$. Furthermore, if $\beta \neq 0$, then α/β is contained in $\mathbf{Q}(\sqrt{d})$. Thus $\mathbf{Q}(\sqrt{d})$ is a field containing the field $\mathbf{Q}$ of rational numbers.

Proof: Let $\alpha = s + t\sqrt{d}$, $\beta = u + v\sqrt{d}$, with s, t, u, v rational. Then

$$\alpha \pm \beta = (s \pm u) + (t \pm v)\sqrt{d}$$
$$\alpha\beta = (su + tvd) + (sv + ut)\sqrt{d},$$

both of which belong to $\mathbf{Q}(\sqrt{d})$. If $\beta \neq 0$, then one of $u, v \neq 0$ (by Lemma 1). Therefore, $u - v\sqrt{d} \neq 0$ (again Lemma 1). Therefore,

$$\frac{1}{\beta} = \frac{1}{u + v\sqrt{d}}\,\frac{u - v\sqrt{d}}{u - v\sqrt{d}} = \frac{u}{u^2 - v^2 d} + \frac{-v}{u^2 - v^2 d}\sqrt{d}$$

is contained in $\mathbf{Q}(\sqrt{d})$, so that $\alpha/\beta = \alpha \cdot (1/\beta)$ is contained in $\mathbf{Q}(\sqrt{d})$ by the first part of the proposition. ■

You should observe that we gave explicit formulas in Proposition 2 for the expressions $\alpha + \beta$, $\alpha - \beta$, $\alpha\beta$, $1/\beta$, and α/β. Thus, for example, if $\alpha =$

$7 - 4\sqrt{3}$ and $\beta = 2 + 3\sqrt{3}$, then

$$\alpha + \beta = 9 - \sqrt{3}$$
$$a - \beta = 5 - 7\sqrt{3}$$
$$\alpha\beta = -22 + 13\sqrt{3}$$
$$\frac{1}{\beta} = -\frac{2}{23} + \frac{3}{23}\sqrt{3}$$
$$\frac{\alpha}{\beta} = -\frac{50}{23} + \frac{29}{23}\sqrt{3}.$$

If $\alpha = s + t\sqrt{d}$ is contained in $\mathbf{Q}(\sqrt{d})$, let us define the *conjugate* α' *of* α to be the number $\alpha' = s - t\sqrt{d}$. Note that if $d < 0$, then the conjugate of α is just the usual complex conjugate of α. However, this is not true if $d > 0$. The properties of conjugates are summarized in the following proposition:

Proposition 3: Let α, β be contained in $\mathbf{Q}(\sqrt{d})$. Then

(i) $(\alpha \pm \beta)' = \alpha' \pm \beta'$.
(ii) $(\alpha\beta)' = \alpha'\beta'$.
(iii) $(\alpha/\beta)' = \alpha'/\beta'$.
(iv) $\alpha' = \alpha$ if and only if α is rational.

Proof: (i)–(iii) Exercises.

(iv) Let $\alpha = s + t\sqrt{d}$. Then $\alpha' = \alpha$ if and only if $s + t\sqrt{d} = s - t\sqrt{d}$. But by Lemma 1, this is equivalent to saying $t = -t$; that is, $t = 0$. Thus, $\alpha' = \alpha$ if and only if $\alpha = s$ is rational. ■

Definition 4: Let $\alpha = s + t\sqrt{d}$. Then the *trace* of α, denoted $Tr(\alpha)$, is defined as

$$Tr(\alpha) = \alpha + \alpha' = 2s.$$

The *norm of* α, denoted $N(\alpha)$, is defined as

$$N(\alpha) = \alpha\alpha' = s^2 - t^2 d.$$

The norm will be one of the primary tools we shall use in the theory of quadratic fields.

The relevant properties of traces and norms are summarized in the following result:

Proposition 5: Let α, β belong to $\mathbf{Q}(\sqrt{d})$. Then

(i) $Tr(\alpha + \beta) = Tr(\alpha) + Tr(\beta)$; $N(\alpha\beta) = N(\alpha)N(\beta)$.
(ii) $Tr(\alpha)$, $N(\alpha)$ are rational numbers.
(iii) $N(\alpha) = 0$ if and only if $\alpha = 0$.
(iv) α is a zero of the polynomial $X^2 - Tr(\alpha)X + N(\alpha)$.

Proof: (i) and (ii) are obvious by Proposition 3.

(iii) $N(\alpha) = 0$ if and only if $s^2 - t^2 d = 0$, where $\alpha = s + t\sqrt{d}$. Assume that $N(\alpha) = 0$. If $t \neq 0$, then $d = (s/t)^2$, which is a contradiction to the fact that d is square-free. Thus, $t = 0$ and $s^2 = 0$, so that $s = 0$ and $\alpha = 0$. Conversely, if $\alpha = 0$, then $N(\alpha) = 0 \cdot 0 = 0$.

(iv) Set $X = \alpha$. Then $\alpha^2 - Tr(\alpha)\alpha + N(\alpha) = \alpha^2 - (\alpha + \alpha')\alpha + \alpha\alpha' = 0$. ■

In addition to the above properties of $\mathbf{Q}(\sqrt{d})$, we need some elementary facts concerning linear algebra in $\mathbf{Q}(\sqrt{d})$. First observe that $\mathbf{Q}(\sqrt{d})$ is a vector space over the field of rational numbers $\mathbf{Q}$.* Note that Lemma 1 simply asserts that $1, \sqrt{d}$ is a basis of $\mathbf{Q}(\sqrt{d})$. Therefore, from linear algebra, we have

Proposition 6: Let α_1 and α_2 be any pair of linearly independent elements of $\mathbf{Q}(\sqrt{d})$. Then, given any α in $\mathbf{Q}(\sqrt{d})$, there exist unique rational numbers r, s such that $\alpha = r\alpha_1 + s\alpha_2$. That is, α is a linear combination of α_1 and α_2.

Example 7: In $\mathbf{Q}(\sqrt{3})$, consider $\alpha_1 = 1 + 2\sqrt{3}, \alpha_2 = 3 + \sqrt{3}$. We show that α_1, α_2 is a basis of $\mathbf{Q}(\sqrt{3})$. Indeed, by Proposition 6, we need only show that α_1, α_2 are linearly independent. If s_1, s_2 are rational, then $s_1\alpha_1 + s_2\alpha_2 = 0$ implies that

$$\begin{aligned} s_1 + 3s_2 &= 0 \\ 2s_1 + s_2 &= 0. \end{aligned}$$

Multiplying the first equation by 2 and substracting it from the second yields $-5s_2 = 0$ or $s_2 = 0$. Thus, $s_1 = 0$, and α_1, α_2 are linearly independent.

Now let $\alpha = 5 - 6\sqrt{3}$. Let us express α as a linear combination of α_1 and α_2. Now $\alpha = s_1\alpha_1 + s_2\alpha_2$ is equivalent to

$$\begin{aligned} s_1 + 3s_2 &= 5 \\ 2s_1 + s_2 &= -6. \end{aligned}$$

Solving these equations, $s_1 = -\frac{23}{5}$, $s_2 = \frac{16}{5}$, and so $5 - 6\sqrt{3} = -\frac{23}{5}(1 + 2\sqrt{3}) + \frac{16}{5}(3 + \sqrt{3})$.

8.2 Exercises

1. Prove Proposition 3, parts (i)–(iii).
2. Compute the following:

   (a) $(3 - 8\sqrt{5})(6 + 3\sqrt{5})$. (b) $(1 - \sqrt{2})^{-1}$.
   (c) $(1 + \sqrt{-1})^3$. (d) $(8 - \sqrt{2})/(4 + \sqrt{2})$.

---

*By Proposition 2, the sum and difference of elements of $\mathbf{Q}(\sqrt{d})$ belong to $\mathbf{Q}(\sqrt{d})$, and the product of an element of $\mathbf{Q}$ and an element of $\mathbf{Q}(\sqrt{d})$ belongs to $\mathbf{Q}(\sqrt{d})$. The remaining vector space axioms are trivially verified.

3. Determine whether the following sets are linearly independent. If not, exhibit a linear dependence relation among them.
   (a) $1, \sqrt{5}$. (b) $\frac{1}{2}, \sqrt{-2}, 2 - 3\sqrt{-2}$. (c) $1, \frac{1}{2}$ (in $\mathbf{Q}(\sqrt{2})$).
   (d) $2 + \sqrt{-2}, 2 - 3\sqrt{-2}$. (e) $1, a + b\sqrt{3}$, a, b rational.
4. Show that α, β in $\mathbf{Q}(\sqrt{d})$ are linearly independent if and only if
   $\begin{vmatrix} \alpha & \alpha' \\ \beta & \beta' \end{vmatrix} \neq 0.$
5. For the following values of D, write D in the form s^2d, s rational, d an integer not divisible by a square > 1:
   (a) $D = 12$. (b) $D = -63$. (c) $D = \frac{34}{9}$. (d) $D = -\frac{15}{2}$.
6. For the following values of α, find $Tr(\alpha)$, $N(\alpha)$, and a quadratic polynomial of which α is a zero:
   (a) $\alpha = -1 + \sqrt{-3}$. (b) $\alpha = 5 + 4\sqrt{2}$. (c) $\alpha = 1 + \sqrt{17}$.
7. (a) Show that $1 + \sqrt{2}, 1 - \sqrt{2}$ is a basis for $\mathbf{Q}(\sqrt{2})$.
   (b) Express $5 + 18\sqrt{2}$ in terms of the basis of part (a).

8.3 The Integers of a Quadratic Field

In Chapter 7, we studied the Gaussian integers, which are contained in the quadratic field $\mathbf{Q}(\sqrt{-1})$. The Gaussian integers arose naturally in connection with the Diophantine equation $x^2 + y^2 = t$. Similarly, we have seen that the Pell-type equation $x^2 - dy^2 = t$ gives rise to the set of numbers

$$x + y\sqrt{d}, \qquad x, y \text{ integers}, \tag{$*$}$$

of the quadratic field $\mathbf{Q}(\sqrt{d})$. In this section, we would like to designate certain elements of $\mathbf{Q}(\sqrt{d})$ as integers and to begin a study of the properties of these integers with an eye toward eventually applying what we learn to the study of Diophantine equations. Certainly the numbers ($*$) should qualify as integers. However, as we shall shortly see, there are other elements of $\mathbf{Q}(\sqrt{d})$ which have good reason to be considered integers as well. To motivate our definition of the integers of $\mathbf{Q}(\sqrt{d})$, let us write down a few properties which we would like integers to have.

1. All elements $x + y\sqrt{d}$, x, y ordinary integers, are integers of $\mathbf{Q}(\sqrt{d})$.
2. If α, β are integers of $\mathbf{Q}(\sqrt{d})$, then $\alpha + \beta$, $\alpha - \beta$, $\alpha \cdot \beta$ are integers. That is, the set of integers of $\mathbf{Q}(\sqrt{d})$ form a ring.
3. If α is an integer of $\mathbf{Q}(\sqrt{d})$, then α' is an integer.
4. If α is an integer of $\mathbf{Q}(\sqrt{d})$ and is a rational number, then α is an ordinary integer.

Requirements 1 and 2 seem reasonable in the light of ($*$) and the properties of the ordinary integers. Requirement 3 merely states that if

$x + y\sqrt{d}$, x, y rational, is an integer of $\mathbf{Q}(\sqrt{d})$, then $x - y\sqrt{d}$ is one also. The last requirement is inserted to prevent numbers such as $\frac{1}{2}$, $\frac{2}{3}$, and $\frac{9}{16}$ from being integers of $\mathbf{Q}(\sqrt{d})$.

Please note that the set of numbers (*) clearly satisfies requirements 1–4. We leave the verification as an exercise. Why then should we not take the set (*) as the set of integers of $\mathbf{Q}(\sqrt{d})$? The point is that sometimes there is a larger set of numbers of $\mathbf{Q}(\sqrt{d})$ which satisfies requirements 1–4 and we would like to take the larger set as our set of integers. How can we go about locating the largest set of numbers of $\mathbf{Q}(\sqrt{d})$ which satisfies requirements 1–4? It turns out to be easy.

Note that if α is any would-be integer of $\mathbf{Q}(\sqrt{d})$, then so is α' by requirement 3. Therefore, by requirement 2, both $Tr(\alpha) = \alpha + \alpha'$ and $N(\alpha) = \alpha\alpha'$ are integers of $\mathbf{Q}(\sqrt{d})$. However, both $Tr(\alpha)$ and $N(\alpha)$ are rational numbers, so that by requirement 4 both are ordinary integers. This suggests that we make the following definition:

Definition 1: Let α be an element of the quadratic field $\mathbf{Q}(\sqrt{d})$. We say that α is an *integer of* $\mathbf{Q}(\sqrt{d})$ provided that $Tr(\alpha)$, $N(\alpha)$ are ordinary integers. We denote the set of integers of $\mathbf{Q}(\sqrt{d})$ by I_d.

Note that Definition 1 allows more integers than just the numbers (*). For example, $(1 + \sqrt{5})/2$ is an integer of $\mathbf{Q}(\sqrt{5})$, since

$$Tr\left(\frac{1+\sqrt{5}}{2}\right) = \frac{1+\sqrt{5}}{2} + \frac{1-\sqrt{5}}{2} = 1,$$

$$N\left(\frac{1+\sqrt{5}}{2}\right) = \frac{1-5}{4} = -1.$$

However, we shall see below that this definition does indeed give us a very reasonable definition of integer. First, however, let us describe precisely the integers of $\mathbf{Q}(\sqrt{d})$ and then prove that requirements 1–4 are satisfied.

Suppose that $\alpha = x + y\sqrt{d}$, x, y rational, is an element of $\mathbf{Q}(\sqrt{d})$. Then α is an integer of $\mathbf{Q}(\sqrt{d})$ if and only if

$$Tr(\alpha) = 2x, \qquad N(\alpha) = x^2 - dy^2 \tag{1}$$

are integers. Set $2x = x_1$, $2y = y_1$. Then the numbers (1) are just x_1 and $(x_1^2 - dy_1^2)/4$, respectively. Thus, α is an integer of $\mathbf{Q}(\sqrt{d})$ if and only if

$$x_1 \quad \text{and} \quad \frac{x_1^2 - dy_1^2}{4}$$

are integers. If this is to occur, y_1 must be an integer also. Indeed, $x_1^2 - dy_1^2$ must be an integer, and since x_1 is an integer, we must have that dy_1^2 is an integer. But since d is square-free, y_1 cannot have any denominator > 1 if dy_1^2 is to be an integer. Thus, if α is an integer of $\mathbf{Q}(\sqrt{d})$, y_1 is an integer and

$$x_1^2 - dy_1^2 \equiv 0 \pmod{4}. \tag{2}$$

Since a perfect square is congruent to 0 or 1(mod 4), we see that if $d \equiv 2$ or 3(mod 4), then x_1 and y_1 must both be even, so that

$$\alpha = x + y\sqrt{d} = \frac{x_1}{2} + \frac{y_1}{2}\sqrt{d}$$

has both $x = x_1/2$ and $y = y_1/2$ integers. If $d \equiv 1 \pmod 4$, then (2) will be satisfied either for x_1 and y_1 both even or x_1 and y_1 both odd. In either case, we have that $(x_1 - y_1)/2$ is an integer, and

$$\alpha = x + y\sqrt{d} = \frac{x_1 - y_1}{2} + y_1\left(\frac{1 + \sqrt{d}}{2}\right).$$

Thus, we see that if $d \equiv 1 \pmod 4$, then the integers of $\mathbf{Q}(\sqrt{d})$ are all of the form

$$a + b\frac{1 + \sqrt{d}}{2}, \qquad a, b \text{ integers.}$$

Conversely, every such number is an integer since

$$Tr\left(a + b\frac{1 + \sqrt{d}}{2}\right) = 2a + b$$

and

$$N\left(a + b\frac{1 + \sqrt{d}}{2}\right) = \left(a + b\frac{1 + \sqrt{d}}{2}\right)\left(a + b\frac{1 - \sqrt{d}}{2}\right)$$
$$= a^2 + ab + b^2\frac{1 - d}{4}$$

are integers (since $d \equiv 1 \pmod 4$). Finally, notice that we can never have $d \equiv 0 \pmod 4$ since d is square-free. Thus, we have proved the following result:

Theorem 2: Let d be a square-free integer. Then the set I_d of integers of $\mathbf{Q}(\sqrt{d})$ consists of the numbers of the form $x + y\omega_d$, where x and y are integers and

$$\omega_d = \begin{cases} \sqrt{d} & \text{if } d \equiv 2 \text{ or } 3 \pmod 4, \\ \dfrac{1 + \sqrt{d}}{2} & \text{if } d \equiv 1 \pmod 4. \end{cases}$$

Example 3:

(i) $I_{-1} = \{x + y\sqrt{-1} \mid x, y \text{ are integers}\}$ = the Gaussian integers, since $-1 \equiv 3 \pmod 4$.

(ii) $I_5 = \left\{x + y\dfrac{1 + \sqrt{5}}{2} \,\middle|\, x, y \text{ are integers}\right\}$, since $5 \equiv 1 \pmod 4$.

On the basis of Theorem 2, it is easy to see that I_d satisfies requirements 1–4. Requirements 1 and 4 are obvious. If α is an integer of $\mathbf{Q}(\sqrt{d})$ and $d \equiv 2$ or 3(mod 4), then α is of the form $x + y\sqrt{d}$, x, y integers, so that

$\alpha' = x - y\sqrt{d}$ is also an integer of $\mathbf{Q}(\sqrt{d})$. If $d \equiv 1 \pmod 4$, then α is of the form $x + y(1 + \sqrt{d})/2$, so that

$$\alpha' = x + y\frac{1-\sqrt{d}}{2} = (x + y) + (-y)\frac{1+\sqrt{d}}{2}$$

is also an integer of $\mathbf{Q}(\sqrt{d})$. Thus, requirement 3 holds.

To prove requirement 2, it suffices to prove that if α and β are integers of $\mathbf{Q}(\sqrt{d})$, then so is $\alpha \cdot \beta$, the case of sums and differences being obvious. However, in case $d \equiv 2$ or $3 \pmod 4$, $\alpha = x + y\sqrt{d}$, $\beta = z + w\sqrt{d}$, x, y, z, w integers, so that

$$\alpha\beta = (xz + ywd) + (yz + xw)\sqrt{d}$$

is also an integer of $\mathbf{Q}(\sqrt{d})$. If $d \equiv 1 \pmod 4$, then $\alpha = x + y(1 + \sqrt{d})/2$, $\beta = z + w(1 + \sqrt{d})/2$, x, y, z, w integers, so that

$$\begin{aligned}\alpha\beta &= xz + yw\left(\frac{1+\sqrt{d}}{2}\right)^2 + (xw + yz)\frac{1+\sqrt{d}}{2} \\ &= xz + yw\frac{d-1}{4} + (xw + yz + yw)\frac{1+\sqrt{d}}{2}\end{aligned}$$

is an integer of $\mathbf{Q}(\sqrt{d})$. Thus, we have proved the following result:

Theorem 4: The set I_d of integers of $\mathbf{Q}(\sqrt{d})$ satisfies requirements 1–4.

We shall refer to the elements of I_d as *integers*. When we wish to speak of what we formerly called integers (e.g., 1, 2, 3, 4, $-5, \ldots$), we shall use the term *rational integers*. Note that by requirement 1, every rational integer belongs to I_d and hence is an integer in our new sense.

A few comments are in order to convince the reader that our definition of the integers of $\mathbf{Q}(\sqrt{d})$ is the correct one. We may rephrase our definition of the integers of $\mathbf{Q}(\sqrt{d})$ as follows: Every number γ of $\mathbf{Q}(\sqrt{d})$ satisfies a quadratic equation with rational coefficients:

$$x^2 + px + q = 0, \tag{3}$$

namely the equation for which $p = -Tr(\gamma)$, $q = N(\gamma)$. Definition 1 states that γ is an integer of $\mathbf{Q}(\sqrt{d})$ if and only if Eq. (3) has rational integer coefficients. To see that this is a reasonable generalization of the usual notion of integers, let us compare the situation in $\mathbf{Q}(\sqrt{d})$ with the situation in the rational numbers. Every rational number r satisfies a linear equation with rational coefficients, namely

$$x - r = 0.$$

And this equation has integer coefficients if and only if r is a rational integer. Thus, we see a clear parallel between the integers of a quadratic field and the rational integers.

Every rational number r has a denominator. That is, there is a rational integer n such that nr is a rational integer. A similar property holds for the integers of $\mathbf{Q}(\sqrt{d})$.

Proposition 5: Let α be any number in $\mathbf{Q}(\sqrt{d})$. Then there is a rational integer n such that $n\alpha$ is an integer of $\mathbf{Q}(\sqrt{d})$.

Proof: If $\alpha = x + y\sqrt{d}$ and n is a common denominator of x and y, then $n\alpha$ is clearly in I_d. ∎

8.3 Exercises

1. Determine the integers of $\mathbf{Q}(\sqrt{d})$, where
   (a) $d = 5$. (b) $d = 11$. (c) $d = -11$. (d) $d = -13$.
2. Let d be a square-free integer $d \equiv 1 \pmod 4$. Show that I_d consists of all numbers of the form $(x + y\sqrt{d})/2$, where x and y are rational integers of the same parity (i.e., both even or odd).
3. Find all integers of norm ≤ 20 in I_{-2}. In I_{-3}.
4. Let $d < 0$. Show that there are only finitely many elements of I_d having given norm.
5. Is the conclusion of Exercise 4 correct if $d > 0$?
6. Verify that the set of numbers (*) satisfies requirements 1–4.
7. Let d be a rational integer, not a perfect cube, and let $\mathbf{Q}(\sqrt[3]{d})$ denote the set of all numbers of the form

$$a + b\sqrt[3]{d} + c(\sqrt[3]{d})^2,$$

   where a, b, c are rationals.
   (a) Show that the sum, product, difference, and quotient of elements of $\mathbf{Q}(\sqrt[3]{d})$ belong to $\mathbf{Q}(\sqrt[3]{d})$.
   (b) Show that every element of $\mathbf{Q}(\sqrt[3]{d})$ is a zero of a cubic polynomial with rational coefficients.
8. (continue Exercise 7) Let us say that an element α of $\mathbf{Q}(\sqrt[3]{d})$ is an *integer* if α is a zero of a cubic polynomial with rational integer coefficients and leading coefficient 1.
   *(a) Show that the sum, difference, and product of two integers of $\mathbf{Q}(\sqrt[3]{d})$ are integers.
   (b) If α is any element of $\mathbf{Q}(\sqrt[3]{d})$, then there exists a nonzero rational integer n such that $n\alpha$ is an integer.
   *(c) Show that the integers of $\mathbf{Q}(\sqrt[3]{2})$ are the numbers $a + b\sqrt[3]{2} + c(\sqrt[3]{2})^2$, where a, b, c are rational integers.

8.4 Binary Quadratic Forms

Now that we have explained the most basic facts concerning quadratic numbers and quadratic integers, let us look at the quadratic Diophantine equation

$$ax^2 + bxy + cy^2 = m. \tag{1}$$

We wish to determine whether rational integers x, y satisfying (1) exist. If so, we would like to give a procedure for determining all such x, y. The polynomial

$$f(x, y) = ax^2 + bxy + cy^2 \tag{2}$$

is called a *binary quadratic form* (or *form* for short).

As we mentioned in Section 1, our approach will be to factor (2) (see Eq. (3)) and work inside the appropriate quadratic field. There are other approaches to the study of Eq. (1). We shall outline one such approach in the exercises.*

Before we start studying Eq. (1) in detail, let us make a minor change in the problem we posed above. Let us allow a, b, c, and m in (1) to be rational numbers. (But we are interested in determining only *rational integers* x, y satisfying (1).) Superficially, this seems to make our problem a more general one, but it really does not, for if t is a common denominator of a, b, c, and m then $a^* = ta$, $b^* = tb$, $c^* = tc$, and $m^* = tm$ are rational integers. Solving (1) is equivalent to solving

$$a^*x^2 + b^*xy + c^*y^2 = m^*.$$

We shall see that there are certain technical advantages in dealing with equations with rational coefficients rather than just rational integral coefficients.

Example 1: Solving the equation $\frac{1}{2}x^2 + 3xy + y^2 = \frac{2}{3}$ for rational integers x, y is the same as solving the equation $3x^2 + 18xy + 6y^2 = 4$.

The binary quadratic form (2), for fixed rational numbers a, b, c, can be factored as follows:

$$f(x, y) = \frac{1}{a}\left(ax + \frac{b + \sqrt{D}}{2}y\right)\left(ax + \frac{b - \sqrt{D}}{2}y\right), \tag{3}$$

where $D = b^2 - 4ac$. (Just multiply it out.)

Definition 2: $D = b^2 - 4ac$ is called the *discriminant* of $f(x, y)$.

---

*In many ways, the approach given in the exercises is simpler than the one we give. However, we chose to present the quadratic fields approach because it is the best method by means of which to consider the deeper parts of the theory (e.g., genera and composition). Also, our method has many applications to other types of Diophantine equations.

If D is the square of a rational number, say $D = s^2$, then the factorization (3) reduces to

$$f(x, y) = \frac{1}{a}\left(ax + \frac{b+s}{2}y\right)\left(ax + \frac{b-s}{2}y\right).$$

In this case, the solutions to (1) can be easily found using the theory of linear Diophantine equations and factorization in the rational integers. We shall outline a procedure for carrying this out in the exercises. Thus, let us henceforth make the following restrictive assumption:

Assumption: D is not the square of a rational number.

This assumption will be in effect whenever binary quadratic forms are discussed.

It is easy to show that D may be written in the form $D = s^2 d$, where d is a square-free rational *integer* $\neq 1$ and s is a positive rational number. Moreover, such s and d are uniquely determined by D. (*Exercise:* For example, $\frac{3}{5} = (\frac{1}{5})^2 \cdot 15$ and $-12 = 2^2 \cdot (-3)$.) Using such a factorization of D, the factorization (3) can be rewritten as

$$f(x, y) = \frac{1}{a}\left(ax + \frac{b+s\sqrt{d}}{2}y\right)\left(ax + \frac{b-s\sqrt{d}}{2}y\right).$$

Let $\alpha = a$ and $\beta = (b + s\sqrt{d})/2$, and notice that α and β are elements of the quadratic field $\mathbf{Q}(\sqrt{d})$. Moreover, $f(x, y)$ can be rewritten in the form

$$\begin{aligned} f(x, y) &= \frac{1}{a}(\alpha x + \beta y)(\alpha' x + \beta' y) \\ &= \frac{1}{a}N(\alpha x + \beta y). \end{aligned} \tag{4}$$

Therefore, our original Diophantine equation (1) can be restated as follows: Determine all rational integers x, y such that $N(\alpha x + \beta y) = am$. This suggests that we consider the set

$$M = \{\alpha x + \beta y \mid x, y \text{ are rational integers.}\}$$

of $\mathbf{Q}(\sqrt{d})$. It is easy to check that α and β are linearly independent (exercise). Thus, each ξ in M has a *unique* expression in the form $\xi = \alpha x + \beta y$, where x and y are rational integers. Thus, we have established a 1-to-1 correspondence between pairs of rational integers x, y and elements $\xi = \alpha x + \beta y$ of M. Moreover, our original problem is equivalent to the problem of determining all ξ in M such that $N(\xi) = am$. Thus, we finally arrive at the following elementary result:

Proposition 3: Let $f(x, y) = ax^2 + bxy + cy^2$ be a binary quadratic form with rational coefficients. Define α and β and M as above. If x, y is a rational integral solution of $f(x, y) = m$, then $\xi = \alpha x + \beta y$ belongs to M and

$N(\xi) = am$. Conversely, if ξ belongs to M and $N(\xi) = am$, then there are unique rational integers x, y such that $\xi = \alpha x + \beta y$ and $f(x, y) = m$.

Proposition 3 suggests the following definition:

Definition 4: Let α, β be any two linearly independent elements of $\mathbf{Q}(\sqrt{d})$. Let

$$M = \{\alpha x + \beta y \mid x, y \text{ are rational integers}\}.$$

Then M is called a *module* of $\mathbf{Q}(\sqrt{d})$ and α, β is called a *basis* of M. If α, β is any basis of M, then we write $M = \{\alpha, \beta\}$.

Example 5:

(i) If $f(x, y) = x^2 + y^2$, then the corresponding module M is $\{1, \sqrt{-1}\}$, the Gaussian integers.

(ii) Suppose that $f(x, y) = x^2 + 4xy + 5y^2$. Using the factorization (3), we see that

$$f(x, y) = (x + (2 + \sqrt{-1})y)(x + (2 - \sqrt{-1})y).$$

Thus, the module corresponding to $f(x, y)$ is $M = \{1, 2 + \sqrt{-1}\}$. But it is easily checked that $\{1, 2 + \sqrt{-1}\} = \{1, \sqrt{-1}\}$ (exercise), so that apparently very different forms give rise to the same module.

Example 6: Theorem 3.2 says that I_d is a module of $\mathbf{Q}(\sqrt{d})$, namely $I_d = \{1, \omega_d\}$.

Let M be a module in a quadratic field $\mathbf{Q}(\sqrt{d})$. Suppose that we are given a specific basis of M, say $M = \{\alpha, \beta\}$. We may then associate to M the form defined by

$$\begin{aligned} f(x, y) &= (\alpha x + \beta y)(\alpha' x + \beta' y) \\ &= N(\alpha)x^2 + Tr(\alpha\beta')xy + N(\beta)y^2. \end{aligned}$$

From Proposition 2.5, we see that $f(x, y)$ has rational coefficients. Also, the discriminant of $f(x, y)$ is

$$Tr(\alpha\beta')^2 - 4N(\alpha)N(\beta) = 4(st - ru)^2 d,$$

where $\alpha = r + s\sqrt{d}$, $\beta = t + u\sqrt{d}$ (exercise). Since α, β is a basis of $\mathbf{Q}(\sqrt{d})$, it is easy to see that $st - ru \neq 0$. Thus, the discriminant of $f(x, y)$ is not the square of a rational number. Therefore, we see that from the basis α, β of M, we can construct a binary quadratic form of the type we restricted ourselves to above. This suggests that modules and forms are, in some sense, opposite faces of the same coin, with every result about forms implying a corresponding result about modules and vice versa. Indeed, this is the case, and it will take us most of the remainder of the book to adequately describe and exploit the connection between modules and forms. The above discussion gives us some motivation for studying the arithmetic of modules in a quadratic field. We shall begin by studying the most elementary properties of modules and then use them to solve

Problem 1: Let $f(x, y)$ and m be given. Give a procedure for determining all rational integers x, y such that $f(x, y) = m$.

The complete solution of Problem 1 will be given at the end of the current chapter. In Chapter 9, we shall return to modules, but at this point we shall delve much deeper into their arithmetic. In particular, we shall develop a theory of factorization of modules which generalizes the theory of factorization in the Gaussian integers. We shall then apply our factorization theory to the study of various Diophantine equations. In Chapter 10, we shall study the Bachet equation and Fermat's last theorem for $n = 3$. Finally, in Chapter 11, we shall return to the study of binary quadratic forms, and we shall consider the following problem:

Problem 2: Let $f(x, y)$ be a given binary quadratic form. Describe the set of rational numbers m for which the Diophantine equation $f(x, y) = m$ is solvable.

Note that Problem 2 is much more difficult than Problem 1. For example, if $f(x, y) = x^2 + y^2$, then the solution to Problem 2 is given by the two-square theorem. However, Problem 1 can be solved for a given integer m by enumeration since there are only finitely many rational integers x, y such that $x^2 + y^2 = m$. We hasten to point out, however, that Problem 1 is not always trivial. For example, if $f(x, y) = x^2 - dy^2$ $(d > 0)$, then we saw in Chapter 6 that if the equation $f(x, y) = m$ has one solution, then it has an infinite number—and determining these solutions was no easy task.

For reference purposes, let us state Problems 1 and 2 in terms of modules.

Problem 1′: Let r be a given rational number and M a given module. Determine all ξ in M such that $N(\xi) = r$.

Problem 2′: Let M be a given module. Determine the set of rational numbers which are norms of elements of M.

8.4 Exercises

1. Find the discriminants of each of the forms
   (a) $x^2 + xy + y^2$. (b) $2x^2 + 3xy - y^2$.
2. Compute the modules associated to the binary quadratic forms of Exercise 1.

3. Let a, b, s be rational integers, and d a square-free rational integer. Show that a, $(b + s\sqrt{d})/2$ is a basis of $\mathbf{Q}(\sqrt{d})$ if and only if $as \neq 0$.
4. Let $D = s^2$ for some rational integer s, and suppose that the binary quadratic form $f(x, y) = ax^2 + bxy + cy^2$ has discriminant D. Let m be a rational integer. Show that $f(x, y) = m$ if and only if $(2ax - (s - b)y)(2ax + (s + b)y) = 4am$. Therefore, deduce that the Diophantine equation $f(x, y) = m$ can be reduced to the case of a system of two linear equations in two variables.
5. Write the rational number D in the form s^2d, d a square-free rational integer, where

   (a) $D = 64$. (b) $D = -\frac{45}{8}$. (c) $D = -\frac{5}{7}$. (d) $D = 57$.
6. Give an example to show that different bases of the same module may give rise to different binary quadratic forms.

*7. Let M be any module of $\mathbf{Q}(\sqrt{d})$. Show that to the module M there always correspond an infinite number of different binary quadratic forms, determined by the different bases of M.

8. Let $f(x, y)$ be the binary quadratic form associated to the module $M = \{\alpha, \beta\}$, where $\alpha = r + s\sqrt{d}$, $\beta = t + u\sqrt{d}$. Show that the discriminant of $f(x, y)$ equals $4(st - ru)^2d$.

8.5 Modules

We shall now begin our investigation of the properties of modules. Throughout, let $\mathbf{Q}(\sqrt{d})$ be a quadratic field, where d is a square-free, rational integer. The only fact we shall assume as known from Section 4 is the definition of a module.

Lemma 1: Let $M = \{\alpha, \beta\}$ be a module and let θ, ζ belong to M. Then $\theta \pm \zeta$ belongs to M. Thus, M is an additive abelian group.

Proof: Let $\theta = \alpha x_1 + \beta y_1$, $\zeta = \alpha x_2 + \beta y_2$, where x_1, x_2, y_1, y_2 are rational integers. Then $\theta \pm \zeta = \alpha(x_1 \pm x_2) + \beta(y_1 \pm y_2)$ belongs to M. ∎

Let $M = \{\alpha, \beta\}$ be a module. Then α and β are by no means uniquely determined by M. To put it another way, M may have many different bases. For example, $M = \{1, \sqrt{2}\}$ is the same module as $\{1, \sqrt{2} + 1\}$. Indeed, if $\gamma = 1\cdot x + \sqrt{2}y$ belongs to M, then $\gamma = 1\cdot(x - y) + (\sqrt{2} + 1)y$ belongs to $\{1, \sqrt{2} + 1\}$. If $\delta = 1\cdot x + (\sqrt{2} + 1)y$ belongs to $\{1, \sqrt{2} + 1\}$, then $\delta = 1\cdot(x + y) + \sqrt{2}\cdot y$ belongs to M. Thus, $\{1, \sqrt{2}\} = \{1, \sqrt{2} + 1\}$. The question we wish to examine now is, when is $\{\alpha, \beta\} = \{\alpha_1, \beta_1\}$?

Thus, let us suppose that α, β and α_1, β_1 are bases of $\mathbf{Q}(\sqrt{d})$. Note that

α and β belong to $\{\alpha, \beta\}$. Thus, if $\{\alpha, \beta\} = \{\alpha_1, \beta_1\}$, we deduce that α, β belong to $\{\alpha_1, \beta_1\}$. Thus, we may write

$$\alpha = \alpha_1 x_1 + \beta_1 y_1$$
$$\beta = \alpha_1 z_1 + \beta_1 w_1,$$

where x_1, y_1, z_1, and w_1 are rational integers. Conversely, reversing the roles of α, β and α_1, β_1, we may write

$$\alpha_1 = \alpha x + \beta y$$
$$\beta_1 = \alpha z + \beta w$$

for rational integers x, y, z, and w. By substituting the second set of equations in the first set and using the fact that α, β is a basis of $\mathbf{Q}(\sqrt{d})$ (this allows us to compare coefficients of α and β) we see that

$$\begin{pmatrix} x_1 & y_1 \\ z_1 & w_1 \end{pmatrix}\begin{pmatrix} x & y \\ z & w \end{pmatrix} = \begin{pmatrix} 1 & 0 \\ 0 & 1 \end{pmatrix}.$$

Taking determinants of both sides, we immediately see that

$$(x_1 w_1 - z_1 y_1)(xw - zy) = 1.$$

Therefore, since both factors in this last equation are rational integers, we see that

$$x_1 w_1 - z_1 y_1 = \pm 1.$$

We may summarize our results in the following proposition:

Proposition 2: Let α, β and α_1, β_1 be bases of $\mathbf{Q}(\sqrt{d})$ and assume that $\{\alpha, \beta\} = \{\alpha_1, \beta_1\}$. Then there exist rational integers x_1, y_1, z_1, and w_1 such that

$$\begin{pmatrix} \alpha \\ \beta \end{pmatrix} = \begin{pmatrix} x_1 & y_1 \\ z_1 & w_1 \end{pmatrix}\begin{pmatrix} \alpha_1 \\ \beta_1 \end{pmatrix}$$

and

$$det \begin{pmatrix} x_1 & y_1 \\ z_1 & w_1 \end{pmatrix} = \pm 1.$$

The converse of Proposition 2 is true, but the proof is left for the exercises.

Our next goal is to give a characterization of modules which will be very useful to us in what follows. The characterization is in terms of the *discriminant,* which is a rational number we shall introduce to measure, in some sense, the "size" of a module. The discriminant will play a fundamental role in our theory.

Let α, β belong to $\mathbf{Q}(\sqrt{d})$. The determinant

$$\Delta(\alpha, \beta) = \begin{vmatrix} \alpha & \alpha' \\ \beta & \beta' \end{vmatrix}^2 = (\alpha\beta' - \beta\alpha')^2 \tag{1}$$

is called the *discriminant* of α, β.

Lemma 3:

(i) $\Delta(\alpha, \beta)$ is a rational number.

(ii) If α, β belong to I_d, then $\Delta(\alpha, \beta)$ is a rational integer.

(iii) α, β is a basis of $\mathbf{Q}(\sqrt{d})$ if and only if $\Delta(\alpha, \beta) \neq 0$.

Proof:

(i) To show that $\Delta(\alpha, \beta)$ is rational, it suffices to show that $\Delta(\alpha, \beta)' = \Delta(\alpha, \beta)$ (Proposition 2.3, part (iv)). But $\Delta(\alpha, \beta)' = ((\alpha\beta' - \alpha'\beta)^2)' = ((\alpha\beta' - \alpha'\beta)')^2 = (\alpha'\beta - \alpha\beta')^2 = \Delta(\alpha, \beta)$, as desired.

(ii) If α, β belong to I_d, then $\Delta(\alpha, \beta) = (\alpha\beta' - \alpha'\beta)^2$ belongs to I_d. But since $\Delta(\alpha, \beta)$ is also a rational number, we see from Theorem 3.4 that $\Delta(\alpha, \beta)$ is a rational integer.

(iii) Let α, β be a basis of $\mathbf{Q}(\sqrt{d})$. To show that $\Delta(\alpha, \beta) \neq 0$, let us reason by contradiction. If $\Delta(\alpha, \beta) = 0$, then $\alpha/\beta = (\alpha/\beta)'$, so that $\alpha/\beta = c$ is a rational number. Therefore, $\alpha \cdot 1 - \beta \cdot c = 0$, which contradicts the fact that α, β is a basis of $\mathbf{Q}(\sqrt{d})$. We leave the converse as an exercise. ∎

We now wish to define the discriminant of a module $M = \{\alpha, \beta\}$ to be $\Delta(\alpha, \beta)$. However, this definition of discriminant might depend on the choice of the module basis α, β. Let us show that it does not. If α_1, β_1 is another module basis of M, then Proposition 2 implies that there exist rational integers x, y, z, w such that

$$\begin{pmatrix} \alpha \\ \beta \end{pmatrix} = \begin{pmatrix} x & y \\ z & w \end{pmatrix} \begin{pmatrix} \alpha_1 \\ \beta_1 \end{pmatrix}$$

and

$$det\begin{pmatrix} x & y \\ z & w \end{pmatrix} = xw - zy = \pm 1.$$

But then*

$$\begin{aligned}\Delta(\alpha, \beta) &= (\alpha\beta' - \alpha'\beta)^2 \\ &= ((x\alpha_1 + y\beta_1)(z\alpha_1 + w\beta_1)' - (x\alpha_1 + y\beta_1)'(z\alpha_1 + w\beta_1))^2 \\ &= (xw - yz)^2(\alpha_1\beta_1' - \alpha_1'\beta_1)^2 \\ &= (\alpha_1\beta_1' - \alpha_1'\beta_1)^2 \\ &= \Delta(\alpha_1, \beta_1).\end{aligned}$$

Thus, the discriminant $\Delta(\alpha, \beta)$ does not depend on the choice of the module basis, and we are entitled to make the following definition:

*This computation reflects the fact that the determinant of the product of two matrices is the product of their determinants.

Definition 4: The *discriminant* Δ_M of the module $M = \{\alpha, \beta\}$ is defined to be the nonzero rational number $\Delta(\alpha, \beta)$ given in (1).

Example 5: The discriminant of $M = \{5, \sqrt{-7} + 3\}$ is just

$$\Delta_M = \begin{vmatrix} 5 & 5 \\ \sqrt{-7} + 3 & -\sqrt{-7} + 3 \end{vmatrix}^2 = (-10\sqrt{-7})^2 = -700.$$

By using discriminants, we can give a very useful characterization of modules.

Theorem 6: Let M be a set of elements of $\mathbf{Q}(\sqrt{d})$. Then M is a module if and only if the following three conditions hold:

(i) M contains a basis of $\mathbf{Q}(\sqrt{d})$.

(ii) If θ, ζ belong to M, then $\theta \pm \zeta$ belongs to M.

(iii) There is a rational integer k such that $kM \subseteq I_d$.

Moreover, when we choose α, β in a module M such that $|\Delta(\alpha, \beta)| > 0$ and is as small as possible, we have $M = \{\alpha, \beta\}$.

Proof: If M is a module, then condition (i) is part of the definition of a module and condition (ii) is Lemma 1. To prove condition (iii) we simply let $M = \{\alpha, \beta\}$ and choose (by Proposition 3.5) n, m such that $n\alpha$, $m\beta$ are in I_d. Then we may set $k = mn$. Thus, a module satisfies conditions (i)–(iii).

Conversely, suppose that M satisfies conditions (i), (ii), and (iii). We know from Lemma 3, part (iii), that α, β in M form a basis of $\mathbf{Q}(\sqrt{d})$ if and only if $|\Delta(\alpha, \beta)| > 0$. Moreover, from Lemma 3, part (ii), we have that $\Delta(k\alpha, k\beta)$ is a rational integer. But $\Delta(k\alpha, k\beta) = k^4\Delta(\alpha, \beta)$, and thus, for every basis α, β of $\mathbf{Q}(\sqrt{d})$ belonging to M, $\Delta(\alpha, \beta)$ is a rational number whose denominator divides k^4. Thus, there is a basis of $\mathbf{Q}(\sqrt{d})$ in M with $|\Delta(\alpha, \beta)|$ least. It remains to show that any such basis of $\mathbf{Q}(\sqrt{d})$ is a basis of M. Let γ belong to M. Since α, β is a basis of $\mathbf{Q}(\sqrt{d})$, we may write $\gamma = \alpha x + \beta y$, where x, y are rational numbers. If x, y are, in fact, rational integers for every γ in M, then we are done. Thus, assume that x is not a rational integer. Then we may write $x = x_1 + x_2$, where x_1 is a rational integer and x_2 is a rational number satisfying $0 < x_2 < 1$. Let us set $\alpha^* = \gamma - x_1\alpha$, $\beta^* = \beta$. We assert that α^*, β^* is a basis of $\mathbf{Q}(\sqrt{d})$ (exercise). Moreover, since α, β, γ all belong to M, it is clear that α^*, β^* belong to M by (ii). Finally,

$$\begin{aligned} |\Delta(\alpha^*, \beta^*)| &= \left| \begin{vmatrix} \alpha^* & \alpha^{*\prime} \\ \beta^* & \beta^{*\prime} \end{vmatrix}^2 \right| = |(\alpha^*\beta^{*\prime} - \beta^*\alpha^{*\prime})^2| \\ &= |((\alpha x_2 + \beta y)\beta' - \beta(\alpha' x_2 + \beta' y))^2| \\ &= |x_2^2(\alpha\beta' - \beta\alpha')^2| \\ &= x_2^2|\Delta(\alpha, \beta)| \\ &< |\Delta(\alpha, \beta)| \end{aligned}$$

since $0 < x_2 < 1$. This contradicts the manner in which α, β were chosen and thus completes the proof of the theorem. ■

Example 7: Let $\alpha_1, \alpha_2, \ldots, \alpha_n$ belong to $\mathbf{Q}(\sqrt{d})$. Suppose that two of the α_i are linearly independent. Set $M = \{x_1\alpha_1 + x_2\alpha_2 + \cdots + x_n\alpha_n \mid x_1, x_2, \ldots, x_n$ are rational integers$\}$. Then the hypotheses of Theorem 6 are easily verified and M is a module.

8.5 Exercises

1. Find Δ_M for $M = \{1, \sqrt{3}\}$; $M = \{2 + \sqrt{3}, 5 - 2\sqrt{3}\}$; and $M = \{\sqrt{-5}/3, \frac{1}{2} - \frac{2}{5}\sqrt{-5}\}$.
2. Which of the following subsets of $\mathbf{Q}(\sqrt{d})$ are modules:
   (a) $\{0\}$.
   (b) The set of all rational integers.
   (c) The set of all numbers of the form $x + (y/2)$, where x, y are rational integers.
   (d) The set of all numbers of the form $x + (y\sqrt{d}/2^n)$, where x, y, n are rational integers.
   (e) The set of numbers $x + y\sqrt{d} + y(3 + 2\sqrt{d}) + w(\frac{1}{2} + \sqrt{d})$, where x, y, z, w are rational integers.
3. Prove the result of Example 7.
4. Find a basis for the module

   $$M = \{(\tfrac{1}{2} + \sqrt{3})x + \sqrt{3}y + (\tfrac{1}{5} + 2\sqrt{3})z \mid x, y, z \text{ rational integers}\}.$$
5. Prove the converse of Proposition 2.
6. Complete the proof of Lemma 3, part (iii).
7. Show that $\{1, \sqrt{3}\} = \{5 + 17\sqrt{3}, 2 + 7\sqrt{3}\}$. Compute the discriminants of these two bases and verify that they are equal.
8. Complete the proof of Theorem 6.
9. Let M be a *subset* of $\mathbf{Q}(\sqrt{d})$. Associate points in the plane to the elements of M as follows: If $\alpha = x + y\sqrt{d}$ is in M, associate the point $\alpha^* = (x, y\sqrt{|d|})$ (if $d < 0$) or the point $\alpha^* = (\alpha, \alpha')$ (if $d > 0$ and hence α, α' are real). Let M^* denote the set of all points α^* for α in M. Show that M is a module if and only if M^* satisfies the following conditions: (i) If α^*, β^* are in M^*, then $\alpha^* \pm \beta^*$ are in M^*. (ii) Not all points of M^* are collinear. (iii) There exists a circle about $(0, 0)$ in the plane containing no points of M^*.
10. Continuing the notation of Exercise 9, assume that $M = \{\alpha, \beta\}$. Let P be the parallelogram whose vertices are $(0, 0)$, α^*, β^*, $\alpha^* + \beta^*$. Show

that

$$|\Delta(\alpha, \beta)|^{1/2} = \begin{cases} \text{area of } P & \text{if } d > 0, \\ 2 \times \text{area of } P & \text{if } d < 0. \end{cases}$$

11. Let $\alpha, \beta, \alpha_1, \beta_1$ belong to $\mathbf{Q}(\sqrt{d})$ and assume that $\alpha_1 = r\alpha + s\beta$ and $\beta_1 = t\alpha + u\beta$ for r, s, t, u rational. Show that $\Delta(\alpha_1, \beta_1) = (ru - st)^2\Delta(\alpha, \beta)$.
12. Let α, β, γ belong to $\mathbf{Q}(\sqrt{d})$. Show that $\Delta(\gamma\alpha, \gamma\beta) = N(\gamma)^2\Delta(\alpha, \beta)$.
13. Let us define the *discriminant* Δ_d of $\mathbf{Q}(\sqrt{d})$ to be the discriminant of I_d. Show that

$$\Delta_d = \begin{cases} d & \text{if } d \equiv 1 (\text{mod } 4), \\ 4d & \text{if } d \equiv 2, 3 (\text{mod } 4). \end{cases}$$

14. (a) Show that the discriminant of a quadratic field is $\equiv 0$ or $1 (\text{mod } 4)$.
    (b) Let Δ be any rational integer $\equiv 0$ or $1 (\text{mod } 4)$. Show Δ may be written in the form $\Delta = s^2\Delta_0$, where s is a rational integer and Δ_0 is the discriminant of a quadratic field.
15. A rational integer D is said to be a *discriminant* if D is the discriminant of some quadratic field $\mathbf{Q}(\sqrt{d})$ (see Exercise 13); D is said to be a *prime discriminant* if D is divisible by only one rational prime. Show that the prime discriminants are the numbers

$$-4, \quad \pm 8, \quad (-1)^{(p-1)/2}p, \quad p \text{ an odd, rational prime.}$$

16. (continue Exercise 15). Show that every discriminant D can be uniquely written as a product of prime discriminants.

8.6 The Coefficient Ring of a Module

In Section 4, we showed that the problem of solving quadratic Diophantine equations of the form $ax^2 + bxy + cy^2 = m$ could be reduced to the problem of determining all numbers in a module having a given rational number as norm. Let us concentrate on a fixed module M and rational number r. Consider the problem of determining all ξ in M such that

$$N(\xi) = r. \tag{1}$$

Let us now observe that if one solution ξ of (1) is given, then we may construct other solutions as follows: Suppose that ϵ is a number of $\mathbf{Q}(\sqrt{d})$ such that* $\epsilon M \subseteq M$ and $N(\epsilon) = 1$. Then if ξ is in M and satisfies (1), we see immediately that $\epsilon\xi$ is in ϵM and hence in M and that, moreover, $N(\epsilon\xi)$

*Here ϵM is just the set of all numbers of $\mathbf{Q}(\sqrt{d})$ of the form $\epsilon\alpha$, where α is an element of M.

$= N(\epsilon)N(\xi) = 1 \cdot r = r$. Thus, $\epsilon\xi$ has norm r, and we have a means of manufacturing new solutions of (1) from known ones. Let us summarize.

Proposition 1: Suppose that ξ is in M and that $N(\xi) = r$, and suppose that ϵ is in $\mathbf{Q}(\sqrt{d})$ and satisfies $\epsilon M \subseteq M$ and $N(\epsilon) = 1$. Then $\epsilon\xi$ is in M and $N(\epsilon\xi) = r$.

Example 2: Let $d = 2$, $M = \{1, \sqrt{2}\}$, and $\xi = 2 + 3\sqrt{2}$. Then $N(\xi) = -14$. Moreover, since $\epsilon = 3 - 2\sqrt{2}$ is such that $\epsilon M \subseteq M$ (why?) and $N(\epsilon) = 1$, we have $N(\epsilon\xi) = -14$. Indeed, $\epsilon\xi = -6 + 5\sqrt{2}$ and $N(\epsilon\xi) = (-6)^2 - 2(5)^2 = -14$.

From Proposition 1, we see that the elements η of $\mathbf{Q}(\sqrt{d})$ such that $\eta M \subseteq M$ (at least those of norm 1) can be used to construct new elements of a given norm from known ones. Therefore, it seems reasonable to study this set of numbers.

Definition 3: The set of elements η of $\mathbf{Q}(\sqrt{d})$ having the property $\eta M \subseteq M$ is called the *ring of coefficients* (or *coefficient ring*) *of M* and will be denoted $\mathcal{O}_M$.

It is clear that every rational integer belongs to $\mathcal{O}_M$, for if $M = \{\alpha, \beta\}$ and n is any rational integer, then any element γ of M is of the form $\alpha x + \beta y$, where x and y are rational integers, so that $n\gamma = \alpha(nx) + \beta(ny)$ belongs to M.

Let us try to locate some other elements of $\mathcal{O}_M$.

Lemma 4: Let γ belong to $\mathbf{Q}(\sqrt{d})$ and let $M = \{\alpha, \beta\}$. Then γ belongs to $\mathcal{O}_M$ if and only if $\gamma\alpha$ and $\gamma\beta$ belong to M.

Proof: If γ belongs to $\mathcal{O}_M$, then $\gamma M \subseteq M$. Thus, since α and β belong to M, we know $\gamma\alpha$, $\gamma\beta$ belong to M. Conversely, suppose that $\gamma\alpha$ and $\gamma\beta$ belong to M. Let μ be any element of M; we must show that $\gamma\mu$ belongs to M. Since $M = \{\alpha, \beta\}$, $\mu = \alpha x + \beta y$ for rational integers x and y. Then $\gamma\mu = (\gamma\alpha)x + (\gamma\beta)y$. Moreover, since $\gamma\alpha$ and $\gamma\beta$ are in M, we may write $\gamma\alpha = \alpha x_1 + \beta y_1$, $\gamma\beta = \alpha x_2 + \beta y_2$, where x_1, x_2, y_1, y_2 are rational integers. Then

$$\begin{aligned}\gamma\mu &= (\alpha x_1 + \beta y_1)x + (\alpha x_2 + \beta y_2)y \\ &= \alpha(xx_1 + yx_2) + \beta(xy_1 + yy_2)\end{aligned}$$

belongs to M. ∎

Lemma 5: Let γ be any element of $\mathbf{Q}(\sqrt{d})$. Then $\mathcal{O}_M$ contains $k\gamma$ for some positive rational integer k.

Proof: By Lemma 4, $k\gamma$ belongs to $\mathcal{O}_M$ if and only if $k\gamma\alpha$ and $k\gamma\beta$ belong to $M = \{\alpha, \beta\}$. Now α, β is a basis of $\mathbf{Q}(\sqrt{d})$, and $\gamma\alpha$, $\gamma\beta$ belong to $\mathbf{Q}(\sqrt{d})$.

Thus, there are rational numbers x, y, z, w such that

$$\begin{aligned} \gamma\alpha &= \alpha x + \beta y \\ \gamma\beta &= \alpha z + \beta w. \end{aligned} \tag{2}$$

Let the rational integer k be a common denominator of x, y, z, w. Multiplying (2) through by k, we see that $k\gamma\alpha$ and $k\gamma\beta$ belong to M, as desired. ■

Before delving any further into the structure of $\mathcal{O}_M$, let us record a few trivial facts.

Proposition 6:

(i) All rational integers are contained in $\mathcal{O}_M$.
(ii) If η, θ belong to $\mathcal{O}_M$, so do $\eta \pm \theta, \eta \cdot \theta$. Thus, $\mathcal{O}_M$ is a ring.

Proof:

(i) This has been observed above.
(ii) If η, θ belong to $\mathcal{O}_M$, then $\eta M \subseteq M$, $\theta M \subseteq M$. Thus, if γ belongs to M, then $\eta\gamma$, $\theta\gamma$ belong to M, so that $(\eta \pm \theta)\gamma = \eta\gamma \pm \theta\gamma$ belongs to M since the sum or difference of two elements of M is again an element of M (Lemma 5.1). Thus, $(\eta \pm \theta)M \subseteq M$ and $\eta \pm \theta$ belongs to $\mathcal{O}_M$. Finally, $(\eta\theta)M \subseteq \eta M \subseteq M$, so that $\eta\theta$ belongs to $\mathcal{O}_M$. ■

Example 7: Let us compute $\mathcal{O}_M$ for $M = \{1, \sqrt{d}\}$. Suppose that $\eta = s + t\sqrt{d}$, where s and t are rational numbers. Then η is in $\mathcal{O}_M$ if and only if $\eta \cdot 1$ and $\eta\sqrt{d}$ belong to M, that is, if and only if

$$\begin{aligned} s + t\sqrt{d} &= x \cdot 1 + y\sqrt{d} \\ td + s\sqrt{d} &= w \cdot 1 + z\sqrt{d}, \end{aligned}$$

where x, y, z, w are rational integers. Since $1, \sqrt{d}$ is a basis of $\mathbf{Q}(\sqrt{d})$, we see from the first equation that $s = x$, $t = y$, so that s and t are rational integers. From the second equation, we see that $w = td$, $z = s$, and w and z are automatically rational integers if s and t are. Thus, $\mathcal{O}_M$ consists of the numbers of the form $s + t\sqrt{d}$, s, t rational integers. That is, $\mathcal{O}_M = M$.

Example 8: Let us do one more example. Consider $M = \{3, 2\sqrt{7}\}$ in $\mathbf{Q}(\sqrt{7})$. Then $\eta = s + t\sqrt{7}$ (s, t rational) belongs to $\mathcal{O}_M$ if and only if

$$\begin{aligned} 3s + 3t\sqrt{7} &= x \cdot 3 + y \cdot 2\sqrt{7} \\ 14t + 2\sqrt{7}s &= z \cdot 3 + w \cdot 2\sqrt{7} \end{aligned}$$

for rational integers x, y, z, w. Comparing coefficients we obtain the four conditions $x = s$, $3t = 2y$, $14t = 3z$, and $s = w$, and we need to determine all rational numbers s, t such that the corresponding x, y, z, w are rational integers. First, $s = x = w$, and s can be any rational integer. Also, $y = 3t/2$

and $z = 14t/3$ must be rational integers, and thus t must be a rational integer divisible by $2 \cdot 3 = 6$. That is, η has the form $\eta = s + 6t'\sqrt{7}$, where s and t' are rational integers. Thus,

$$\mathcal{O}_M = \{1, 6\sqrt{7}\}.$$

Note that $\mathcal{O}_M$ in Example 8 has a basis of the form $1, \mathcal{f}\omega_7$, where $\mathcal{f}$ is a rational integer ($\omega_7 = \sqrt{7}$). In Theorem 10 we shall show that Example 8 is typical. As a first step, we need

Lemma 9: If γ is in $\mathcal{O}_M$, then γ is in I_d. Thus, $\mathcal{O}_M$ is a subring of I_d.

Proof: If γ is in $\mathcal{O}_M$, and $M = \{\alpha, \beta\}$, then

$$\gamma\alpha = \alpha x + \beta y, \qquad \gamma\beta = \alpha w + \beta z,$$

where x, y, z, w are rational integers. These last equations may be rewritten in the form

$$\alpha(x - \gamma) + \beta y = 0$$
$$\alpha w + \beta(z - \gamma) = 0.$$

Solving the first equation for β and substituting the result in the second equation, we easily see that (since $\alpha \neq 0$)

$$(x - \gamma)(z - \gamma) - wy = 0.$$

In other words, γ satisfies the equation

$$\gamma^2 - (x + z)\gamma + (xz - wy) = 0.$$

The other root of this equation is easily seen to be γ' (use the quadratic formula), so that

$$(X - \gamma)(X - \gamma') = X^2 - (x + z)X + (xz - wy).$$

In other words, $Tr(\gamma) = x + z$ and $N(\gamma) = xz - wy$ are rational integers. Thus, we are done by Definition 3.1. ∎

Note that we may immediately deduce from our above discussion and Theorem 5.6 that $\mathcal{O}_M$ is a module. We can completely describe $\mathcal{O}_M$:

Theorem 10: There exists a positive rational integer $\mathcal{f}$ such that $\mathcal{O}_M = \{1, \mathcal{f}\omega_d\}$. The rational integer* $\mathcal{f}$ is characterized as the least positive rational integer such that $\mathcal{f}\omega_d$ is in $\mathcal{O}_M$.

Proof: We have $I_d = \{1, \omega_d\}$, where ω_d is defined in Theorem 3.2. From Lemma 5, there is a positive rational integer f such that $f\omega_d$ belongs to $\mathcal{O}_M$. Let us choose $\mathcal{f}$ to be the smallest such f. Since 1 and $\mathcal{f}\omega_d$ are in $\mathcal{O}_M$, Proposition 6 implies that every number of the form $x + y\mathcal{f}\omega_d$, x, y rational integers, is

---

*$\mathcal{f}$ is called the *conductor* of $\mathcal{O}_M$.

contained in $\mathcal{O}_M$. Thus, we see that

$$\{1, \mathcal{f}\omega_d\} \subseteq \mathcal{O}_M. \tag{3}$$

Conversely, let γ be in $\mathcal{O}_M$. Then, since $\mathcal{O}_M \subseteq I_d = \{1, \omega_d\}$ (Lemma 9), we have $\gamma = s + t\omega_d$ for some rational integers s, t. Since s is in $\mathcal{O}_M$, we see that $\gamma - s = t\omega_d$ is in $\mathcal{O}_M$. By the division algorithm, we may write $t = q\mathcal{f} + r$, where $0 \leq r < \mathcal{f}$. Then, $t\omega_d = q\mathcal{f}\omega_d + r\omega_d$ is in $\mathcal{O}_M$ as is $q\mathcal{f}\omega_d$, so that $t\omega_d - q\mathcal{f}\omega_d = r\omega_d$ is in $\mathcal{O}_M$. But by the definition of $\mathcal{f}$, we see that $r = 0$. Therefore, $t\omega_d = q\mathcal{f}\omega_d$, and $\gamma = s + q\mathcal{f}\omega_d$ is in $\{1, \mathcal{f}\omega_d\}$. Thus, we see that

$$\mathcal{O}_M \subseteq \{1, \mathcal{f}\omega_d\}. \tag{4}$$

From Eqs. (3) and (4), we conclude the proof. ■

Example 11: Let M be a module and $\mathcal{O}_M$ be its coefficient ring. Then by Theorem 10, $\mathcal{O}_M$ is a module. What is its coefficient ring? We know that it is $\{1, g\omega_d\}$, where g is the least rational integer such that $g\omega_d\mathcal{O}_M \subseteq \mathcal{O}_M$. Writing $\mathcal{O}_M = \{1, \mathcal{f}\omega_d\}$ it is then trivially seen that $g = \mathcal{f}$. Thus, $\mathcal{O}_M$ is its own coefficient ring.

We shall conclude this section by giving a procedure for computing the coefficient ring of any module $M = \{\alpha, \beta\}$. By Theorem 10, this amounts to determining $\mathcal{f}$ from α and β.

Let us first make a useful reduction. Observe that $M = \alpha\{1, \beta/\alpha\}$. The following lemma implies that the coefficient ring of M is the same as the coefficient ring of $\{1, \beta/\alpha\}$. Therefore, in computing coefficient rings, it suffices to consider modules of the form $\{1, \gamma\}$.

Lemma 12: Let M_1 and M_2 be modules such that $M_1 = \alpha M_2$. Then $\mathcal{O}_{M_1} = \mathcal{O}_{M_2}$.

Proof: If γ is an element of $\mathbf{Q}(\sqrt{d})$, then $\gamma M_2 \subseteq M_2$ if and only if $\gamma\alpha M_2 \subseteq \gamma\alpha M_2$. Thus, γ belongs to $\mathcal{O}_{M_1}$ if and only if γ belongs to $\mathcal{O}_{M_2}$. ■

Let us now calculate the ring of coefficients of a module of the form $\{1, \gamma\}$. We know that γ satisfies the quadratic equation $X^2 - Tr(\gamma)X + N(\gamma) = 0$. Since $Tr(\gamma)$ and $N(\gamma)$ are rational numbers, we may write this equation in the form

$$aX^2 - bX + c = 0, \tag{5}$$

where a, b, c are rational integers, $a > 0$, and a, b, c have no common factor > 1.

Lemma 13: Let $M = \{1, \gamma\}$ be a module. Then $\mathcal{O}_M = \{1, a\gamma\}$, where a is given in Eq. (5).

Proof: Note that $1, \gamma$ is a basis of $\mathbf{Q}(\sqrt{d})$. Therefore, if θ is any element of $\mathbf{Q}(\sqrt{d})$, we may write $\theta = x + y\gamma$ for rational x, y. Now θ belongs to $\mathcal{O}_M$ if and only if $\theta M \subseteq M$, which is equivalent to $\theta \cdot 1, \theta \cdot \gamma$ in M. In other words, θ belongs to $\mathfrak{O}_M$ if and only if x, y are rational integers and

$$\theta\gamma = x\gamma + y\gamma^2 = \left(x + \frac{b}{a}y\right)\gamma - \frac{c}{a}y$$

is in M. This, in turn, is equivalent to the requirement that x, y, by/a, cy/a are rational integers. Since a, b, c have no common factor > 1, the only way by/a and cy/a can be rational integers is if $a \mid y$. Thus, $x + y\gamma$ belongs to $\mathfrak{O}_M$ if and only if x, y are rational integers and $a \mid y$. Thus, $\mathfrak{O}_M = \{1, a\gamma\}$. ∎

With γ given in (5), we know that $\gamma = (b \pm \sqrt{D})/2a$, where $D = b^2 - 4ac$. Moreover, if $M = \{\alpha, \beta\}$ and $M' = \{\alpha', \beta'\}$, it is trivial to check that $\mathfrak{O}_M = \mathfrak{O}_{M'}$. The following theorem completes our method for calculating the coefficient ring of any given module M, since we can always replace M by M' if need be:

Theorem 14: Let $M = \{a, (b + \sqrt{D})/2\}$, where $D = b^2 - 4ac$ is not a perfect square and a, b, c are rational integers having no common factor greater than 1. Let $D = \mathcal{f}^2 d$, where d is square-free. Then we have

$$\mathfrak{O}_M = \begin{cases} \left\{1, \dfrac{\mathcal{f}}{2}\omega_d\right\}, & d \equiv 2 \text{ or } 3 \pmod 4, \\ \{1, \mathcal{f}\omega_d\}, & d \equiv 1 \pmod 4. \end{cases}$$

Proof: Set $\alpha = (b + \sqrt{D})/2a$. Define the module $M_1 = \{1, \alpha\}$. We know from Lemma 12 that $\mathfrak{O}_M = \mathfrak{O}_{M_1}$. Moreover, by Lemma 13 we know that $\mathfrak{O}_M = \{1, a\alpha\}$. Thus,

$$\mathfrak{O}_M = \left\{1, \frac{b + \mathcal{f}\sqrt{d}}{2}\right\}.$$

Since $D = \mathcal{f}^2 d = b^2 - 4ac$, we see that

$$\mathcal{f}^2 d \equiv b^2 \pmod 4.$$

If $d \equiv 3 \pmod 4$, then $-\mathcal{f}^2 \equiv b^2 \pmod 4$, and thus $\mathcal{f}^2 \equiv b^2 \equiv 0 \pmod 4$ since a perfect square is congruent to 0 or $1 \pmod 4$. Thus, if $d \equiv 3 \pmod 4$, both $\mathcal{f}$ and b are even. If $d \equiv 2 \pmod 4$, then $2\mathcal{f}^2 \equiv b^2 \pmod 4$, so that b is even; hence, $\mathcal{f}$ is also even. Thus, if $d \equiv 2$ or $3 \pmod 4$,

$$\frac{b + \mathcal{f}\sqrt{d}}{2} = \frac{b}{2} + \frac{\mathcal{f}}{2}\sqrt{d},$$

with $b/2$ and $\mathcal{f}/2$ rational integers. Thus, if $d \equiv 2$ or $3 \pmod 4$,

$$\mathfrak{O}_M = \left\{1, \frac{b}{2}\sqrt{d}\right\} = \left\{1, \frac{\mathcal{f}}{2}\omega_d\right\}.$$

Finally, if $d \equiv 1 \pmod 4$, $\mathcal{f}^2 \equiv b^2 \pmod 4$, so that $\mathcal{f}$ and b are either both even or both odd. Thus,

$$\frac{b + f\sqrt{d}}{2} = \frac{b - f}{2} + f\frac{1+\sqrt{d}}{2}$$

with $(b - f)/2$ and f rational integers, so that

$$\mathcal{O}_M = \left\{1, f\frac{1+\sqrt{d}}{2}\right\} = \{1, f\omega_d\}. \qquad \blacksquare$$

Example 15: We shall compute $\mathcal{O}_M$ for $M = \{3, 2\sqrt{5}\}$. Here $\gamma = 2\sqrt{5}/3$ and γ satisfies $9X^2 - 20 = 0$, so that $D = 4 \cdot 9 \cdot 20 = 12^2 \cdot 5$. Since $5 \equiv 1 \pmod 4$, we have $\omega_5 = (1 + \sqrt{5})/2$ and $\mathcal{O}_M = \{1, 12\omega_5\}$.

8.6 Exercises

1. Determine $\mathcal{O}_M$ if $M = \{1, \sqrt{3}\}$; $M = \{2 + \sqrt{3}, 5 - 2\sqrt{3}\}$; $M = \{7, 4\sqrt{2}\}$.
2. Let M_1, M be modules such that $M \subseteq M_1 \subseteq I_d$. Suppose that $\mathcal{O}_M = \{1, f\omega_d\}$, $\mathcal{O}_{M_1} = \{1, f_1\omega_d\}$. Show that $f_1 \mid f$.
3. Let M be a module and suppose that $\alpha M = M$ for all α in I_d. Show that $\mathcal{O}_M = I_d$.
4. Let $M = \{1, f\omega_d\}$, where f is a positive rational integer. Show that $\mathcal{O}_M = M$. Thus, the possible coefficient rings in $\mathbf{Q}(\sqrt{d})$ are precisely the modules $M = \{1, f\omega_d\}$ for $f = 1, 2, 3, \ldots$. (These subrings of I_d are often called *orders*.)
5. Let a, b, d be pairwise relatively prime rational integers. Let $M = \{a, b\sqrt{d}\}$. Show that $\mathcal{O}_M = \{1, ab\sqrt{d}\}$.

8.7 The Unit Theorem

Proposition 6.1 tells us that if we know those elements ϵ of $\mathcal{O}_M$ such that $N(\epsilon) = 1$, then from one element of M of norm r we may manufacture others in a straightforward way. This procedure will be one of the two crucial steps in solving our first problem concerning binary quadratic forms, which in terms of modules amounts to determining all elements of a given module M having a given rational number as norm. In this section, we shall give a very precise description of all those ϵ of $\mathcal{O}_M$ such that $N(\epsilon) = 1$.

Recall the following definition from abstract algebra:

Definition 1: A *unit* of the ring $\mathcal{O}_M$ is an element ϵ of $\mathcal{O}_M$ such that ϵ^{-1} belongs to $\mathcal{O}_M$.

The relevance of units to the discussion of this section is seen in

Lemma 2: Let ϵ belong to $\mathcal{O}_M$. Then ϵ is a unit of $\mathcal{O}_M$ if and only if $N(\epsilon) = \pm 1$.

Proof: If ϵ is a unit, then ϵ^{-1} is in $\mathcal{O}_M \subseteq I_d$, and thus $N(\epsilon)$, $N(\epsilon^{-1})$ are both rational integers. However,

$$1 = N(1) = N(\epsilon\epsilon^{-1}) = N(\epsilon)N(\epsilon^{-1}),$$

so that $N(\epsilon) = \pm 1$. Conversely, if $N(\epsilon) = \pm 1$, then $\epsilon\epsilon' = \pm 1$ and $\epsilon^{-1} = \pm \epsilon'$. But it is easy to see (from Theorem 6.10) that if ϵ belongs to $\mathcal{O}_M$, then so does ϵ'. Thus, ϵ^{-1} is in $\mathcal{O}_M$. ■

Example 3: Let $\mathcal{O}_M = I_2 = \{1, \sqrt{2}\}$. Then $\epsilon = 1 + \sqrt{2}$ is a unit since $\epsilon^{-1} = -1 + \sqrt{2}$ belongs to $\mathcal{O}_M = I_2$. However, $\lambda = 1 + 2\sqrt{2}$ is not a unit of $\mathcal{O}_M$ since $\lambda^{-1} = -\frac{1}{7} + \frac{2}{7}\sqrt{2}$ does not belong to $\mathcal{O}_M = I_2$. Alternatively, $N(1 + \sqrt{2}) = -1$ and $N(1 + 2\sqrt{2}) = -7$, so that $1 + \sqrt{2}$ is a unit and $1 + 2\sqrt{2}$ is not by, Lemma 2.

In this section, we shall first describe all the units of $\mathcal{O}_M$. We shall then determine which of the units have norm $+1$.

By Theorem 6.10, $\mathcal{O}_M = \{1, f\omega_d\}$ for some rational integer $f \geq 1$, where $\omega_d = \sqrt{d}$ if $d \equiv 2$ or $3 \pmod 4$ and $= (1 + \sqrt{d})/2$ if $d \equiv 1 \pmod 4$. Thus, every element ϵ in $\mathcal{O}_M$ has the form $\epsilon = x + f\omega_d y$ for unique rational integers x and y. Thus, ϵ is a unit of $\mathcal{O}_M$ if and only if

$$N(x + f\omega_d y) = \pm 1.$$

However, since

$$\begin{aligned} N(x + f\omega_d y) &= (x + f\omega_d y)(x + f\omega_d' y) \\ &= x^2 - f^2 d y^2 && \text{if } d \equiv 2 \text{ or } 3 \pmod 4, \\ &= x^2 + f xy + \frac{1-d}{4} f^2 y^2 && \text{if } d \equiv 1 \pmod 4, \end{aligned}$$

we have the following result:

Proposition 4: The number $\epsilon = x + f\omega_d y$, x, y rational integers, is a unit of $\mathcal{O}_M = \{1, f\omega_d\}$ if and only if x and y satisfy

$$\begin{aligned} x^2 - f^2 d y^2 &= \pm 1 && \text{if } d \equiv 2 \text{ or } 3 \pmod 4, \\ x^2 + f xy + \frac{1-d}{4} f^2 y^2 &= \pm 1 && \text{if } d \equiv 1 \pmod 4. \end{aligned}$$

As is very common in this theory, we must consider the cases $d < 0$ and $d > 0$ separately. The answers are very different.

Case 1: $d < 0$. First assume that $d \equiv 2$ or $3 \pmod 4$. Since $x^2 - d f^2 y^2 = x^2 + |d| f^2 y^2 \geq 0$, Proposition 4 implies that we must study the equation

$$x^2 + |d| f^2 y^2 = 1. \tag{1}$$

If $|d| > 1$ or $f^2 > 1$ and $y \neq 0$, then $x^2 + |d| y^2 f^2 \geq |d| f^2 y^2 > 1$, and there are no solutions to (1). Thus, if $|d| > 1$ or $f^2 > 1$, then $y = 0$, and so

$x = \pm 1$, and the only units are $\epsilon = \pm 1$. If $|d| = 1$ and $\mathscr{f}^2 = 1$, then (1) becomes $x^2 + y^2 = 1$, which has the solutions $(x, y) = (\pm 1, 0), (0, \pm 1)$, and so in this case, the units are $\epsilon = \pm 1, \pm\sqrt{-1}$. This completes the case $d < 0$, $d \equiv 2$ or $3 (\text{mod } 4)$.

Now suppose that $d < 0$ and $d \equiv 1 (\text{mod } 4)$. Since

$$x^2 + \mathscr{f}xy + \frac{1-d}{4}\mathscr{f}^2y^2 = \left(x + \frac{\mathscr{f}}{2}y\right)^2 + \frac{|d|}{4}\mathscr{f}^2y^2 \geq 0,$$

we must study the equation

$$\left(x + \frac{\mathscr{f}}{2}y\right)^2 + \frac{|d|}{4}\mathscr{f}^2y^2 = 1. \tag{2}$$

If $|d| > 4$ and $y \neq 0$, then $(|d|/4)\mathscr{f}^2y^2 > 1$, and there are no solutions to (2). Thus, $|d| > 4$ implies that $y = 0$ and $x = \pm 1$. Thus, the units are $\epsilon = \pm 1$. If $|d| \leq 4$, then $d = -3$ since $d < 0$ and $d \equiv 1 (\text{mod } 4)$. If $\mathscr{f}^2 > 1$ and $y \neq 0$, then $\frac{3}{4}\mathscr{f}^2y^2 > 1$, and there are no solutions to (2). Thus, $\mathscr{f}^2 > 1$ implies that $y = 0$, and again $\epsilon = \pm 1$ are the only units. Finally, if $\mathscr{f} = 1$ and $d = -3$, then (2) becomes $(x + \frac{1}{2}y)^2 + \frac{3}{4}y^2 = 1$. If $|y| \geq 2$, then $\frac{3}{4}y^2 > 1$, and there are no solutions. In the other three cases for y one trivially checks that $y = 1$ implies that $x = 0$ or -1, $y = 0$ implies that $x = \pm 1$, and $y = -1$ implies that $x = 0$ or 1. A simple calculation then shows that if $d = -3$ and $\mathscr{f} = 1$, the units are $\epsilon = \pm 1, (\pm 1 \pm \sqrt{-3})/2$, where all combinations of signs are permitted.

We summarize in

Theorem 5: Suppose that $\mathcal{O}_M = \{1, \mathscr{f}\omega_d\} = \mathcal{O}_{\mathscr{f}}$, where $d < 0$. Then the units of $\mathcal{O}_M$ are $\epsilon = \pm 1$ except in the following two cases:

(i) $d = -1$, $\mathscr{f} = 1$, in which case the units are $\epsilon = 1, -1, \sqrt{-1}, -\sqrt{-1}$.

(ii) $d = -3$, $\mathscr{f} = 1$, in which case the units are $\epsilon = 1, -1, (1 + \sqrt{-3})/2, (1 - \sqrt{-3})/2, (-1 + \sqrt{-3})/2, (-1 - \sqrt{-3})/2$.

Case 2: $d > 0$. This case is much more intricate than the case of negative d. In particular there will always be an infinite number of units. We shall avoid most of the difficulties by appealing to the solution of Pell's equation given in Chapter 6. Specifically we shall use the following result (Theorem 6.6.3):

If $d_0 > 0$ is a rational integer which is not a perfect square, then the Diophantine equation

$$x^2 - d_0y^2 = 1 \tag{3}$$

has an infinite number of solutions.

To determine the units of $\mathcal{O}_{\mathscr{f}} = \{1, \mathscr{f}\omega_d\}$ we require three lemmas.

Lemma 6: There are an infinite number of units in $\mathcal{O}_{\mathscr{f}}$.

Proof: We make the simple observation that $\epsilon = x + y\mathcal{f}\sqrt{d}$ belongs to $\mathcal{O}_{\mathcal{f}}$ for all rational integers x, y. (This is just the definition of $\mathcal{O}_{\mathcal{f}}$ if $d \equiv 2, 3 \pmod 4$, since then $\omega_d = \sqrt{d}$. If $d \equiv 1 \pmod 4$ and $\omega_d = (1 + \sqrt{d})/2$, then

$$x + \mathcal{f}\sqrt{d}\,y = (x - \mathcal{f}y) + (2y)\mathcal{f}\omega_d$$

belongs to $\mathcal{O}_{\mathcal{f}}$.) Now $N(\epsilon) = N(x + \mathcal{f}\sqrt{d}\,y) = x^2 - (\mathcal{f}^2 d)y^2$. Setting $d_0 = \mathcal{f}^2 d$, we see that d_0 is not a perfect square. There are an infinite number of solutions (x, y) of (3). If (x, y) is a solution, then $\epsilon = x + y\mathcal{f}\sqrt{d}$ is in $\mathcal{O}_{\mathcal{f}}$, and $N(\epsilon) = 1$, so that ϵ is a unit of $\mathcal{O}_{\mathcal{f}}$. ■

Lemma 7: If ϵ is a unit of $\mathcal{O}_{\mathcal{f}}$, then so are $-\epsilon$ and $1/\epsilon$. If ϵ_1, ϵ_2 are units of $\mathcal{O}_{\mathcal{f}}$, then so is $\epsilon_1 \cdot \epsilon_2$. In particular, the units of $\mathcal{O}_{\mathcal{f}}$ form a group.

Proof: This fact holds for arbitrary rings and is an easy exercise. ■

Lemma 8: Let $B > 1$ be a given real number. Then there exists only a finite number of units ϵ of $\mathcal{O}_{\mathcal{f}}$ such that $1 < \epsilon < B$.

Proof: Let ϵ be a unit of $\mathcal{O}_{\mathcal{f}}$. From Proposition 2.5 we know that ϵ is a root of the equation

$$X^2 - Tr(\epsilon)X + N(\epsilon) = 0$$

with rational integer coefficients. Since ϵ is a unit, we have $N(\epsilon) = \pm 1$, and thus ϵ is a root of the equation

$$X^2 - Tr(\epsilon)X \pm 1 = 0.$$

Now $\epsilon > 1$ and $N(\epsilon) = \epsilon\epsilon' = \pm 1$, and so $|\epsilon'| = (1/\epsilon) < 1$. Thus,

$$|Tr(\epsilon)| = |\epsilon + \epsilon'| \leq |\epsilon| + |\epsilon'| < B + 1.$$

Hence, ϵ must be a root of one of the equations $X^2 + nX \pm 1 = 0$, where n is a rational integer such that $|n| \leq B + 1$. There are only a finite number of such equations, and each equation can have at most two roots. Thus, there are only a finite number of possibilities for ϵ. ■

By Lemma 6, there is a unit ϵ in $\mathcal{O}_{\mathcal{f}}$ such that $\epsilon \neq \pm 1$. If ϵ is negative, replace ϵ by $-\epsilon$, and we still have a unit of $\mathcal{O}_{\mathcal{f}}$ by Lemma 7. Thus, assume that $\epsilon > 0$. If $\epsilon < 1$, replace ϵ by ϵ^{-1} (again a unit of $\mathcal{O}_{\mathcal{f}}$ by Lemma 7). Thus, there is a unit ϵ of $\mathcal{O}_{\mathcal{f}}$ such that $\epsilon > 1$. By Lemma 8 there is a unique smallest unit $\epsilon_{\mathcal{f}}$ of $\mathcal{O}_{\mathcal{f}}$ such that $1 < \epsilon_{\mathcal{f}} \leq \epsilon$. In fact, $\epsilon_{\mathcal{f}}$ is the smallest unit of $\mathcal{O}_{\mathcal{f}}$ larger than 1.

Definition 9: The smallest unit of $\mathcal{O}_{\mathcal{f}}$ which is larger than 1 is called the *fundamental unit* of $\mathcal{O}_{\mathcal{f}}$, denoted $\epsilon_{\mathcal{f}}$.

Theorem 10: Suppose that $\mathcal{O}_{\mathcal{f}} = \{1, \mathcal{f}\omega_d\}$, where $d > 0$. Then the units of $\mathcal{O}_{\mathcal{f}}$ are the numbers $\pm\epsilon_{\mathcal{f}}^n$ ($n = 0, \pm 1, \pm 2, \ldots$).

Proof: Let $\epsilon \geq 1$ be a unit of $\mathcal{O}_{\mathcal{f}}$. Since $1 < \epsilon_{\mathcal{f}}$, we see that $\epsilon_{\mathcal{f}}^n \longrightarrow \infty$ $(n \longrightarrow \infty)$, and so there is an integer $n \geq 0$ such that

$$\epsilon_{\mathcal{f}}^n \leq \epsilon < \epsilon_{\mathcal{f}}^{n+1}.$$

Thus,

$$1 \leq \epsilon\epsilon_{\mathcal{f}}^{-n} < \epsilon_{\mathcal{f}}.$$

By Lemma 7, $\epsilon\epsilon_{\mathcal{f}}^{-n}$ is a unit, and it is smaller than $\epsilon_{\mathcal{f}}$, the smallest unit larger than 1. Thus, $1 = \epsilon\epsilon_{\mathcal{f}}^{-n}$, and so $\epsilon = \epsilon_{\mathcal{f}}^n$. In general, if ϵ is any unit of $\mathcal{O}_{\mathcal{f}}$, then one of ϵ, ϵ^{-1}, $-\epsilon$, or $(-\epsilon)^{-1}$ is a unit ≥ 1 and thus of the form $\epsilon_{\mathcal{f}}^n$ for some integer $n \geq 0$. ∎

You should observe the similarity between the proof of Theorem 10 and the proof given describing the method for obtaining the general solution to Pell's equation (Theorem 6.6.7). You should also note that in case $d \equiv 1 \pmod 4$ we managed to avoid considering the more complicated Diophantine equation given in Proposition 4 (by virtue of Lemma 6). However, we may now conclude that if $d \equiv 1 \pmod 4$, then $x^2 + \mathcal{f}xy + ((1-d)/4)\mathcal{f}^2y^2 = 1$ has an infinite number of solutions in rational integers x, y.

Example 11: In case $d \equiv 2, 3 \pmod 4$, the problem of determining the units of $\{1, \mathcal{f}\sqrt{d}\}$ is equivalent to solving the Diophantine equations

$$x^2 - \mathcal{f}^2dy^2 = \pm 1.$$

If $d \not\equiv 2 \pmod 8$, then reasoning with congruences modulo 8 shows that the equation

$$x^2 - \mathcal{f}^2dy^2 = -1$$

has no solution. Thus, determining the units is equivalent to solving Pell's equation $x^2 - \mathcal{f}^2dy^2 = 1$. The fundamental unit $\epsilon_{\mathcal{f}}$ is then just $x_0 + y_0\mathcal{f}\sqrt{d}$, where (x_0, y_0) is the fundamental solution of the Pell equation. We shall take up the general problem of computing $\epsilon_{\mathcal{f}}$ in a finite number of steps after we prove Theorem 12.

Using our knowledge of the units of $\mathcal{O}_M$, let us now determine all units of norm 1. This is easy from what we have already proved.

Theorem 12: Let $\mathcal{O}_M = \{1, \mathcal{f}\omega_d\} = \mathcal{O}_{\mathcal{f}}$, and in case $d > 0$, let $\epsilon_{\mathcal{f}}$ be the fundamental unit of $\mathcal{O}_M$. Further, let $\mathbf{U}_+$ denote the set of all units of norm $+1$. Then, if $d > 0$,

$$\mathbf{U}_+ = \begin{cases} \{\pm\epsilon_{\mathcal{f}}^n \mid n = 0, \pm 1, \pm 2, \ldots\} & \text{if } N(\epsilon_{\mathcal{f}}) = 1, \\ \{\pm\epsilon_{\mathcal{f}}^{2n} \mid n = 0, \pm 1, \pm 2, \ldots\} & \text{if } N(\epsilon_{\mathcal{f}}) = -1. \end{cases}$$

If $d < 0$, all units have norm 1.

Proof: $N(\pm\epsilon_{\mathcal{f}}^k) = N(\epsilon_{\mathcal{f}})^k$ for any integer k. ∎

Let us finally turn to the problem of explicitly computing the fundamental unit $\epsilon_{\mathcal{f}}$ in the case $d > 0$. We require three lemmas to aid in the calculations.

Lemma 13: Let $d > 0$ and let $\epsilon = x + \mathcal{f}\omega_d y$ be a unit of $\mathcal{O}_{\mathcal{f}}$ such that $\epsilon > 1$. Then $x > 0$ and $y > 0$, except for the case $d = 5$, $\mathcal{f} = 1$, $\epsilon = (1 + \sqrt{5})/2$.

Proof: First note that $\omega_d - \omega_d' > 0$. Moreover, $N(\epsilon) = \epsilon\epsilon' = \pm 1$ implies that $\epsilon^{-1} = \pm\epsilon'$, and thus we have

$$\epsilon - \epsilon' = \epsilon \mp \frac{1}{\epsilon} > 0.$$

Thus, $\epsilon - \epsilon' = \mathcal{f}(\omega_d - \omega_d')y > 0$, so that $y > 0$. Also, the facts $\epsilon > 1$ and $\epsilon\epsilon' = \pm 1$ imply that

$$|\epsilon'| = |x + \mathcal{f}\omega_d' y| < 1.$$

Moreover, $\omega_d' < 0$, so that if $x \neq 0$, we must have $x > 0$. If $x = 0$, then, since $N(\epsilon) = \pm 1$, we have $y^2\mathcal{f}^2\omega_d\omega_d' = \pm 1$. But, $\omega_d\omega_d'$ is either $-d$ (if $d \equiv 2$ or $3 \pmod 4$) or $(1 - d)/4$ (if $d \equiv 1 \pmod 4$). Since $d > 1$, the first case cannot occur. Thus, $d \equiv 1 \pmod 4$. If $d > 5$, then $|y^2\mathcal{f}^2\omega_d\omega_d'| > 1$. Thus, we see that $d = 5$ and $y^2\mathcal{f}^2 = 1$, so that $\mathcal{f} = 1$, $y = \pm 1$. Thus, since $\epsilon > 1$, we must have $\epsilon = (1 + \sqrt{5})/2$. ■

Lemma 14: Let $d > 0$ and let $\epsilon = x + \mathcal{f}\omega_d y$ be a unit of $\mathcal{O}_{\mathcal{f}}$. Assume that $\epsilon > 1$ and $x \neq 0$. Then x is the integer closest to the irrational number $-y\mathcal{f}\omega_d'$.

Proof: By Lemma 13, $x > 0$ and $y > 0$, so that $x \geq 1$ and $y \geq 1$ (since x and y are rational integers). Moreover, $\omega_d > 1$ and $\mathcal{f} \geq 1$, and thus $\epsilon = x + \mathcal{f}\omega_d y > 2$. Since $N(\epsilon) = \epsilon\epsilon' = \pm 1$, we see that

$$|\epsilon'| = \frac{1}{\epsilon} < \frac{1}{2}.$$

But $\epsilon' = x + \mathcal{f}\omega_d' y$, and therefore

$$|x + \mathcal{f}\omega_d' y| < \tfrac{1}{2}.$$

Thus, the assertion follows immediately. ■

Note, for reference, that

$$-\mathcal{f}\omega_d' y = \begin{cases} \mathcal{f}\sqrt{d}\,y & \text{if } d \equiv 2, 3 \pmod 4, \\ \mathcal{f}\dfrac{\sqrt{d} - 1}{2}y & \text{if } d \equiv 1 \pmod 4. \end{cases}$$

Lemma 15: Let $d > 0$ and let $\epsilon = x + \mathcal{f}\omega_d y$ be a unit of $\mathcal{O}_{\mathcal{f}}$ such that $\epsilon > 1$. For each rational integer $n \geq 1$, set

$$\epsilon^n = x_n + \mathcal{f}\omega_d y_n.$$

Then $0 < x = x_1 < x_2 < x_3 < \cdots$ and $0 < y = y_1 < y_2 < y_3 < \cdots$.

Proof: Note that

$$\epsilon^n = \epsilon\epsilon^{n-1} = (x_1 + \mathcal{f}\omega_d y_1)(x_{n-1} + \mathcal{f}\omega_d y_{n-1})$$

$$= \begin{cases} (x_1 x_{n-1} + \mathcal{f}^2 d y_1 y_{n-1}) \\ \quad + \mathcal{f}\omega_d(x_1 y_{n-1} + y_1 x_{n-1}) & \text{if } d \equiv 2 \text{ or } 3 (\text{mod } 4), \\ \left(x_1 x_{n-1} + \dfrac{d-1}{4}\mathcal{f}^2 y_1 y_{n-1}\right) \\ \quad + \mathcal{f}\omega_d(x_1 y_{n-1} + y_1 x_{n-1} + \mathcal{f} y_1 y_{n-1}) & \text{if } d \equiv 1 (\text{mod } 4). \end{cases}$$

Thus, we have explicit formulas for x_n and y_n in terms of $x = x_1$, $y = y_1$ and x_{n-1}, y_{n-1}. The assertion follows immediately by induction on n. ■

We may now give our procedure for computing $\epsilon_{\mathcal{f}}$. Let us exclude the case $\mathcal{f} = 1$, $d = 5$. By Lemmas 13, 14 and 15, we have the following:

Algorithm: Inspect each of the numbers $x + \mathcal{f}\omega_d y$, $y = 1, 2, 3, \ldots$, $x =$ the integer closest to $-\mathcal{f}\omega'_d y$. The first one which is a unit is the fundamental unit $\epsilon_{\mathcal{f}}$.

Example 16: Let $\mathcal{f} = 1$, $d = 23$. In this case $\epsilon_{\mathcal{f}}$ is the fundamental unit of I_{23}. Since $23 \equiv 3 (\text{mod } 4)$, we are looking for a solution to $x^2 - 23y^2 = \pm 1$. We tabulate our data as follows:

y	$-y\mathcal{f}\omega'_d = \sqrt{23}y$	x	$N(x + \sqrt{23}y) = x^2 - 23y^2$
1	4.8	5	$25 - 23 = 2$
2	9.6	10	$100 - 92 = 8$
3	14.4	14	$196 - 207 = -11$
4	19.2	19	$361 - 368 = -7$
5	24.0	24	$576 - 575 = 1$

Thus, $\epsilon_1 = 24 + 5\sqrt{23}$ is the fundamental unit.

Example 17: Let us compute the fundamental unit for $\mathcal{f} = 2$, $d = 3$. Again we tabulate the data:

y	$-\mathcal{f}\omega'_d y = 2\sqrt{3}\,y$	x	$N(x + 2\sqrt{3}\,y) = x^2 - 12y^2$
1	3.4	3	$9 - 12 = -3$
2	6.9	7	$49 - 48 = 1$

Thus, $\epsilon_2 = 7 + 2 \cdot 2\sqrt{3} = 7 + 4\sqrt{3}$ is the fundamental unit of $\{1, 2\sqrt{3}\}$.

There is another way of proceeding in the case $f > 1$. Suppose that the fundamental unit ϵ_1 of I_d has been determined. Then the fundamental unit ϵ_f of $\mathcal{O}_f = \{1, f\omega_d\}$ can be determined as follows. Since $\mathcal{O}_f$ is contained in I_d, we see that ϵ_f is a unit of I_d. Therefore, $\epsilon_f = \pm\epsilon_1^n$ for some integer n. Moreover, since $\epsilon_1 > 1$, $\epsilon_f > 1$, we see that $\epsilon_f = \epsilon_1^n$ for some $n \geq 1$. In general, ϵ_f is the first power of ϵ_1 which belongs to $\mathcal{O}_f$.

In Example 17, we easily compute that $\epsilon_1 = 2 + \sqrt{3}$ is the fundamental unit of I_3. Now $\epsilon_1^2 = 7 + 2 \cdot 2\sqrt{3}$ is clearly in $\{1, 2\sqrt{3}\}$. Since ϵ_1 does not belong to $\{1, 2\sqrt{3}\}$, we see that $\epsilon_2 = 7 + 4\sqrt{3}$. Also, $\epsilon_4 = 7 + 4\sqrt{3}$.

In closing, we point out that although the above method for determining the fundamental unit seems very efficient, we have given no upper bound on how big the first value of y which gives a unit will be. Unfortunately, y may be very large, even for small d. For example, for $f = 1$, $d = 31$, $\epsilon_f = 1520 + 273\sqrt{31}$; if $f = 1$, $d = 94$, $\epsilon_f = 2143295 + 221064\sqrt{94}$. We have included a table of the fundamental units of I_d for $1 < d < 200$ at the end of the book (see Table 2).

There is a much more efficient method for determining the fundamental unit. The method relies on the theory of continued fractions, a topic often covered in an elementary course in number theory. The interested reader should consult G. H. Hardy and E. M. Wright, *An Introduction to the Theory of Numbers*, 3rd ed., Oxford University Press, Inc., New York, 1960 and Y. Borevich and I. Shafarevich, *Number Theory*, Academic Press, New York, 1966, Chapter 2, Section 7.3.

8.7 Exercises

1. Show that $(1 + \sqrt{5})/2$ is the fundamental unit of I_5.
2. Compute the fundamental unit of I_d where
   (a) $d = 3$. (b) $d = 6$. (c) $d = 17$. (d) $d = 21$. (e) $d = 35$.
3. Write a computer program to compute ϵ_d, the fundamental unit of I_d. Use your program to check as much of Table 2 as you can.
4. Compute the fundamental unit of $\mathcal{O}_f = \{1, f\omega_d\}$, where
   (a) $d = 3$, $f = 1$. (b) $d = 3$, $f = 2$.
   (c) $d = 3$, $f = 4$. (d) $d = 5$, $f = 1$.
   (e) $d = 5$, $f = 5$. (f) $d = 5$, $f = 7$.
   Use both the algorithm given in the text and the method which locates the first power of ϵ_1 which lies in $\mathcal{O}_f$. Compare the efficiency of these two computations.
5. Let $\mathbf{R}_+$ denote the set of positive real numbers and let G be a subgroup of $\mathbf{R}_+$ under multiplication such that there exists an interval (a, b) about 1 in $\mathbf{R}_+$ which contains no element of G other than 1.

(a) Show that G consists of all powers (positive and negative) of some element of G.
(b) Can part (a) be used to simplify the proof of the unit theorem?

6. Let ϵ_1 denote the fundamental unit of I_d, $d > 0$.
(a) Show that $\epsilon_1 \geq 1 + \sqrt{d}$ if $d \equiv 2$ or $3 \pmod 4$.
(b) Show that $\epsilon_1 \geq (3 + \sqrt{d})/2$ if $d \equiv 1 \pmod 4$, $d \neq 5$.
(c) Show that $\epsilon_1 \longrightarrow \infty$ as $d \longrightarrow \infty$.

8.8 Computing Elements of a Given Norm in a Module

We are finally ready to give a complete solution of Problem 1′ posed in Section 4. Namely, we shall give a procedure for determining all elements of given norm in a given module. When translated into the language of forms, this solution will allow us to solve completely any quadratic Diophantine equation in two variables. Let M be a module in $\mathbf{Q}(\sqrt{d})$ and let r be a rational number. Our object in this section is to determine all ξ in M such that $N(\xi) = r$.

First, let us note that the case $d < 0$ is trivial, for if $M = \{\alpha, \beta\}$, with $\alpha = a + b\sqrt{d}$, $\beta = c + e\sqrt{d}$, then $N(\alpha x + \beta y) = (ax + cy)^2 + (bx + ey)^2|d|$ for x, y rational. If x and y are rational integers and g is a common denominator of a, b, c, e, then $ax + cy$ and $bx + ey$ are rational numbers with denominators dividing g. Thus, if $N(\alpha x + \beta y) = r$, we see that $r \geq 0$ and $|ax + cy| \leq \sqrt{r}$, $|bx + ey| \leq \sqrt{r/|d|}$. In particular, there are only finitely many choices for the rational numbers $ax + cy$, $bx + ey$. Thus, there are only finitely many choices for $\alpha x + \beta y$, and it is easy to determine which of these have norm r. *Henceforth, let us assume that* $d > 0$.

Let us now transform Problem 1′ into two other problems which we shall be able to solve. Motivated by Proposition 6.1, we make the following Definition:

Definition 1: We say that two elements α, β of the module M are *associates* (or *are associated with one another*) if there exists a unit ϵ of $\mathcal{O}_M$ such that $\alpha = \epsilon\beta$.

Example 2: $2 + 3\sqrt{2}$ and $4 - \sqrt{2}$ are associates in $M = \{1, \sqrt{2}\}$, since $4 - \sqrt{2} = (-1 + \sqrt{2})(2 + 3\sqrt{2})$, and $-1 + \sqrt{2}$ is a unit of $\mathcal{O}_M = M$.

From Lemma 7.2, we see that the numbers ϵ of $\mathbf{Q}(\sqrt{d})$ which satisfy the hypotheses of Proposition 6.1 are just the units of $\mathcal{O}_M$ which have norm 1. Thus, if ξ belongs to M, then $\epsilon\xi$ is an associate of ξ. Therefore, we can break our problem of determining all elements ξ of M having norm r into two subproblems, namely

Problem 1-A: Determine all units of $\mathcal{O}_M$ of norm 1.

Problem 1-B: Find a set S_r of elements of M such that

(i) No two elements of S_r are associates.
(ii) Each elements of S_r has norm r.
(iii) If ξ in M has norm r, then ξ is an associate of some element of S_r.

Of course, Problem 1-A is solved by Theorem 7.12. We shall solve Problem 1-B in the remainder of this section.

A set S_r satisfying (i)–(iii) in Problem 1-B is called a *complete set of nonassociate elements of norm r*. We shall specify a procedure for finding such a set. Moreover, it is somewhat surprising that S_r is always finite.

First, let us note that if Problems 1-A and 1-B are solved, then we can determine all ξ in M having norm r. Indeed, if $S_r = \{\xi_1, \ldots, \xi_t\}$ and ξ is any element of M of norm r, then ξ is an associate of ξ_i for some i $(1 \leq i \leq t)$. Thus, there exists a unit ϵ of $\mathcal{O}_M$ such that $\xi = \epsilon\xi_i$. However, since $N(\xi) = N(\xi_i) = r$, we see that

$$r = N(\xi) = N(\epsilon\xi_i) = N(\epsilon)N(\xi_i) = N(\epsilon)r.$$

Thus, if $r \neq 0$, we see that $N(\epsilon) = 1$. Thus, we have the following result:

Theorem 3: Let M be a module and let r be a nonzero rational number. Suppose that $\xi_1, \ldots, \xi_t$ is a complete set of nonassociated elements of M having norm r. If ξ is any element of M of norm r, then there exists a unit ϵ of $\mathcal{O}_M$ of norm 1 and an i such that $\xi = \epsilon\xi_i$. That is, the set of all ξ in M such that $N(\xi) = r$ is precisely the set of all $\epsilon\xi_i$, where $1 \leq i \leq t$ and ϵ is a unit of $\mathcal{O}_M$ of norm 1.

We now turn to the solution of Problem 1-B.

Lemma 4: Let ξ be an element of M such that $N(\xi) = r$. Then there is an associate ξ_1 of ξ satisfying

$$1 \leq \xi_1 < \epsilon_{\mathcal{f}} \quad \text{and} \quad \frac{|r|}{\epsilon_{\mathcal{f}}} < |\xi_1'| \leq |r|,$$

where $\epsilon_{\mathcal{f}}$ is the fundamental unit of $\mathcal{O}_M = \{1, \mathcal{f}\omega_d\}$.

Proof: Since $N(\xi) = r$, we have $N(-\xi) = r$. And since an associate of $-\xi$ is also an associate of ξ, we may assume, without loss of generality, that $\xi > 0$. Since $\epsilon_{\mathcal{f}} > 1$, we see that $\epsilon_{\mathcal{f}}^n \longrightarrow \infty$ as $n \longrightarrow \infty$ and $\epsilon_{\mathcal{f}}^n \longrightarrow 0$ as $n \longrightarrow -\infty$. Thus, there is a rational integer n such that

$$\epsilon_{\mathcal{f}}^n \leq \xi < \epsilon_{\mathcal{f}}^{n+1}.$$

Equivalently,

$$1 \leq \xi\epsilon_{\mathcal{f}}^{-n} < \epsilon_{\mathcal{f}}.$$

Set $\xi_1 = \xi\epsilon_{\mathcal{f}}^{-n}$. Then ξ_1 is an associate of ξ and

$$1 \leq \xi_1 < \epsilon_{\mathcal{f}}. \tag{1}$$

Moreover, since

$$r = N(\xi_1) = \xi_1 \xi_1',$$

we see that

$$\xi_1' = \frac{r}{\xi_1},$$

so that from Eq. (1),

$$\frac{|r|}{\epsilon_{\mathcal{L}}} < |\xi_1'| \leq |r|. \qquad \blacksquare$$

Suppose that ξ_1 is the associate of ξ constructed in Lemma 4, and suppose that $\xi_1 = a + \sqrt{d}\,b$, where a and b are rational numbers. Then

$$a = \frac{\xi_1 + \xi_1'}{2}, \qquad b = \frac{\xi_1 - \xi_1'}{2\sqrt{d}},$$

so that

$$|a| \leq \frac{1}{2}(|\xi_1| + |\xi_1'|) \leq \frac{1}{2}(\epsilon_{\mathcal{L}} + |r|), \tag{2}$$

$$|b| \leq \frac{1}{2\sqrt{d}}(\epsilon_{\mathcal{L}} + |r|). \tag{3}$$

Note that a and b are rational numbers. What can we say about their denominators? Quite a good deal, as the following lemma shows:

Lemma 5: Let M be a module. Then there is a positive rational integer g such that for all λ in M, we have $g\lambda = a + \sqrt{d}\,b$, where a and b are rational integers.

Proof: Exercise. See Theorem 5.6, (iii). ■

Let g be the rational integer of Lemma 5 associated to the module M we have been considering. Then a and b are rational numbers with denominators dividing g (since $\xi_1 = a + \sqrt{d}\,b$ belongs to M). Therefore, a is one of the rational numbers

$$0, \pm\frac{1}{g}, \pm\frac{2}{g}, \ldots, \pm\frac{P}{g},$$

where P is the largest integer less than $g(\epsilon_{\mathcal{L}} + |r|)/2$. Similarly, b is one of the rational numbers

$$0, \pm\frac{1}{g}, \pm\frac{2}{g}, \ldots, \pm\frac{Q}{g},$$

where Q is the largest rational integer less than $g(\epsilon_{\mathcal{L}} + |r|)/2\sqrt{d}$. Note that these two sets of rational numbers are finite and depend only on r and the module M. In particular, we see that there are only finitely many choices for ξ_1 and that these choices can be enumerated in a simple way. To find which of the choices for ξ_1 have norm r, it suffices to test each of the finite number of choices. We shall give some examples of such calculations below. But first, let us summarize our conclusions.

Theorem 6: Let M be a given module and let r be a given rational number. Then a complete set of nonassociated elements of M is finite.* In other words, there exists a finite set of elements $\{\xi_1, \ldots, \xi_t\}$ of M such that (i) $N(\xi_i) = r$ $(1 \leq i \leq t)$ and (ii) every element ξ of M such that $N(\xi) = r$ is an associate of some ξ_i and (iii) no two ξ_i are associates.

The calculation we have outlined above is a viable solution to Problem 1-B. Let us illustrate this by means of two examples.

Example 7: Let $M = I_2 = \{1, \sqrt{2}\}$, and let us determine a complete set of nonassociated elements of M of norm 7. It is easy to see that $\epsilon_1 = 1 + \sqrt{2}$ is the fundamental unit of $\mathfrak{O}_M = I_2$. Moreover, since every element of M is of the form $x + \sqrt{2}y$, x, y rational integers, we may take $g = 1$ in Lemma 5. Thus, we may select a complete set of nonassociated elements of I_2 from among the elements $\xi_1 = a + \sqrt{2}b$, where from Eqs. (2) and (3)

$$|a| \leq \frac{1}{2}(\epsilon_1 + 7) = \frac{8 + \sqrt{2}}{2} = 4.707\ldots$$

$$|b| \leq \frac{1}{2\sqrt{2}}(\epsilon_1 + 7) = \frac{1}{2\sqrt{2}}(8 + \sqrt{2}) = 3.328\ldots.$$

Thus, $a = 0, \pm 1, \pm 2, \pm 3, \pm 4$ and $b = 0, \pm 1, \pm 2, \pm 3$, giving rise to a list of 63 elements: $0, \pm\sqrt{2}, \pm 2\sqrt{2}, \pm 3\sqrt{2}, \pm 1, \pm 1 \pm \sqrt{2}, \pm 1 \pm 2\sqrt{2}, \pm 1 \pm 3\sqrt{2}, \pm 2, \pm 2 \pm \sqrt{2}, \pm 2 \pm 2\sqrt{2}, \pm 2 \pm 3\sqrt{2}, \pm 3, \pm 3 \pm \sqrt{2}, \pm 3 \pm 2\sqrt{2}, \pm 3 \pm 3\sqrt{2}, \pm 4, \pm 4 \pm \sqrt{2}, \pm 4 \pm 2\sqrt{2}, \pm 4 \pm 3\sqrt{2}$. These have norms 0, -2, -8, -18, 1, -1, -7, -17, 4, 2, -4, -14, 9, 7, 1, -9, 16, 14, 8, -2. Thus, every element of norm 7 is an associate of one of $\pm 3 \pm \sqrt{2}$. But $-3 - \sqrt{2} = (-1)\cdot(3 + \sqrt{2})$ and $-3 + \sqrt{2} = (-1)(3 - \sqrt{2})$, so that $-3 - \sqrt{2}$ and $3 + \sqrt{2}$ are associates and $-3 + \sqrt{2}$ and $3 - \sqrt{2}$ are associates. Thus, every element of M of norm 7 is an associate of $3 + \sqrt{2}$ or $3 - \sqrt{2}$. Moreover, $3 + \sqrt{2}$ and $3 - \sqrt{2}$ are not associates, for if they were,

$$\frac{3 + \sqrt{2}}{3 - \sqrt{2}} = \frac{11}{7} + \frac{6}{7}\sqrt{2}$$

would be a unit of $\mathfrak{O}_M = \{1, \sqrt{2}\}$, whereas it does not even belong to $\mathfrak{O}_M$. Thus, every element of M of norm 7 is an associate of precisely one of $3 + \sqrt{2}, 3 - \sqrt{2}$. Finally, since $N(\epsilon_1) = N(1 + \sqrt{2}) = -1$, we see that the units of $\mathfrak{O}_M$ of norm $+1$ are precisely the numbers $\pm\epsilon_1^{2n}$ $(n = 0, \pm 1, \pm 2, \ldots)$. Thus, the elements ξ of M of norm 7 are precisely given by

$$\pm(1 + \sqrt{2})^{2n}(3 + \sqrt{2}), \qquad \pm(1 + \sqrt{2})^{2n}(3 - \sqrt{2})$$
$$(n = 0, \pm 1, \pm 2, \ldots).$$

Now that we have, at least in principle, solved the problem of computing all elements of given norm in a given module, let us return to the problem

*For given r, this set may be empty.

which initiated our discussion of modules, namely the problem of computing all solutions in rational integers x, y of the Diophantine equation

$$ax^2 + bxy + cy^2 = m, \tag{4}$$

where a, b, c, m are given rational integers. As we saw in Section 4, we can reduce all considerations to the case where $D = b^2 - 4ac$ is not a perfect square. Write $D = \mathcal{f}^2 d$, where $\mathcal{f}$, d are rational integers and d is square-free. Then, Eq. (4) is equivalent to

$$N\left(ax + \frac{b + \mathcal{f}\sqrt{d}}{2}y\right) = am.$$

Thus, we must find all elements of norm am in the module $M = \{a, (b + \mathcal{f}\sqrt{d})/2\}$. Since we computed the coefficient ring of this M in Theorem 6.14, we may directly apply the above method.

Example 8: Let us put together all that we have learned above in order to determine all solutions x, y to the Diophantine equation

$$x^2 + 3xy - 5y^2 = 65. \tag{5}$$

In this case $D = 3^2 + 4 \cdot 5 = 29$, $d = 29$, $\mathcal{f} = 1$. Moreover, by Theorem 6.14 we have

$$M = \left\{1, \frac{3 + \sqrt{29}}{2}\right\} = \left\{1, \frac{1 + \sqrt{29}}{2}\right\}$$

$$\mathcal{O}_M = \left\{1, \frac{1 + \sqrt{29}}{2}\right\} = I_{29}.$$

By the algorithm presented in Section 7, we have $\epsilon_1 =$ the fundamental unit of $I_{29} = 2 + \omega_{29} = (5 + \sqrt{29})/2$. We must determine all elements ξ of M of norm 65. If we set $\xi = a + b\sqrt{29}$, where a and b are rational numbers, then as in Lemma 5, we may take a and b to have denominator 2. Thus, if $N(\xi) = 65$ and $a = a_1/2$, $b = b_1/2$ with a_1, b_1 rational integers, we see that a_1 and b_1 satisfy $a_1^2 - 29b_1^2 = 4 \cdot 65 = 260$. Moreover, possibly replacing ξ with an associate, by Eqs. (2) and (3) we see that

$$|a| \leq \frac{1}{2}\left(\frac{5 + \sqrt{29}}{2} + 65\right) = 35.10\ldots$$

$$|b| \leq \frac{1}{2\sqrt{29}}\left(\frac{5 + \sqrt{29}}{2} + 65\right) = 6.52\ldots.$$

Thus, we must have

$$|a_1| \leq 70$$

$$|b_1| \leq 13.$$

But computing $260 + 29b_1^2$ for $b_1 = 0, 1, \ldots, 13$ and determining which are perfect squares, we find that the only choices for a_1 and b_1 are

$$(a_1, b_1) = (\pm 17, \pm 1), (\pm 41, \pm 7), (\pm 46, \pm 8),$$

so that if ξ belongs to M and $N(\xi) = 65$, then* ξ is an associate of one of $(\pm 17 \pm \sqrt{29})/2$, $(\pm 41 \pm 7\sqrt{29})/2$, or $(\pm 23 \pm 4\sqrt{29})$. But $(17 + \sqrt{29})/2$ and $-(17 + \sqrt{29})/2$ are associates, as are $(17 - \sqrt{29})/2$ and $-(17 - \sqrt{29})/2$, etc. Moreover,

$$\frac{23 + 4\sqrt{29}}{(41 - 7\sqrt{29})/2} = \frac{27 + 5\sqrt{29}}{2} = \epsilon_1^2,$$

and thus $23 + 4\sqrt{29}$ and $(41 - 7\sqrt{29})/2$ are associates. Similarly, $23 - 4\sqrt{29}$ and $(41 + 7\sqrt{29})/2$ are associates. Thus, if ξ belongs to M and $N(\xi) = 65$, then ξ is an associate of one of $(17 \pm \sqrt{29})/2$, $(41 \pm 7\sqrt{29})/2$. None of these four numbers are associates. For example,

$$\frac{(17 + \sqrt{29})/2}{(17 - \sqrt{29})/2} = \frac{(17 + \sqrt{29})^2}{N(17 - \sqrt{29})} = \frac{318 + 34\sqrt{29}}{260}$$

is not in $\mathfrak{O}_M$, and so $(17 + \sqrt{29})/2$ and $(17 - \sqrt{29})/2$ are not associates. Our factorization of (5) was

$$x^2 + 3xy - 5y^2 = N\left(x + \frac{3 + \sqrt{29}}{2}y\right).$$

Since

$$\frac{17 + \sqrt{29}}{2} = 7 + \frac{3 + \sqrt{29}}{2}, \quad \frac{17 - \sqrt{29}}{2} = 10 - \frac{3 + \sqrt{29}}{2},$$

$$\frac{41 + 7\sqrt{29}}{2} = 10 + 7\frac{3 + \sqrt{29}}{2}, \quad \frac{41 - 7\sqrt{29}}{2} = 31 - 7\frac{3 + \sqrt{29}}{2},$$

a complete set of nonassociate solutions to the Diophantine equation (5) is given by

$$(x, y) = (7, 1), \quad (10, -1), \quad (10, 7), \quad (31, -7).$$

Since $\epsilon_1 = (5 + \sqrt{29})/2$ and $N(\epsilon_1) = -1$, we see that the elements of M having norm 65 are just given by

$$\pm\left(\frac{17 \pm \sqrt{29}}{2}\right)\left(\frac{5 + \sqrt{29}}{2}\right)^{2n}, \qquad \pm\left(\frac{41 \pm 7\sqrt{29}}{2}\right)\left(\frac{5 + \sqrt{29}}{2}\right)^{2n}$$

for $n = 0, \pm 1, \pm 2, \ldots$, and all possible combinations of signs, Thus, we see that if x_0, y_0 is one solution of Eq. (5), then other solutions x, y may be computed from

$$\pm\left(x_0 + y_0\frac{3 + \sqrt{29}}{2}\right)\epsilon_1^2 = x + y\frac{3 + \sqrt{29}}{2} \tag{*}$$

$$\pm\left(x_0 + y_0\frac{3 + \sqrt{29}}{2}\right)\epsilon_1^{-2} = x + y\frac{3 + \sqrt{29}}{2}. \tag{**}$$

---

*Note that the numbers listed belong to M. It is possible in some examples to arrive at (a_1, b_1), which gives rise to ξ not belonging to M.

Moreover, starting from the solutions $(x_0, y_0) = (7, 1)$, $(10, -1)$, $(10, 7)$, $(31, -7)$, all solutions may be computed from these formulas. Since $\epsilon_1^2 = (27 + 5\sqrt{29})/2$ and $\epsilon_1^{-2} = (27 - 5\sqrt{29})/2$, we see that Eqs. (*) and (**) may be written

$$x + y\frac{3 + \sqrt{29}}{2} = \pm\left(6x_0 + 25y_0 + (5x_0 + 21y_0)\frac{3 + \sqrt{29}}{2}\right)$$

$$x + y\frac{3 + \sqrt{29}}{2} = \pm\left(21x_0 - 25y_0 + (6y_0 - 5x_0)\frac{3 + \sqrt{29}}{2}\right).$$

Thus,

$$x = \pm(6x_0 + 25y_0)$$
$$y = \pm(5x_0 + 21y_0)$$

or

$$x = \pm(21x_0 - 25y_0)$$
$$y = \pm(6y_0 - 5x_0).$$

By putting in $(x_0, y_0) = (7, 1)$, $(10, -1)$, $(10, 7)$, $(31, -7)$, we get two more solutions. Using, these, we get two more, and so forth. In this manner, we generate all solutions to our Diophantine equation (5).

8.8 Exercises

1. Determine whether the following elements of I_2 are associates:
   (a) $5 + 7\sqrt{2}$ and $-13 + 11\sqrt{2}$.
   (b) $3 + \sqrt{2}$ and $13 + 9\sqrt{2}$.
   (c) $8 + 15\sqrt{2}$ and 3.
   (d) $5 + \sqrt{2}$ and $5 - \sqrt{2}$.
2. Suppose that α_1 and α_2 are associates in I_d. Show that $|N(\alpha_1)| = |N(\alpha_2)|$.
3. Calculate a complete set of nonassociates of norm m in I_5, where
   (a) $m = 2$. (b) $m = 3$. (c) $m = 11$.
   (d) $m = 44$. (e) $m = 55$.
4. Determine the coefficient ring associated to the binary quadratic form $x^2 - xy + y^2$.
5. Determine all solutions of the following Diophantine equations:
   (a) $3x^2 + 3xy - 5y^2 = 55$.
   (b) $x^2 + xy + y^2 = 3$.
   (c) $x^2 + y^2 = 3$.
   (d) $x^2 - 3y^2 = -8$.

9

Factorization Theory in Quadratic Fields

9.1 The Failure of Unique Factorization

In Section 8.3, we constructed the integers of $\mathbf{Q}(\sqrt{d})$. As we have already indicated, we would like to build a theory of factorization for the integers of $\mathbf{Q}(\sqrt{d})$ which would yield information about Diophantine equations. We constructed such a theory for the Gaussian integers in Chapter 7. One might hope that a similar theory could be constructed for I_d. However, as we shall show in this section, any such hopes are far too optimistic.

Let us first recall the basic definitions from abstract algebra concerning factorization in the particular case of the integral domain I_d. If α and β belong to I_d, then we say that α *divides* β (denoted $\alpha \mid \beta$) provided that there exists γ in I_d such that $\beta = \alpha\gamma$. We call ϵ in I_d a *unit* provided that $1/\epsilon$ belongs to I_d. We say that α, β in I_d are *associates* if there is a unit ϵ of I_d such that $\alpha = \beta\epsilon$. We call α in I_d *irreducible** if α is a nonzero, nonunit and, whenever we have $\alpha = \beta\gamma$ with β, γ in I_d, one of β or γ must be a unit of I_d. We say that I_d is a *unique factorization domain* (UFD) if (1) every nonzero, nonunit α of I_d can be written as the product of irreducible elements of I_d and (2) if $\alpha = \pi_1 \cdots \pi_t = \theta_1 \cdots \theta_s$, where $\pi_1, \ldots, \pi_t, \theta_1, \ldots, \theta_s$, are irreducible, then $s = t$ and we can renumber $\pi_1, \ldots, \pi_t$, so that π_1 and θ_1 are associates, π_2 and θ_2 are associates, and so forth. Further, recall that an *ideal* of I_d is a subset A such that (1) if α, β are in A, then $\alpha \pm \beta$ are in A, and (2) if α is in A and θ is in I_d, then $\alpha\theta$ is in A.

---

*We prefer to use the term *irreducible* instead of *prime* for technical reasons (see Example 7).

A typical example of a UFD is the Gaussian integers, as we proved in Chapter 7. We shall show below that I_d is not always a UFD. However, it is true that every nonzero nonunit α of I_d is the product of irreducible elements. To prove this we need two lemmas.

Lemma 1: Let α, β belong to I_d and assume that $\alpha \mid \beta$. Then $N(\alpha) \mid N(\beta)$.

Proof: If $\alpha \mid \beta$, then $\beta = \alpha\gamma$ for some γ in I_d. But then $N(\beta) = N(\alpha\gamma) = N(\alpha)N(\gamma)$, so that $N(\alpha) \mid N(\beta)$ since $N(\alpha)$, $N(\beta)$, $N(\gamma)$ are all rational integers. ■

Lemma 2: Let α belong to I_d and assume that $|N(\alpha)| = p$, where p is a rational prime. Then α is irreducible.

Proof: Since $p \geq 2$, we see that α is nonzero and a nonunit by Lemma 8.7.2. If $\alpha = \beta\gamma$ with β, γ in I_d, then $p = |N(\alpha)| = |N(\beta\gamma)| = |N(\beta)||N(\gamma)|$. However, since p is a rational prime, we see that either $|N(\beta)| = 1$ or $|N(\gamma)| = 1$. Thus, by Lemma 8.7.2, either β or γ is a unit. ■

Theorem 3: Let α be a nonzero, nonunit of I_d. Then α can be written in the form $\alpha = \pi_1 \cdots \pi_t$, where $\pi_1, \ldots, \pi_t$ are irreducible elements of I_d.

Proof: Since $\alpha \neq 0$ is a nonunit, Lemma 8.7.2 implies that $|N(\alpha)| \geq 2$. If $|N(\alpha)| = 2$, then α is irreducible by Lemma 2, and we may take $t = 1$, $\pi_1 = \alpha$. Thus, assume that $|N(\alpha)| > 2$, and let us proceed by induction on $|N(\alpha)|$. If α is irreducible, we are done. Thus, assume that α is not irreducible. Then we may write $\alpha = \beta\gamma$, where neither β nor γ is a unit. Thus, by Lemma 8.7.2, $|N(\beta)| \geq 2, |N(\gamma)| \geq 2$. Therefore, since

$$|N(\alpha)| = |N(\beta)||N(\gamma)|,$$

we have

$$1 < |N(\beta)|, |N(\gamma)| < |N(\alpha)|.$$

Thus, by induction, β and γ can be written in the form

$$\beta = \theta_1 \cdots \theta_s$$

$$\gamma = \eta_1 \cdots \eta_u,$$

where $\theta_1, \ldots, \theta_s, \eta_1, \ldots, \eta_u$ are irreducible elements of I_d. Finally,

$$\alpha = \beta\gamma = \theta_1 \cdots \theta_s\eta_1 \cdots \eta_u$$

is an expression of α as a product of irreducible elements. ■

Let us now give an example to show that I_d is not usually a UFD.

Example 4: Let $d = -5$. Then I_d consists of all numbers of the form $x + y\sqrt{-5}$, x, y rational integers. In particular, 6 belongs to I_d. Let us factor 6. A moment's calculation shows that

$$6 = 3 \cdot 2 = (1 + \sqrt{-5})(1 - \sqrt{-5}). \tag{1}$$

We assert that 3, 2, $1 + \sqrt{-5}$, and $1 - \sqrt{-5}$ are all irreducible in I_d. For example, if $\alpha | 3$, then $N(\alpha) | N(3)$, and so $N(\alpha) | 9$ by Lemma 1. Moreover, if $\alpha = x + y\sqrt{-5}$, then $N(\alpha) = x^2 + 5y^2 \geq 0$, so that $N(\alpha)$ is one of 1, 3, or 9. If $N(\alpha) = 1$, then α is a unit by Lemma 8.7.2. If $N(\alpha) = 3$, then the Diophantine equation $x^2 + 5y^2 = 3$ has a solution in rational integers x, y, which it clearly does not. Finally, if $N(\alpha) = 9$, then $3/\alpha$ belongs to I_d (since $\alpha | 3$), and $N(3/\alpha) = N(3)/N(\alpha) = 1$, so that $3/\alpha$ is a unit. Thus, if $3 = \alpha\beta$, with α and β in I_{-5}, then either α is a unit or $\beta = 3/\alpha$ is a unit, and hence 3 is irreducible. Similar arguments may be given for 2, $1 + \sqrt{-5}$, and $1 - \sqrt{-5}$. Thus, Eq. (1) gives us two factorizations of 6 into products of irreducible elements. If factorization were unique, 3 would be an associate of either $1 + \sqrt{-5}$ or $1 - \sqrt{-5}$. But it is not, since $(1 + \sqrt{-5})/3$ and $(1 - \sqrt{-5})/3$ do not even belong to I_{-5}, let alone are they units. Thus, factorization is not unique in I_{-5}.

There are other pathologies, in addition to the failure of unique factorization, which can occur in I_d. We shall content ourselves with exhibiting two.

Definition 5: Given α and β in I_d, we call γ in I_d a *greatest common divisor* of α and β provided that (i) $\gamma | \alpha$ and $\gamma | \beta$, and (ii) if $\delta | \alpha$ and $\delta | \beta$, then $\delta | \gamma$. We write $\gamma = \gcd(\alpha, \beta)$ to mean that γ is a greatest common divisor of α and β.

Example 6: Let us show that two elements of I_d need not have a greatest common divisor. For example, let us again take $d = -5$ and let us set $\alpha = 9$, $\beta = 6 + 3\sqrt{-5}$. Let us prove that α and β have no greatest common divisor. Note that $N(\alpha) = 81 = N(\beta)$. Moreover, as we have seen in Example 4, all norms of elements in I_{-5} are ≥ 0. Therefore, if $\gamma | \alpha$ and $\gamma | \beta$, we must have $N(\gamma) = 1, 3, 9, 27$, or 81 (by Lemma 1). If $\gamma = x + y\sqrt{-5}$, then we must have $x^2 + 5y^2 = 1, 3, 9, 27$, or 81. These five Diophantine equations are trivially solved. The only solutions are

$$(x, y) = (\pm 1, 0), \quad (\pm 3, 0), \quad (\pm 2, \pm 1), \quad (\pm 9, 0), \quad (\pm 6, \pm 3), \quad (\pm 1, \pm 4),$$

where all possible combinations of the signs are permitted. The corresponding γ's are

$$\gamma = \pm 1, \quad \pm 3, \quad \pm 9, \quad \pm 2 \pm \sqrt{-5}, \quad \pm 6 \pm 3\sqrt{-5}, \quad \pm 1 \pm 4\sqrt{-5}.$$

Of these, the only ones which actually divide both α and β (calculate α/γ, β/γ and see whether they lie in I_{-5}) are

$$\gamma = \pm 1, \quad \pm 3, \quad \pm(2 + \sqrt{-5}),$$

so that these are the only common divisors of α and β. But none is a greatest common divisor, since ± 3 does not divide $\pm(2 + \sqrt{-5})$ and vice versa. (Do the division.) Thus, 9 and $6 + 3\sqrt{-5}$ do not have a greatest common divisor.

Example 7: In Chapter 2 of this book, we emphasized that the most basic property of rational primes was Euclid's lemma (Lemma 2.4.3), which says that if p is a prime and $p \mid ab$, then $p \mid a$ or $p \mid b$. Note, however, that the analogue of Euclid's lemma is generally false for the irreducible elements of I_d. For example, in Example 4 we showed that 3, 2, $1 + \sqrt{-5}$, and $1 - \sqrt{-5}$ are irreducible elements in I_{-5}. Moreover, by Eq. (1), we see that $3 \mid (1 + \sqrt{-5})(1 - \sqrt{-5})$, but clearly 3 does not divide either $1 + \sqrt{-5}$ or $1 - \sqrt{-5}$. (If $3 \mid 1 \pm \sqrt{-5}$, then $(1 \pm \sqrt{-5})/3$ would belong to I_{-5}, which is not true.) Thus, the analogue of Euclid's lemma is false in I_{-5}.

In Examples 4, 6, and 7, we seem to have destroyed any hope of building a theory of factorization for I_d. All the customary tools for factoring in the rational integers and the Gaussian integers have been taken from us. Not only is factorization not always unique, but numbers do not always have greatest common divisors, and Euclid's lemma is not always true. But is the situation totally hopeless? Is there nothing at all that we can say about factorization in I_d? Clearly we must look for some new concepts to replace the concepts of gcd and irreducible element. These new concepts, originally proposed by Kummer, will now be introduced.

Let us reexamine closely the two situations in which greatest common divisors were shown to exist. We showed that the greatest common divisor of two rational integers a and b (not both zero) exists by considering the set of all rational integers of the form $ax + by$, x, y rational integers (Theorem 2.3.3). Let us denote this set by (a,b). We showed that all elements of this set are multiples of a single positive element d. We then showed that d is a gcd of a and b.

The procedure for finding a gcd of two Gaussian integers α and β (not both zero) was similar to that used for the rational integers. We considered the set $(\alpha, \beta) = \{\alpha\theta + \beta\zeta \mid \theta \text{ and } \zeta \text{ are Gaussian integers}\}$. We then showed that (α, β) consists of all multiples of a single element γ, that is $(\alpha,\beta) = \gamma I_{-1}$.

We can try to repeat this procedure for the integral domain I_d. Let α and β be elements of I_d, not both zero, and let $(\alpha, \beta) = \{\alpha\theta + \beta\zeta \mid \theta, \zeta \text{ in } I_d\}$. This set will not generally consist of the multiples of a single element of I_d. Nevertheless, the set will serve as a replacement for the greatest common divisor of α and β, as we shall see later in this chapter. Let us first determine some properties of this set.

Proposition 8: (α, β) is a nonzero ideal of I_d.

Proof: Since $\alpha = \alpha \cdot 1 + \beta \cdot 0$ and $\beta = \alpha \cdot 0 + \beta \cdot 1$ belong to (α,β) and α and β are not both zero, we see that (α,β) contains nonzero elements. Let $\lambda = \alpha\theta_1 + \beta\zeta_1$, $\omega = \alpha\theta_2 + \beta\zeta_2$ belong to (α,β). Then $\lambda \pm \omega = \alpha(\theta_1 \pm \theta_2) + \beta(\zeta_1 \pm \zeta_2)$ belongs to (α, β). If δ is any element of I_d, then $\lambda\delta = \alpha(\delta\theta_1) + \beta(\delta\zeta_1)$ belongs to (α, β). Thus, (α, β) is an ideal of I_d. ■

We now note that $\{0\}$ is also an ideal of I_d. This trivial ideal will play no role in what follows. Therefore, we make the following

Convention: Henceforth, all ideals of I_d will be assumed to be nonzero.

Proposition 9: α and β have a gcd in I_d if and only if $(\alpha, \beta) = \gamma I_d$ for some γ in I_d. In this case, γ is a gcd of α, β.

Proof: If $(\alpha, \beta) = \gamma I_d$, then since α and β belong to (α,β) (see the proof of Proposition 8), we have that α and β are multiples of γ. If $\delta \mid \alpha$ and $\delta \mid \beta$, then $\alpha = \delta\theta$, $\beta = \delta\zeta$ for some θ, ζ in I_d. Since $\gamma = 1 \cdot \gamma$ belongs to $(\alpha,\beta) = \gamma I_d$, we have $\gamma = \alpha\eta + \beta\lambda$ for some η, λ in I_d. Thus, $\gamma = \delta\theta\eta + \delta\zeta\lambda = \delta(\theta\eta + \zeta\lambda)$, so that $\delta \mid \gamma$. Thus, γ is a gcd of α and β. We leave the proof of the converse as an exercise. ∎

Let us recall from abstract algebra that an ideal of the form γI_d is called a *principal ideal.* We have seen above that not every pair of elements α, β of I_d has a greatest common divisor. Therefore, by Proposition 9 not every ideal of I_d is principal. However, from Proposition 9 it is clear that there is a connection between factorization theory and ideal theory in the integral domain I_d. The ingenious idea of Kummer was to develop a factorization for the ideals of I_d instead of the numbers in I_d. To give some idea of what this means, let us look at the example of the Gaussian integers again. Before we can do this, let us recall the following easily proved fact concerning ideals:

Lemma 10: Let $A = \alpha I_d$, $B = \beta I_d$ be two principal ideals of I_d. Then $A = B$ if and only if α and β are associates.

Example 11: Let us first determine the ideals of I_{-1}. Let A be any ideal. Choose α in A such that $N(\alpha)$ is positive and as small as possible. Using the division algorithm in I_{-1} we may show that $A = \alpha I_{-1}$. *Thus, every ideal of I_{-1} is principal.* Let us *define* the product of two ideals αI_{-1}, βI_{-1} by $(\alpha I_{-1}) \cdot (\beta I_{-1}) = (\alpha\beta)I_{-1}$. (Check that this definition makes sense; that is, if $\alpha I_{-1} = \alpha_1 I_{-1}$, $\beta I_{-1} = \beta_1 I_{-1}$, then $(\alpha\beta)I_{-1} = (\alpha_1\beta_1)I_{-1}$.) Let $A = \alpha I_{-1}$ be any ideal. Lemma 10 implies that α is not a unit of I_{-1} if and only if $A \neq I_{-1}$. In case α is not a unit, we may factor α as $\alpha = \pi_1\pi_2 \cdots \pi_t$, where $\pi_1, \pi_2, \ldots, \pi_t$ are irreducible elements of I_{-1}. Set $P_i = \pi_i I_{-1}$ $(i = 1, 2, \ldots, t)$. It is easy to check that P_i is irreducible. That is, if $P_i = CD$ for ideals C, D, then either $C = I_{-1}$ or $D = I_{-1}$. Moreover, $A = P_1P_2 \cdots P_t$. Thus, we see that every ideal $\neq I_{-1}$ can be written as a product of irreducible ideals. Moreover, by Lemma 10 and the fact that I_{-1} is a unique factorization domain, we see that the factorization of A into a product of irreducible ideals is unique up to a rearrangement of factors. Thus, in the special case of I_{-1}, the unique factorization of ideals is a direct consequence of the unique factorization for elements.

Now we may outline our approach for the remainder of the chapter. We shall define the product of two arbitrary ideals of I_d. Since not every ideal is principal, the definition given in Example 11 will not suffice. The appropriate definition will be given in Section 9.3. We shall then develop the properties of this multiplication. In Section 9.4, we shall define the concept of an irreducible ideal. Finally, we shall prove that every ideal of I_d (other than I_d) can be written uniquely as a product of irreducible ideals.

One of our primary purposes in developing the theory of quadratic fields was to study the general quadratic Diophantine equation $ax^2 + bxy + cy^2 = t$. We saw in Chapter 8 that there is a natural connection between modules in $\mathbf{Q}(\sqrt{d})$ and these equations. On the other hand, we saw in Chapter 7 how the factorization theory in I_{-1} can be used to study the special quadratic Diophantine equation $x^2 + y^2 = t$. It turns out that it is possible to develop a general theory of factorization of modules, not just ideals. Such a general theory has important applications to the study of quadratic Diophantine equations. But what is the connection between ideals and modules? The answer is given in Corollary 14, which summarizes the next two results.

Theorem 12: Every ideal A of I_d is a module.

Proof: We must show that there are α, β in A such that every γ in A can be written uniquely in the form $\gamma = \alpha x + \beta y$, for rational integers x, y. This is not easy to carry out explicitly, and so we shall use the characterization of a module given in Theorem 8.5.6. Let γ be any nonzero element of A. Since $1, \sqrt{d}$ belong to I_d and form a basis of $\mathbf{Q}(\sqrt{d})$, we see that $\gamma, \gamma\sqrt{d}$ belong to A and are a basis of $\mathbf{Q}(\sqrt{d})$. This is hypothesis (i) of Theorem 8.5.6. Hypothesis (ii) of Theorem 8.5.6 is justified by the fact that A is an ideal. Finally, hypothesis (iii) is verified since $A \subseteq I_d$. Thus, by Theorem 8.5.6, A is a module. ∎

Proposition 13: Let A be an ideal of I_d. Then $\mathcal{O}_A = I_d$.

Proof: By Lemma 8.6.9, $\mathcal{O}_A \subseteq I_d$. However, if $\gamma \in I_d$, then $\gamma A \subseteq A$ (since A is an ideal). Thus, $\gamma \in \mathcal{O}_A$ and $I_d \subseteq \mathcal{O}_A$. ∎

Corollary 14: A subset A of I_d is an ideal of I_d if and only if A is a module and $\mathcal{O}_A = I_d$.

Theorem 12 and Corollary 14 suggest the following definition:

Definition 15: Let $\mathcal{O}$ be a ring of coefficients, and M a module. If M has $\mathcal{O}$ as its ring of coefficients, then we say that M *belongs to* $\mathcal{O}$. If $M \subseteq \mathcal{O}$ and M belongs to $\mathcal{O}$, then we say that M is an *integral module* (for $\mathcal{O}$). That is, M is an integral module if and only if $M \subseteq \mathcal{O}_M$.

In the rest of this chapter, we shall consider the following situation. Let $\mathcal{O}$ be a fixed coefficient ring. We shall consider the set of all modules belonging to $\mathcal{O}$. We shall develop a factorization theory for all integral modules belonging to $\mathcal{O}$. The modules play the role of *generalized rational numbers*, whereas the integral modules play the role of the *generalized integers*. Factoring principal modules (i.e., modules of the form $\gamma\mathcal{O}$) will allow us to draw conclusions about the factorization of individual elements γ of $\mathbf{Q}(\sqrt{d})$. In the special case $\mathcal{O} = I_d$, our factorization theory will concern the ideals of I_d. This general form of the theory is most convenient for applications to quadratic Diophantine equations.

9.1 Exercises

1. Determine which of the following statements is true:
   (a) $2 - \sqrt{-3} \mid 5 - 2\sqrt{-3}$. (b) $2 + \sqrt{2} \mid 16 + 13\sqrt{2}$.
2. Show that $2 + 3\sqrt{2}$ and $8 + 5\sqrt{2}$ are associates.
3. Which of the following elements of I_2 are irreducible?
   (a) $3 + 5\sqrt{2}$. (b) $1 + \sqrt{2}$.
   (c) $3 + 6\sqrt{2}$. (d) $36 - \sqrt{2}$.
4. Factor the following elements of I_{-5} into irreducible elements:
   (a) 17. (b) 6. (c) 35.
5. Show that the division algorithm holds in I_{-3}. That is, show that if α, β are integers of I_{-3}, $\beta \neq 0$, then there exist integers γ, δ of I_{-3} such that $\alpha = \beta\gamma + \delta, 0 \leq |N(\delta)| < |N(\beta)|$.
6. Use Exercise 5 to show that I_{-3} is a unique factorization domain.
7. Do Exercises 5 and 6 with I_{-3} replaced by I_3.
8. Show that the division algorithm does not hold in I_{-5}.
9. Give an example to demonstrate that I_{-6} is not a unique factorization domain.
10. Give an example of two elements of I_{-6} which do not have a greatest common divisor.
11. Give an example to show that the analogue of Euclid's lemma is false for I_{-6}.
12. Complete the proof of Proposition 9.
13. Prove Lemma 10.
14. Suppose that every ideal of I_d is principal. Show that I_d is a UFD.

15. Let $\alpha_1, \ldots, \alpha_n$ be nonzero elements of I_d. Let $(\alpha_1, \ldots, \alpha_n) = \{\alpha_1\beta_1 + \cdots + \alpha_n\beta_n \mid \beta_i \text{ belong to } I_d\}$. Show that $(\alpha_1, \ldots, \alpha_n)$ is an ideal of I_d.

16. Find bases for the following ideals of I_3:
    (a) (2). (b) $(3 + 5\sqrt{3})$. (c) $(2\sqrt{3})$. (d) $(1, 2\sqrt{3})$.
    (e) $(2, 3)$. (f) $(5, 10 + 5\sqrt{3}, \sqrt{3})$.

17. Which of the following ideals of I_3 are principal?
    (a) $(2, \sqrt{3})$. (b) $(5, 1 + \sqrt{3})$.
    (c) $(7, 7 + 2\sqrt{3})$. (d) $(3, 3\sqrt{3})$.

18. Compute the discriminants of each of the ideals of Exercise 16.

19. Suppose that α, β belong to I_d and that $\Delta(\alpha, \beta)$ is a square-free integer $\neq 0$. Prove that $(\alpha, \beta) = \{\alpha, \beta\}$.

20. Give an example of an ideal $\{\alpha, \beta\}$ such that $\Delta(\alpha, \beta)$ is *not* square-free.

21. Give an example of a module $M \subseteq I_d$ which is not an ideal of I_d.

9.2 Generalized Congruences and the Norm of a Module

In this section, we shall introduce the notion of the norm of a module, which is a rational integer measuring the "size" of a module. The notion of the norm of a module will be very important in deriving our theory of factorization.

Recall that in Chapter 3 we introduced the notion of a congruence modulo a positive rational integer n. Then the number of elements in a complete residue system modulo n was shown to be n. In this section, we shall generalize the notion of congruence by replacing n by a module M. We shall define the notion of a complete residue system modulo M. Then the norm of M will be defined as the number of elements in a complete residue system modulo M.

Let us begin by defining our generalized congruences. Let M and A be modules and assume that $M \subseteq A$. We say that two elements α and β of A are *congruent* to one another modulo* M, denoted $\alpha \equiv \beta (\text{mod } M)$, provided that $\alpha - \beta$ belongs to M. For example, if $M = \gamma I_d$, where γ is an element of I_d, and $A = \mathfrak{O}_M = I_d$, and if α, β belong to I_d, then $\alpha \equiv \beta (\text{mod } M)$ if and only if $\alpha - \beta$ is of the form $\gamma\delta$ for some δ in I_d. Thus, $\alpha \equiv \beta (\text{mod } M)$, in this case, is equivalent to $\gamma \mid \alpha - \beta$. Thus, our new notion of congruence generalizes what we are accustomed to think of as congruences.

Example 1: Let $M = \{4, 3\sqrt{5}\}$, $A = I_5$. Then $5 + 3\sqrt{5} \equiv -3 (\text{mod } M)$ since $5 + 3\sqrt{5} - (-3) = 4 \cdot 2 + 3\sqrt{5} \cdot 1$ belongs to M.

---

*Another way of viewing this is as follows: A and M are abelian groups with M a subgroup of A. Then α and β are congruent mod M provided that they are in the same coset of A modulo M. Thus, we are really considering A/M.

Let us construct a complete residue system modulo M in a manner similar to that used when we discussed congruences among rational integers.

Definition 2: Let M, A be modules such that $M \subseteq A$. A *complete residue system of A modulo M* is a set S of elements of A such that every element of A is congruent modulo M to precisely one element of S.

Let us give a concrete method for constructing a complete residue system of A modulo M. The key to our method is to construct a special basis for the module M.

Lemma 3: Let $A = \{\alpha, \beta\}$ be a module in $\mathbf{Q}(\sqrt{d})$ and let γ be any element of $\mathbf{Q}(\sqrt{d})$. Then there exists a rational integer $a > 0$ such that $a\gamma$ belongs to A.

Proof: α, β is a basis of $\mathbf{Q}(\sqrt{d})$, so that $\gamma = r\alpha + s\beta$ for rational numbers r, s. We may choose a to be a common denominator of r and s. ∎

Let A and M be modules, $M \subseteq A$, and suppose that $A = \{\alpha, \beta\}$. Let a be the least positive rational integer such that $a\alpha$ belongs to M (Lemma 3). Since $M \subseteq A$, every γ in M is of the form $\gamma = r\alpha + s\beta$ for rational integers r and s. There must be a γ in M for which $s \neq 0$ since M contains a basis of $\mathbf{Q}(\sqrt{d})$. Possibly replacing such a γ by $-\gamma$, we can find a γ in M such that $s > 0$. Among all elements $r\alpha + s\beta$ in M with $s > 0$, we may choose one with s least. The two numbers so constructed give a basis of M.

Theorem 4: Let A and M be modules, $M \subseteq A$, and let $A = \{\alpha, \beta\}$. Let $a > 0$ be the least rational integer such that $a\alpha$ is in M and let $b\alpha + c\beta$ be that element of M for which $c > 0$ and is as small as possible. Then $M = \{a\alpha, b\alpha + c\beta\}$. Moreover, we may always choose $b\alpha + c\beta$ so that $0 \leq b < a$.

Proof: First observe that the division algorithm for rational integers implies that if k is a rational integer such that $k\alpha$ belongs to M, then $a \mid k$. Similarly, observe that if $\gamma = x\alpha + y\beta$ belongs to M, then $c \mid y$. Let $\gamma = x\alpha + y\beta$ belong to M. Then $c \mid y$, say $y = sc$. Then

$$\gamma - s(b\alpha + c\beta) = (x - sb)\alpha$$

belongs to M. Therefore, $a \mid x - sb$, say $x - sb = ra$. Then

$$\gamma = r(a\alpha) + s(b\alpha + c\beta). \tag{1}$$

Conversely, any number of the form (1) belongs to M since $a\alpha$ and $b\alpha + c\beta$ belong to M. Thus, $M = \{a\alpha, b\alpha + c\beta\}$. Note that if we write $b = aq + t$, $0 \leq t < a$, then $M = \{a\alpha, (b\alpha + c\beta) - q(a\alpha)\}$ and $(b\alpha + c\beta) - q(a\alpha) = t\alpha + c\beta$, $0 \leq t < a$. Therefore, we may choose $b\alpha + c\beta$ so that $0 \leq b < a$. ∎

We shall apply Theorem 4 most frequently in the case M is an integral module and $A = \mathfrak{O}_M$ (for example, whenever M is an ideal of I_d). In this case we have the following result:

Corollary 5: Let M be a module contained in its coefficient ring $\mathfrak{O}_M = \{1, f\omega_d\}$. Let $a > 0$ be the least rational integer in M and let $b + cf\omega_d$ be an element of M for which $c > 0$ is as small as possible. Then $M = \{a, b + cf\omega_d\}$ and we may assume that $0 \leq b < a$.

We may now exhibit a complete system of residues of A modulo M.

Theorem 6: Let M, A be modules, $M \subseteq A$, $A = \{\alpha, \beta\}$ and let $M = \{a\alpha, b\alpha + c\beta\}$, where a, b, c are as in Theorem 4. Then a complete residue system of A modulo M is given by

$$x\alpha + y\beta, \qquad 0 \leq x < a, \quad 0 \leq y < c. \tag{2}$$

In particular, such a system contains ac elements.

Proof: All the numbers listed clearly belong to A. First we show that no two can be congruent modulo M: If $0 \leq x, x_1 < a$, $0 \leq y, y_1 < c$, and

$$x\alpha + y\beta \equiv x_1\alpha + y_1\beta (\text{mod } M),$$

then $(x - x_1)\alpha + (y - y_1)\beta$ belongs to M, so that

$$(x - x_1)\alpha + (y - y_1)\beta = ua\alpha + v(b\alpha + c\beta)$$

for some rational integers u, v. But then, $y - y_1 = vc$, $x - x_1 = ua + vb$. Since $0 \leq y, y_1 < c$, the first equation shows that $v = 0$ and $y = y_1$. Then the second equation shows that $x = x_1$ since $0 \leq x, x_1 < a$.

Now we show that any number in A is congruent to one of the numbers (2): If $\gamma = x_1\alpha + y_1\beta$ belongs to A, write $y_1 = qc + y$ with $0 \leq y < c$ and $x_1 - qb = pa + x$ with $0 \leq x < a$. Then $\gamma - (x\alpha + y\beta) = p(a\alpha) + q(b\alpha + c\beta)$ belongs to M and thus

$$\gamma \equiv x\alpha + y\beta (\text{mod } M). \qquad \blacksquare$$

Definition 7: Let M, A be modules, $M \subseteq A$. We define the *index of M in A*, denoted $(A:M)$, to be the number of elements in any complete residue system of A modulo M. If M is an integral module (that is, $M \subseteq \mathfrak{O}_M$), then we define the *norm of M*, denoted $N(M)$, to be the positive integer $(\mathfrak{O}_M:M)$.

We leave it to the reader to show that any two complete residue systems of A modulo M have the same number of elements, so that the index $(A:M)$ is well defined.*

Using Theorem 6, we can obtain a neat formula for $(A:M)$: Let $A = \{\alpha, \beta\}$, $M = \{a\alpha, b\alpha + c\beta\}$, where a, b, c are chosen as in Theorem 4. Then

*Of course, this number is just the number of cosets of M in A.

$$\Delta_A = \begin{vmatrix} \alpha & \alpha' \\ \beta & \beta' \end{vmatrix}^2 = (\alpha\beta' - \alpha'\beta)^2$$

$$\Delta_M = \begin{vmatrix} a\alpha & a\alpha' \\ b\alpha + c\beta & b\alpha' + c\beta' \end{vmatrix}^2$$
$$= (ac)^2(\alpha\beta' - \alpha'\beta)^2 = (A:M)^2\Delta_A,$$

by Theorem 6. Thus, we have derived the following formula:

Theorem 8: Let M, A be modules, $M \subseteq A$. Then the number of elements in a complete residue system of A modulo M is just

$$(A:M) = \sqrt{\frac{\Delta_M}{\Delta_A}}.$$

In particular, if M is an integral module, then

$$N(M) = \sqrt{\frac{\Delta_M}{\Delta_{\mathcal{O}_M}}}.$$

From Theorem 8, we get the following useful corollary:

Corollary 9: Let M, A, B be modules, $M \subseteq A \subseteq B$. Then

$$(B:M) = (B:A)(A:M).$$

Proof: Exercise. ■

Our theory of factorization will be based on the properties of the norm. It is very important for us to define the notion of the norm of a module for all modules, not only integral modules. Note that the second formula of Theorem 8 makes sense for any M. This suggests the following definition:

Definition 10: Let M be any module with coefficient ring $\mathcal{O}_M$. Set

$$N(M) = \sqrt{\frac{\Delta_M}{\Delta_{\mathcal{O}_M}}}.$$

Then $N(M)$ is called the *norm of M*.

Theorem 11: Let γ be a nonzero element of $\mathbf{Q}(\sqrt{d})$ and let $\mathcal{O}$ be a coefficient ring in $\mathbf{Q}(\sqrt{d})$. Then the norm of the module $\gamma\mathcal{O}$ is just $|N(\gamma)|$.

Proof: The coefficient ring of $\gamma\mathcal{O}$ is $\mathcal{O}$ (Lemma 8.6.12). Moreover, by Theorem 8.6.10, $\mathcal{O} = \{1, f\omega_d\}$ for some rational integer $f > 0$. Therefore, $\gamma\mathcal{O} = \{\gamma, \gamma f\omega_d\}$, so that

$$\Delta_{\gamma\mathcal{O}} = \begin{vmatrix} \gamma & \gamma' \\ \gamma f\omega_d & \gamma' f\omega_d' \end{vmatrix}^2$$
$$= (\gamma\gamma' f(\omega_d - \omega_d'))^2$$

$$\Delta_{\mathcal{O}} = \begin{vmatrix} 1 & 1 \\ f\omega_d & f\omega_d' \end{vmatrix}^2 = (f(\omega_d - \omega_d'))^2.$$

Therefore, $N(M) = (\Delta_{\gamma\mathcal{O}}/\Delta_{\mathcal{O}})^{1/2} = |\gamma\gamma'| = |N(\gamma)|$. ■

Example 12:

(i) Let $M = \{2 + 2\sqrt{3}, 6 + 2\sqrt{3}\} = (2 + 2\sqrt{3})I_3$. Then $N(M) = |N(2 + 2\sqrt{3})| = 8$. Since M is contained in $\mathfrak{O}_M = I_3$, we know from our above discussion that the number of elements of I_3 in a complete residue system modulo M is 8.

(ii) Let $M = \{\frac{1}{2}, 2 + \sqrt{3}\}$. Then, by Lemma 8.6.13, we have $\mathfrak{O}_M = \{1, 2\sqrt{3}\}$. Then, from Definition 10,

$$N(M) = \sqrt{\frac{\begin{vmatrix} \frac{1}{2} & \frac{1}{2} \\ 2 + \sqrt{3} & 2 - \sqrt{3} \end{vmatrix}^2}{\begin{vmatrix} 1 & 1 \\ 2\sqrt{3} & -2\sqrt{3} \end{vmatrix}^2}} = \frac{1}{4}.$$

Note that $N(M)$ is not an integer in this case. However, M is not contained in $\mathfrak{O}_M$.

Let us observe a very useful property of the index $(A : M)$.

Proposition 13: Let M, A be modules, $M \subseteq A$. Then $(A : M) = 1$ if and only if $M = A$. In particular, if M is an integral module, then $N(M) = 1$ if and only if $M = \mathfrak{O}_M$.

Proof: $(A : M) = 1$ if and only if a complete residue system of A modulo M contains one element, which is equivalent to saying that 0 is a complete residue system modulo M. Thus, $(A : M) = 1$ if and only if every γ in $\mathfrak{O}_M$ satisfies $\gamma \equiv 0 (\text{mod } M)$; that is, γ belongs to M. Thus, $(A : M) = 1$ if and only if $A = M$. ■

Throughout this section, we have made extensive use of the special basis for the module M which was described in Theorem 4. Note that it is fairly easy to find such a special basis for a given integral module M. Let us assume that $M \subseteq \mathfrak{O}_M = \{1, f\omega_d\}$ and that $M = \{\alpha, \beta\}$. Then we have $\alpha = a_0 + b_0 f\omega_d$, $\beta = c_0 + e_0 f\omega_d$ for rational integers a_0, b_0, c_0, e_0. Moreover, a typical element of M is of the form

$$\alpha x + \beta y = (a_0 x + c_0 y) + (b_0 x + e_0 y) f\omega_d$$

for rational integers x, y. We have seen in Chapter 2 that the least positive rational integer of the form $b_0 x + e_0 y$ is just $r = \gcd(b_0, e_0)$. Then, by determining r and solving the linear Diophantine equation $b_0 x + e_0 y = r$, we can find values x_0 and y_0 for x and y. In this way, we may determine the b and c of Corollary 5, namely $c = r$ and $b = a_0 x_0 + c_0 y_0$. Moreover, we may compute a using Theorem 6, since $ac = N(M)$, and thus

$$a = \frac{N(M)}{c} = \frac{1}{c}\sqrt{\frac{\Delta_M}{\Delta_{\mathfrak{O}_M}}}.$$

Then a basis for M is $\{a, b + cf\omega_d\}$.

9.2 Exercises

1. Determine whether the following congruences modulo $M = \{2, 1 + \sqrt{3}\}$ hold (here take $A = I_3$):
   (a) $3 \equiv 5 + \sqrt{3} \pmod{M}$.
   (b) $3 \equiv 5 - \sqrt{3} \pmod{M}$.
   (c) $17 + 2\sqrt{3} \equiv 13 + 8\sqrt{3} \pmod{M}$.
2. Determine $N(M)$ for the following modules M:
   (a) $M = \{2, \sqrt{3}\}$.
   (b) $M = \{1 + \sqrt{7}, 1 - \sqrt{7}\}$.
   (c) $M = \{11, 3 + 2\sqrt{2}\}$.
   (*Note:* In each example you must first determine $\mathfrak{O}_M$.)
3. Determine bases of the form $\{a, b + c\mathfrak{f}\omega_d\}$ for the following modules M:
   (a) $M = \{1 + \sqrt{7}, 1 - \sqrt{7}\}$.
   (b) $M = \{105 + 20\sqrt{6}, 137 + 25\sqrt{6}\}$.
   (*Note:* In each case, determine $\mathfrak{O}_M$ and observe that $M \subseteq \mathfrak{O}_M$.)
4. Prove Corollary 9.
5. Determine complete residue systems of $\mathfrak{O}_M$ modulo M for M one of the modules of Exercise 3.
6. Determine all integral modules M of norm 15 belonging to $\mathfrak{O} = \{1, 3\omega_7\}$.
7. Prove that any two complete residue systems have the same cardinality.
8. Let A be a given module. Prove that there exist only finitely many modules $M \subseteq A$ with a given index $(A:M)$. Conclude that there are only finitely many integral modules of given norm belonging to a given coefficient ring $\mathfrak{O}$.
9. Let $\mathfrak{O}$ be a given coefficient ring and let γ be in $\mathfrak{O}$. Find a complete residue system of $\gamma\mathfrak{O}$ in $\mathfrak{O}$.
10. Let $M \subseteq A$ be modules. Let $n = (A:M)$. Show that $nA \subseteq M$.

9.3 Products and Sums of Modules

We are now ready to define and prove the basic properties of products of modules.

Definition 1: Let M_1 and M_2 be two modules of $\mathbf{Q}(\sqrt{d})$. The *product* M_1M_2 of M_1 and M_2 is the set of all sums of the form

$$\alpha_1\beta_1 + \alpha_2\beta_2 + \cdots + \alpha_n\beta_n, \tag{1}$$

where $\alpha_1, \ldots, \alpha_n$, belong to M_1 and $\beta_1, \ldots, \beta_n$ belong to M_2.

Example 2: Let $M_1 = \gamma I_d$, $M_2 = \delta I_d$. Then it is clear that $M_1M_2 = (\gamma\delta)I_d$, so that multiplication of principal ideals corresponds to multiplication of elements of I_d.

Let us express M_1M_2 in terms of bases for M_1 and M_2.

Proposition 3: Let $M_1 = \{\alpha_1, \beta_1\}$, $M_2 = \{\alpha_2, \beta_2\}$ be modules in $\mathbf{Q}(\sqrt{d})$. Then M_1M_2 is the set of all numbers of $\mathbf{Q}(\sqrt{d})$ of the form

$$x_1\alpha_1\alpha_2 + x_2\alpha_1\beta_2 + x_3\alpha_2\beta_1 + x_4\beta_1\beta_2, \tag{2}$$

where x_1, x_2, x_3, x_4 are rational integers.

Proof: Exercise. ■

So far, we have not verified that the product of two modules is, again, a module. Let us rectify this omission now.

Corollary 4: Let M_1, M_2 be any modules of $\mathbf{Q}(\sqrt{d})$. Then M_1M_2 is a module in $\mathbf{Q}(\sqrt{d})$.

Proof: Let $M_1 = \{\alpha_1, \beta_1\}$, $M_2 = \{\alpha_2, \beta_2\}$, and let us apply Theorem 8.5.6. There are three conditions to check:

(i) M_1M_2 contains a basis of $\mathbf{Q}(\sqrt{d})$: By Proposition 3, M_1M_2 contains $\alpha_1\alpha_2$, $\beta_1\alpha_2$. And since $\alpha_2 \neq 0$, these two elements are a basis of $\mathbf{Q}(\sqrt{d})$.

(ii) The sum or difference of two elements of the form (2) is again of the form (2) and hence belongs to M_1M_2.

(iii) Let $n > 0$ be a rational integer such that $n\alpha_1, n\alpha_2, n\beta_1, n\beta_2$ belong to I_d (Lemma 2.3). From (2), we see that $n^2M_1M_2 \subseteq I_d$.

Thus, M_1M_2 is a module. ■

Example 5: Set $M_1 = \{2, 21\sqrt{5}\}$, $M_2 = \{6, 7\sqrt{5}\}$. Then M_1M_2 consists of all sums of the form

$$\begin{aligned} 12x_1 + 21\cdot 6\sqrt{5}\,x_2 + 2\cdot 7\sqrt{5}\,x_3 + 21\cdot 7\cdot 5x_4 \\ = (12x_1 + 21\cdot 7\cdot 5x_4) + (21\cdot 6x_2 + 2\cdot 7x_3)\sqrt{5}, \end{aligned}$$

where x_1, x_2, x_3, x_4 run over all rational integers. Since $\gcd(12, 21\cdot 35) = 3$, $12x_1 + 21\cdot 35x_4$ runs over all multiples of 3. (See the proof of the existence of a gcd.) Similarly, $21\cdot 6x_2 + 2\cdot 7x_3$ runs over all multiples of 14. Thus, $M_1M_2 = \{3, 14\sqrt{5}\}$.

Proposition 6: Let M, M_1, M_2, M_3 be modules of $\mathbf{Q}(\sqrt{d})$. Then

(i) $M_1M_2 = M_2M_1$.

(ii) $(M_1M_2)M_3 = M_1(M_2M_3)$.

(iii) $M\mathfrak{O}_M = \mathfrak{O}_M M = M$.

Proof: Parts (i) and (ii): Exercises.

Part (iii): $\mathfrak{O}_M M$ consists of all sums of the form $\alpha_1\beta_1 + \cdots + \alpha_n\beta_n$, where $\alpha_1, \ldots, \alpha_n$ belong to $\mathfrak{O}_M$ and $\beta_1, \ldots, \beta_n$ belong to M. From the definition of $\mathfrak{O}_M$, we have $\alpha_i\beta_i$ belongs to M $(1 \le i \le n)$, so that every element of $\mathfrak{O}_M M$ is in M. But 1 belongs to $\mathfrak{O}_M$, so that any $\alpha = 1 \cdot \alpha$ in M belongs to $\mathfrak{O}_M M$. Thus, $\mathfrak{O}_M M = M$. By part (i), $M\mathfrak{O}_M = M$. ■

Definition 7: Let M be a module of $\mathbf{Q}(\sqrt{d})$. The *conjugate module* M' of M is the set of all α' such that α is in M. Equivalently, if $M = \{\alpha, \beta\}$, then

$$M' = \{\alpha', \beta'\}.$$

We leave it to the reader to show that M' is a module having the same coefficient ring as M (i.e., $\mathfrak{O}_{M'} = \mathfrak{O}_M$).

Our whole subsequent development hinges on the following fundamental fact:

Theorem 8: Let M be a module, and $\mathfrak{O}_M$ its ring of coefficients. Then

$$MM' = N(M)\mathfrak{O}_M.$$

To prove Theorem 8, it is convenient to make the following

Definition 9: Let M_1, M_2 be modules of $\mathbf{Q}(\sqrt{d})$. We say that M_1 and M_2 are *similar* if there exists an $\alpha \neq 0$ in $\mathbf{Q}(\sqrt{d})$ such that $M_1 = \alpha M_2$.

Often, computations can be made simpler by replacing a module by one similar to it. Fortunately, in doing so, the coefficient ring does not change, as we have already observed in Lemma 8.6.12.

Proof of Theorem 8*:* Assume that we know Theorem 8 for modules of the form $\{1, \gamma\}$. Then we may deduce it in general as follows: Suppose that $M = \{\alpha, \beta\}$, $M_1 = \{1, \beta/\alpha\}$. Then $M = \alpha M_1$ so that M, M_1 are similar. By the assumed special case of Theorem 8, we have

$$M_1 M_1' = N(M_1)\mathfrak{O}_{M_1} = N(M_1)\mathfrak{O}_M,$$

since M, M_1 are similar. Therefore,

$$\begin{aligned} MM' &= \alpha M_1 \alpha' M_1' = \alpha\alpha' N(M_1)\mathfrak{O}_M \\ &= |N(\alpha)|\, N(M_1)\mathfrak{O}_M \qquad \text{(since } \pm 1 \text{ belongs to } \mathfrak{O}_M). \end{aligned} \tag{3}$$

However, by Definition 2.10

$$N(M) = \left(\frac{\Delta_M}{\Delta_{\mathfrak{O}_M}}\right)^{1/2}$$

and $\Delta_M = (\alpha\beta' - \beta\alpha')^2 = (\alpha\alpha')^2((\beta/\alpha) - (\beta'/\alpha'))^2 = N(\alpha)^2\Delta_{M_1}$, so that

$$\begin{aligned} N(M) &= |N(\alpha)| \left(\frac{\Delta_{M_1}}{\Delta_{\mathfrak{O}_M}}\right)^{1/2} \\ &= |N(\alpha)| \left(\frac{\Delta_{M_1}}{\Delta_{\mathfrak{O}_{M_1}}}\right)^{1/2} \qquad \text{(since } \mathfrak{O}_M = \mathfrak{O}_{M_1}) \\ &= |N(\alpha)|\, N(M_1). \end{aligned}$$

Thus, by (3), $MM' = N(M)\mathcal{O}_M$, and the general case of Theorem 8 is proved from the special case.

Let us now prove Theorem 8 for modules of the form $M = \{1, \gamma\}$. We know that γ satisfies an equation of the form

$$a\gamma^2 + b\gamma + c = 0, \tag{4}$$

where a, b, c are rational integers with no common factor > 1 and $a > 0$. Now $M' = \{1, \gamma'\}$, and hence MM' coincides with the set of all sums of the form

$$x + y\gamma + z\gamma' + w\gamma\gamma' = x + y\gamma + z\left(-\gamma - \frac{b}{a}\right) + w\left(\frac{c}{a}\right),$$

where x, y, z, w are rational integers (Proposition 3) and where a, b, c are defined in Eq. (4). Thus, MM' consists of all numbers of the form

$$\frac{1}{a}(ax - bz + cw + ua\gamma),$$

where $u = y - z$. As x, z, w run over all rational integers, so does $ax - bz + cw$, since a, b, c have no common factor > 1. Thus, $MM' = (1/a)\{1, a\gamma\} = (1/a)\mathcal{O}_M$ by Lemma 8.6.13. Finally,

$$N(M) = \left(\frac{\Delta_M}{\Delta_{\mathcal{O}_M}}\right)^{1/2} = \left(\frac{\begin{vmatrix} 1 & 1 \\ \gamma & \gamma' \end{vmatrix}^2}{\begin{vmatrix} 1 & 1 \\ a\gamma & a\gamma' \end{vmatrix}^2}\right)^{1/2} = \left(\frac{(\gamma - \gamma')^2}{a^2(\gamma - \gamma')^2}\right)^{1/2} = \frac{1}{a}.$$

Thus, $MM' = N(M)\mathcal{O}_M$, and we are done. ■

Let us observe that in the course of the above proof, we established the following useful fact:

Corollary 10: Let $M = \{1, \gamma\}$ be a module and suppose that γ satisfies the quadratic equation $a\gamma^2 + b\gamma + c = 0$, where a, b, c are rational integers with no common factor > 1 and $a > 0$. Then $N(M) = 1/a$.

Let us draw some more or less immediate consequences of Theorem 8.

Corollary 11: If M_1 and M_2 both belong to the coefficient ring $\mathcal{O}$, then so does M_1M_2.

Proof: Suppose that M_1M_2 has coefficient ring $\mathcal{O}^*$. Then by Theorem 8, we have

$$(M_1M_2)(M_1M_2)' = N(M_1M_2)\mathcal{O}^*.$$

However, it is easy to see that $(M_1M_2)' = M'_1M'_2$ (exercise), so that

$$\begin{aligned}(M_1M_2)(M_1M_2)' &= (M_1M'_1)(M_2M'_2)\\ &= N(M_1)\mathfrak{O}N(M_2)\mathfrak{O}\\ &= N(M_1)N(M_2)\mathfrak{O}.\end{aligned}$$

Thus, we see that $\mathfrak{O} = \alpha\mathfrak{O}^*$, where $\alpha = N(M_1M_2)/N(M_1)N(M_2)$ belongs to $\mathbf{Q}(\sqrt{d})$. Thus, $\mathfrak{O}$ and $\mathfrak{O}^*$ are similar. But $\mathfrak{O} = \{1, \mathcal{f}\omega_d\}$, $\mathfrak{O}^* = \{1, \mathcal{f}^*\omega_d\}$ for appropriate rational integers $\mathcal{f}$, $\mathcal{f}^*$, and it is easy to see that if $\mathcal{f} \neq \mathcal{f}^*$, then $\mathfrak{O}$ and $\mathfrak{O}^*$ cannot be similar (exercise). Thus, $\mathcal{f} = \mathcal{f}^*$ and $\mathfrak{O} = \mathfrak{O}^*$. ■

Corollary 12: Let M_1 and M_2 belong to the same coefficient ring $\mathfrak{O}$. Then $N(M_1M_2) = N(M_1)N(M_2)$.

Proof: From the proof of the preceding result, we have that

$$N(M_1M_2)\mathfrak{O} = N(M_1)N(M_2)\mathfrak{O}.$$

Thus, $\mathfrak{O} = r\mathfrak{O}$, where $r = N(M_1)N(M_2)/N(M_1M_2)$ is a rational number. Write $\mathfrak{O} = \{1, \mathcal{f}\omega_d\}$ for some rational integer $\mathcal{f}$. Then, since $r\cdot 1 = r$ is in $\mathfrak{O}$, we see that $r = a + b\mathcal{f}\omega_d$ for some rational integers a, b. Since 1, $\mathcal{f}\omega_d$ is a basis of $\mathbf{Q}(\sqrt{d})$, we see that $b = 0$ and $r = a$, and so r is a rational integer. Similarly, $\mathfrak{O} = (1/r)\mathfrak{O}$ implies that $1/r$ is a rational integer. Thus, since both r and $1/r$ are rational integers, we have $r = \pm 1$. But $r > 0$, so that $r = 1$ and $N(M_1M_2) = N(M_1)N(M_2)$. ■

Remark 13: It is easy to find a basis for the product of two modules M_1 and M_2 belonging to the same ring of coefficients $\mathfrak{O} = \{1, \mathcal{f}\omega_d\}$. Without loss of generality,* assume that M_1 and M_2 are integral and thus, by Corollary 2.5, can be written in the form $M_i = \{a_i, b_i + c_i\mathcal{f}\omega_d\}$ $(i = 1, 2)$, a_i, b_i, c_i rational integers. In Section 2, we described a procedure for writing a module basis in this form. Let us suppose that $M_1M_2 = \{a, b + c\mathcal{f}\omega_d\}$. Our purpose now is to give a procedure for determining a, b, c. By the definition of the product of modules, M_1M_2 consists of all the numbers of the form

$$xa_1a_2 + ya_1(b_2 + c_2\mathcal{f}\omega_d) + za_2(b_1 + c_1\mathcal{f}\omega_d) + w(b_1 + c_1\mathcal{f}\omega_d)(b_2 + c_2\mathcal{f}\omega_d),$$

where x, y, z, w are rational integers. Assume, for the sake of simplicity, that $d \equiv 2$ or $3 \pmod 4$, so that $\omega_d = \sqrt{d}$. Then the typical element of M_1M_2 is just

$$(xa_1a_2 + ya_1b_2 + za_2b_1 + wb_1b_2 + wc_1c_2\mathcal{f}^2d) + (ya_1c_2 + za_2c_1 + w(b_1c_2 + c_1b_2))\mathcal{f}\omega_d,$$

where x, y, z, w are rational integers. By Corollary 2.5, we may calculate c as the smallest positive integer of the form $ya_1c_2 + za_2c_1 + w(b_1c_2 + c_1b_2)$. But

*For if d_i are rational integers such that $d_iM_i \subseteq \mathfrak{O}$, then $(d_1M_1)(d_2M_2) = \{a, b + c\mathcal{f}\omega_d\}$ implies that $M_1M_2 = \{ad_1^{-1}d_2^{-1}, bd_1^{-1}d_2^{-1} + cd_1^{-1}d_2^{-1}\mathcal{f}\omega_d\}$.

this integer is just the greatest common divisor of a_1c_2, a_2c_1, $b_1c_2 + c_1b_2$, which can be calculated using, say, the fundamental theorem of arithmetic or the Euclidean algorithm twice. Thus, c can be calculated. Moreover, y, z, w can be calculated as solutions to the linear Diophantine equation

$$ya_1c_2 + za_2c_1 + w(b_1c_2 + c_1b_2) = c.$$

(see Exercise 16 in Section 2.3). Therefore, we may calculate b from

$$b = xa_1a_2 + ya_1b_2 + za_2b_1 + wb_1b_2 + wc_1c_2\mathcal{f}^2d,$$

where x can be taken to be any rational integer. Finally, since $N(M_1M_2) = ac$ by Theorem 2.6, we may compute a from

$$a = \frac{1}{c}N(M_1M_2) = \frac{1}{c}N(M_1)N(M_2)$$

$$= \frac{1}{c}\left(\frac{\Delta_{M_1}}{\Delta_{\Theta}}\right)^{1/2}\left(\frac{\Delta_{M_2}}{\Delta_{\Theta}}\right)^{1/2}$$

and

$$\Delta_{M_i} = \begin{vmatrix} a_i & a_i \\ b_i + c_i\mathcal{f}\omega_d & b_i + c_i\mathcal{f}\omega_d' \end{vmatrix}^2, \qquad \Delta_{\Theta} = \begin{vmatrix} 1 & 1 \\ \mathcal{f}\omega_d & \mathcal{f}\omega_d' \end{vmatrix}^2.$$

For example, let $M_1 = \{2, 1 + \sqrt{7}\}$, $M_2 = \{3, 1 + \sqrt{7}\}$, both modules belonging to the coefficient ring I_7 (exercise). In this case $a_1 = 2$, $b_1 = 1$, $c_1 = 1$, $a_2 = 3$, $b_2 = 1$, $c_2 = 1$. We immediately see that $c = \gcd(2, 3, 2) = 1$. Therefore, we may set $x = 0$, $y = -1$, $z = 1$, $w = 0$. Thus, an immediate calculation shows that $b = 1$. Since $N(M_1) = 2$, $N(M_2) = 3$, we know that $N(M_1M_2) = 6$, so that $a = c^{-1}N(M_1M_2) = 6$. Thus, $M_1M_2 = \{6, 1 + \sqrt{7}\}$.

In the above discussion, we developed the properties of multiplication of modules. Let us now introduce the sum of two modules and develop some of the fundamental properties of this operation.

Definition 14: Let M_1 and M_2 be modules in $\mathbf{Q}(\sqrt{d})$. The *sum* $M_1 + M_2$ is the set of all sums $\alpha + \beta$, where α belongs to M_1 and β belongs to M_2.

Example 15: Let $M_1 = \{2, 3\sqrt{7}\}$, $M_2 = \{3, 6\sqrt{7}\}$. Then $M_1 + M_2$ consists of all sums of the form

$$2x_1 + 3\sqrt{7}x_2 + 3x_3 + 6\sqrt{7}x_4 = (2x_1 + 3x_3) + (3x_2 + 6x_4)\sqrt{7},$$

x_1, x_2, x_3, x_4 rational integers. But since $\gcd(2, 3) = 1$, $\gcd(3, 6) = 3$, we see that $2x_1 + 3x_3$ runs over all integers and that $3x_2 + 6x_4$ runs over all multiples of 3, so that $M_1 + M_2 = \{1, 3\sqrt{7}\}$.

It will turn out that the sum of M_1 and M_2 is usually their greatest common divisor (see Section 4). Therefore, it is a good idea to establish some of the properties of addition of modules.

Theorem 16: Let M, M_1, M_2, M_3 be modules in $\mathbf{Q}(\sqrt{d})$.

(i) $M_1 + M_2$ is a module in $\mathbf{Q}(\sqrt{d})$.
(ii) $M_1 \subseteq M_1 + M_2$, $M_2 \subseteq M_1 + M_2$.
(iii) $M_1 + M_2 = M_2 + M_1$.
(iv) $M_1 + (M_2 + M_3) = (M_1 + M_2) + M_3$.
(v) $M(M_1 + M_2) = MM_1 + MM_2$.
(vi) If M_1 and M_2 belong to $\mathcal{O}$, then $M_1 + M_2$ has coefficient ring containing $\mathcal{O}$.
(vii) If M_1 and M_2 belong to I_d, then so does $M_1 + M_2$ (i.e., if M_1 and M_2 are ideals of I_d, then so is $M_1 + M_2$).

Proof: Parts (ii), (iii), and (iv) are completely trivial.

Part (i): We may verify that the conditions of Theorem 8.5.6 are satisfied. The details are similar to the arguments given in Corollary 4 and are left to the reader.

Part (v): Exercise.

Part (vi): Let γ belong to $\mathcal{O}$. Then $\gamma M_1 \subseteq M_1$, $\gamma M_2 \subseteq M_2$. Let $\alpha + \beta$ be an element of $M_1 + M_2$, α in M_1, β in M_2. Then $\gamma\alpha$ and $\gamma\beta$ are, respectively, in M_1 and M_2. Thus, $\gamma\cdot(\alpha + \beta) = \gamma\alpha + \gamma\beta$ is in $M_1 + M_2$. Thus, $\gamma(M_1 + M_2) \subseteq M_1 + M_2$, and γ is in the coefficient ring of $M_1 + M_2$.

Part (vii): By part (vi), $\mathcal{O}_{M_1+M_2} \supseteq I_d$. However, by Lemma 8.6.9 $I_d \supseteq \mathcal{O}_{M_1+M_2}$, so that $\mathcal{O}_{M_1+M_2} = I_d$. ■

It is not generally true that if M_1 and M_2 belong to $\mathcal{O}$, then $M_1 + M_2$ belongs to $\mathcal{O}$. We shall give a counterexample in the exercises. However, if $\mathcal{O} = \{1, \mathcal{f}\omega_d\}$, M_1, M_2 integral, and if $\gcd(N(M_1), \mathcal{f}) = 1$, $\gcd(N(M_2), \mathcal{f}) = 1$, then $M_1 + M_2$ has the coefficient ring $\mathcal{O}$ (see Section 9.4).

9.3 Exercises

1. Calculate M_1M_2 and $M_1 + M_2$ (that is, give a basis) if
   (a) $M_1 = \{2, \sqrt{5}\}$, $M_2 = \{3, 1 + \sqrt{5}\}$.
   (b) $M_1 = \{1 - \sqrt{-5}, 1 + \sqrt{-5}\}$, $M_2 = \{12, 3 + \sqrt{-5}\}$.
2. Let $M = \{\alpha, \beta\}$. Show that $M' = \{\alpha', \beta'\}$.
3. Let M_1 and M_2 be modules. Show that $(M_1M_2)' = M_1'M_2'$.
4. Prove that $\mathcal{O}_M = \mathcal{O}_{M'}$ for any module M.
5. Prove Proposition 3.
6. Prove Proposition 6, parts (i) and (ii).
7. Prove Theorem 16, part (v).

8. Show that if $M_1 = \gamma I_d$ and $M_2 = \delta I_d$ for γ, δ nonzero elements in $\mathbf{Q}(\sqrt{d})$, then $M_1 M_2 = \gamma\delta I_d$.
9. Check that $MM' = N(M)\mathcal{O}_M$ in the case $M = \{5, 2 + \sqrt{-1}\}$.
10. Determine $\mathcal{O}_M$ for M one of the modules of Exercise 1.
11. Use Corollary 10 to compute $N(M)$ in the case M is one of the modules of Exercise 1.
12. Suppose that A and B are modules belonging to $\mathcal{O}$. Does $A + B$ necessarily belong to $\mathcal{O}$? (*Hint:* Try $A = \{4, 1 + \sqrt{2}\}$, $B = \{2, 1 + 2\sqrt{2}\}$.)
13. Let A, B be modules belonging to $\mathcal{O}$, such that $A, B \subseteq \mathcal{O}$. Show that $A \cap B \supseteq AB$.
14. Let A and B be modules belonging to $\mathcal{O}$ such that $A + B = \mathcal{O}$. Show that $A \cap B = AB$.
15. Show that if $\{1, f\omega_d\}$ is similar to $\{1, f^*\omega_d\}$, then $f = f^*$.
16. Extend the comments of Remark 13 to provide a method for computing the product of two modules M_1, M_2 of $\mathbf{Q}(\sqrt{d})$, where $\omega_d = (1 + \sqrt{d})/2$.
17. Show that $\{1, \sqrt{5}\}\{3, \sqrt{5}\} = \{1, \sqrt{5}\}$. Since $\{1, \sqrt{5}\}$ is a coefficient ring, why does this not violate Proposition 6, part (iii)?

9.4 The Fundamental Factorization Theorem

At last we are ready to state and prove our unique factorization theorem for modules belonging to a given coefficient ring $\mathcal{O}$ in $\mathbf{Q}(\sqrt{d})$.

Definition 1: Let M be an integral module belonging to $\mathcal{O}$. We say that M is *prime* if and only if, whenever $M = M_1 M_2$, with M_1 and M_2 integral modules belonging to $\mathcal{O}$, we have either $M_1 = \mathcal{O}$ or $M_2 = \mathcal{O}$.

Since $\mathcal{O}$ plays the role of 1 under multiplication (Proposition 3.6, (iii)), we see that M is a prime module if and only if the only integral modules belonging to $\mathcal{O}$ which are factors of M are $\mathcal{O}$ and M.

Theorem 2: Let p be a rational prime and let M be an integral module belonging to $\mathcal{O}$. If $N(M) = p$, then M is prime.

Proof: Assume that $M = M_1 M_2$, where M_1, M_2 are integral modules belonging to $\mathcal{O}$. Then $p = N(M) = N(M_1 M_2) = N(M_1)N(M_2)$ (Corollary 3.12). Since M_1 and M_2 are contained in their ring of coefficients $\mathcal{O}$, $N(M_1)$ and $N(M_2)$ are positive integers. Thus, since p is a rational prime, either $N(M_1) = 1$ or $N(M_2) = 1$. Thus, by Proposition 2.13, either $M_1 = \mathcal{O}$ or $M_2 = \mathcal{O}$. ■

Corollary 3: Let $M \neq \mathcal{O}$ be an integral module belonging to $\mathcal{O}$. Then M can be written as a product of prime modules.

Proof: By Proposition 2.13, $N(M)$ is a rational integer greater than 1. Let us proceed by induction on $N(M)$. If $N(M) = 2$, then M is prime by Theorem 2. Thus, we may assume that $N(M) > 2$ and that every integral module of norm $< N(M)$ can be factored into a product of prime modules. If M is a prime module, we are done, so assume that M is not prime. Then we may write $M = M_1 M_2$, where neither M_1 nor M_2 equals $\mathcal{O}$. By Proposition 2.13, we then have $N(M_1) > 1$, $N(M_2) > 1$, so that $1 < N(M_1), N(M_2) < N(M)$. By induction, we may write M_1 and M_2 as products of prime modules. Therefore, we may write $M = M_1 M_2$ as a product of prime modules. This completes the induction. ■

Note: You should compare the proof of Corollary 3 with the proofs of Lemma 2.4.2 and Theorem 1.3.

Thus, we see that prime modules belonging to $\mathcal{O}$ are the building blocks from which all integral modules belonging to $\mathcal{O}$ can be constructed. Our main result will be that the analogue of the fundamental theorem of arithmetic holds for integral modules M belonging to $\mathcal{O} = \{1, \mathcal{f}\omega_d\}$ and which satisfy the additional condition $\gcd(N(M), \mathcal{f}) = 1$. We shall prove in this case that the factorization of M into a product of prime modules is unique.

Theorem 4 (Fundamental Theorem): Let $M \neq \mathcal{O}$ be an integral module belonging to $\mathcal{O} = \{1, \mathcal{f}\omega_d\}$. Assume that $\gcd(N(M), \mathcal{f}) = 1$. Then M can be written in the form

$$M = P_1 \cdots P_t,$$

where $P_1, \ldots, P_t$ are prime modules belonging to $\mathcal{O}$. Moreover, this representation is unique up to the order of $P_1, \ldots, P_t$.

Note that if $\mathcal{O} = I_d$, then $\mathcal{f} = 1$, and the above theorem applies to *all* integral modules belonging to I_d, namely to all ideals of I_d. Before we can prove Theorem 4, we shall need some auxiliary facts. Our proof of the fundamental theorem will be modeled on the proof of the fundamental theorem of arithmetic given in Chapter 2. Recall that a central role in that proof was played by Euclid's lemma, which asserted that if a prime p divides ab, then $p \mid a$ or $p \mid b$. This result, in turn, rested on the concept of and the existence of greatest common divisors. Our proof of Theorem 4 will be derived from analogous facts concerning integral modules belonging to $\mathcal{O}$.

Definition 5: Let A, B be integral modules belong to $\mathcal{O}$. Then we say that A *divides* B, written $A \mid B$, if there is an integral module C belonging to $\mathcal{O}$ such that $B = AC$.

Theorem 6: Let A and B be integral modules belonging to $\mathfrak{O}$. Then $A \mid B$ if and only if $A \supseteq B$.

Proof:

(i) Suppose that $A \mid B$. Then $B = AC$ for some integral module C belonging to $\mathfrak{O}$. Since A belongs to $\mathfrak{O}$ and C is integral, $B = AC \subseteq A\mathfrak{O} \subseteq A$. Thus, $B \subseteq A$.

(ii) Suppose that $B \subseteq A$. By Theorem 3.8, we see that $AA' = N(A)\mathfrak{O}$, so that if $\alpha = N(A)^{-1}$, we have

$$A \cdot (\alpha A') = \mathfrak{O}.$$

Therefore, since $A \supseteq B$, we see that

$$\mathfrak{O} = A \cdot (\alpha A') \supseteq B \cdot (\alpha A'). \tag{$*$}$$

Let $C = B \cdot (\alpha A')$. By Lemma 8.6.12 $\alpha A'$ belongs to $\mathfrak{O}$, and thus by Corollary 3.11, C belongs to $\mathfrak{O}$. Then by ($*$), C is an integral module. Finally, since $AA' = N(A)\mathfrak{O}$, we have

$$AC = AB(\alpha A') = B\frac{1}{N(A)}(AA') = B\mathfrak{O} = B,$$

so that $A \mid B$. ■

Definition 7: Let A and B be integral modules belonging to $\mathfrak{O}$. We call an integral module D belonging to $\mathfrak{O}$ a *greatest common divisor* of A and B provided that

(i) $D \mid A$ and $D \mid B$.

(ii) Whenever C is an integral module belonging to $\mathfrak{O}$ such that $C \mid A$ and $C \mid B$, we have $C \mid D$.

As was the case with the rational integers, we must prove the existence and uniqueness of the greatest common divisor. The uniqueness will be left for the exercises. The existence is considerably more delicate than in the case of the rational integers, owing to difficulties which can arise if $\mathcal{f} > 1$. It is because of these difficulties that we shall prove the existence of a greatest common divisor only under an additional hypothesis, namely $\gcd(N(A), \mathcal{f}) = 1$. The following lemma is a technical device to be used in our proof of the existence of a greatest common divisor:

Lemma 8: Let $\mathfrak{O} = \{1, \mathcal{f}\omega_d\}$ be a ring of coefficients of $\mathbf{Q}(\sqrt{d})$, and let M be an integral module belonging to $\mathfrak{O}$ such that $\gcd(N(M), \mathcal{f}) = 1$. Let M_1 be a module of $\mathbf{Q}(\sqrt{d})$ such that $M \subseteq M_1 \subseteq \mathfrak{O}$ and $M_1\mathfrak{O} \subseteq M_1$. Then M_1 belongs to $\mathfrak{O}$.

Proof: Let $\mathfrak{O}_1$ denote the ring of coefficients of M_1. Since $M_1\mathfrak{O} \subseteq M_1$, we see that $\mathfrak{O} \subseteq \mathfrak{O}_1$. Let $\mathfrak{O}_1 = \{1, \mathcal{f}_1\omega_d\}$. Since $\mathcal{f}\omega_d$ belongs to $\mathfrak{O}_1$, we see that $\mathcal{f}_1 \mid \mathcal{f}$.

It suffices to show that $\mathcal{f} \mid \mathcal{f}_1$. For then $\mathcal{f} = \mathcal{f}_1$ and $\mathfrak{O} = \mathfrak{O}_1$. Note that $M \subseteq M_1 \subseteq \mathfrak{O}$, so that by Corollary 2.9, we have

$$N(M) = (\mathfrak{O} : M) = (\mathfrak{O} : M_1)(M_1 : M).$$

Set $a = (\mathfrak{O} : M_1)$, $b = (M_1 : M)$. Then $N(M) = ab$, and the hypothesis $\gcd(N(M), \mathcal{f}) = 1$ implies that $\gcd(a, \mathcal{f}) = 1$, $\gcd(b, \mathcal{f}) = 1$. Since $(\mathfrak{O} : M_1) = a$ and since 1 belongs to $\mathfrak{O}$, we see that $a \cdot 1 = a$ belongs to M_1 (Exercise 10 in Section 2). But then, since $\mathcal{f}_1\omega_d$ belongs to $\mathfrak{O}_1$ and since $\mathfrak{O}_1$ is the ring of coefficients of M_1, we see that $a\mathcal{f}_1\omega_d$ belongs to M_1 and hence to $\mathfrak{O}$. Therefore, since $\mathfrak{O} = \{1, \mathcal{f}\omega_d\}$, we see that $\mathcal{f} \mid a\mathcal{f}_1$. But since $\gcd(a, \mathcal{f}) = 1$, we have $\mathcal{f} \mid \mathcal{f}_1$ and we are done. ∎

Theorem 9: Let A and B be integral modules belonging to $\mathfrak{O} = \{1, \mathcal{f}\omega_d\}$ and assume that $\gcd(N(A), \mathcal{f}) = 1$. Then $A + B$ is a greatest common divisor of A and B.

Proof: First, observe that $M_1 = A + B$ is a module such that $A \subseteq M_1 \subseteq \mathfrak{O}$. Moreover, by Theorem 3.16 part (vi), we know that the ring of coefficients of M_1 contains $\mathfrak{O}$, so that $M_1\mathfrak{O} \subseteq M_1$. Thus, since $\gcd(N(A), \mathcal{f}) = 1$, Lemma 8 can be applied with $M = A$ to yield that $M_1 = A + B$ belongs to $\mathfrak{O}$. Since $A + B \supseteq A, B$, we know from Theorem 6 that $A + B \mid A$, $A + B \mid B$. Moreover, if $C \mid A$ and $C \mid B$, for some C belonging to $\mathfrak{O}$, then again by Theorem 6, we have $C \supseteq A$, $C \supseteq B$. Therefore, if α is in A and β is in B, then α and β are both in C, and hence $\alpha + \beta$ is in C. Thus, $A + B \subseteq C$, so that by Theorem 6, we have $C \mid A + B$. Thus, $A + B$ is a greatest common divisor of A and B. ∎

From the preceding result, we can deduce our analogue of Euclid's lemma.

Corollary 10 (Euclid's lemma): Let P, A, B be integral modules belonging to $\mathfrak{O} = \{1, \mathcal{f}\omega_d\}$. Assume that P is prime and that $\gcd(N(P), \mathcal{f}) = 1$. If $P \mid AB$, then $P \mid A$ or $P \mid B$.

Proof: Suppose that P does not divide A. Let M be a greatest common divisor of P and A (from Theorem 9). Since P is prime and $M \mid P$, we have $M = P$ or $M = \mathfrak{O}$. But since $M \mid A$, and P does not divide A, we see that $M = \mathfrak{O}$. Therefore, by Theorem 9,

$$\mathfrak{O} = P + A.$$

Therefore (Theorem 3.16, (v)),

$$B = \mathfrak{O}B = (P + A)B = PB + AB.$$

Since $P \mid AB$, we have $AB = CP$, so that*

$$B = PB + PC = (B + C)P. \tag{1}$$

**Note:* We are not done yet. We must show that $B + C$ belongs to $\mathfrak{O}$.

Since $PP' = N(P)\mathfrak{O}$, we have

$$BP' = PP'(B + C) = N(P)(B + C),$$

so that by Lemma 8.6.12 the coefficient ring of $B + C$ is the same as the coefficient ring of BP', which equals $\mathfrak{O}$ by Corollary 3.11. Thus, $B + C$ belongs to $\mathfrak{O}$. Moreover, Eq. (1) implies that $P \mid B$. ∎

Now that we have the analogue of Euclid's lemma, we may prove our fundamental theorem with great ease.

Proof of Theorem 4: Let us reason by contradiction. Suppose that there exist modules whose norm is relatively prime to $\mathcal{f}$, whose factorization into prime modules is not unique. By the well-ordering principle, there exists such a module M, the number of whose prime module factors is as small as possible. Suppose that

$$M = P_1 \cdots P_t = Q_1 \cdots Q_s$$

are two factorizations of M with t as small as possible. If $t = 1$, then M is a prime and we immediately see that $s = 1$ and $Q_1 = P_1$. Thus, $t \geq 2$. Note that $P_1 \mid M$. Therefore, $P_1 \mid Q_1 \cdots Q_s$. Since $\gcd(N(M), \mathcal{f}) = 1$, we have $\gcd(N(P_1), \mathcal{f}) = 1$. By Corollary 10, we see that $P_1 \mid Q_i$ for some i $(1 \leq i \leq s)$. Therefore, if, say $P_1 \mid Q_1$, then $Q_1 = P_1 A$ for some integral module A belonging to $\mathfrak{O}$. But since P_1 and Q_1 are prime modules, $A = \mathfrak{O}$, and thus $P_1 = Q_1$. Thus, we have

$$P_1 P_2 \cdots P_t = P_1 Q_2 \cdots Q_s,$$

so that

$$N(P_1)^{-1} P_1' P_1 P_2 \cdots P_t = N(P_1)^{-1} P_1' P_1 Q_2 \cdots Q_s.$$

However, by Theorem 3.8, we have $N(P_1)^{-1} P_1' P_1 = \mathfrak{O}$, so that

$$\mathfrak{O} P_2 \cdots P_t = \mathfrak{O} Q_2 \cdots Q_s.$$

Thus,

$$M_1 = P_2 \cdots P_t = Q_2 \cdots Q_s.$$

Since $t \geq 2$, M_1 is a module $\neq \mathfrak{O}$ with two distinct factorizations into prime modules belonging to $\mathfrak{O}$, a contradiction to the way in which M was chosen. ∎

In the case $\mathfrak{O} = I_d$, the modules belonging to $\mathfrak{O}$ are precisely the ideals of I_d. Also, $\mathcal{f} = 1$, and so the fundamental theorem applies to all integral ideals. Therefore, as a consequence of the fundamental theorem, we deduce the following result for ideals:

Corollary 11: Let A be an ideal of I_d, $A \neq I_d$. Then A can be written uniquely as a product of prime ideals, where uniqueness is meant up to rearrangement of the factors and where a prime ideal is interpreted to mean a prime module* belonging to I_d.

This, at last, is our theory of unique factorization in I_d. If I_d is a unique factorization domain, then the unique factorization of ideals is equivalent to the unique factorization of the elements. Indeed, I_d has unique factorization if and only if all ideals of I_d are of the form γI_d (γ in I_d) (exercise). Therefore, the uniqueness of factorization of integers in I_d is the same as the unique factorization of ideals. If I_d does not have unique factorization, the factorization of ideals acts as a substitute for the factorization of elements.

9.4 Exercises

1. Are the following ideals prime (considered as modules belonging to I_d)?
   (a) $3I_{-5}$. (b) $7I_{-3}$. (c) $6I_2$. (d) $\{2, \sqrt{2}\}$.

*2. Write the following modules as products of prime modules:
   (a) $M = \{5, 2 + \sqrt{3}\}$. (b) $M = \{3, 8 - 7\sqrt{6}\}$.

3. Let $\mathfrak{O}$ be a ring of coefficients, α, β elements of $\mathfrak{O}$. Show that $\alpha\mathfrak{O} \mid \beta\mathfrak{O}$ if and only if $\alpha \mid \beta$.

4. Find the gcd of the two modules $M_1 = \{161, 74 + \sqrt{2}\}$, $M_2 = \{35, 20 + 5\sqrt{2}\}$. (*Ans.*: $\{7, 4 + \sqrt{2}\}$.)

5. Suppose that A, B are integral modules belonging to $\mathfrak{O}$ and that $\gcd(\mathscr{f}, N(A)N(B)) = 1$, where $\mathfrak{O} = \{1, \mathscr{f}\omega_d\}$. Further, suppose that
   $$A = \prod P^{a_P}, \qquad B = \prod P^{b_P}$$
   are expressions of A, B, respectively, as products of prime modules. Show that
   $$A + B = \gcd(A, B) = \prod P^{c_P},$$
   where $c_P =$ the smaller of a_P and b_P.

6. Show that the gcd of two modules is unique.

7. Let A, B be integral modules belonging to $\mathfrak{O}$ such that $A + B$ belongs to $\mathfrak{O}$. Show that A and B have a gcd and that $\gcd(A,B) = A + B$.

8. Let P be a prime module. Show that $N(P) = p$ or p^2 for some rational prime p. (*Hint: $PP' = N(P)\mathfrak{O}$.*)

9. Let M be any integral module belonging to $\mathfrak{O} = \{1, \mathscr{f}\omega_d\}$. Show that M can be written uniquely in the following form: $M = P_1 \cdots P_t A$, where

*See Proposition 5.1 to check that this use of the word *prime* ideal is consistent with the terminology of abstract algebra.

P_i is a prime module belonging to $\mathfrak{O}$ such that $\gcd(N(P_i), \mathscr{f}) = 1$ and A is an integral module belonging to $\mathfrak{O}$ such that the only primes dividing $N(A)$ are primes dividing $\mathscr{f}$.

10. Let $\mathfrak{O} = \{1, \mathscr{f}\omega_d\}$ be a coefficient ring of $\mathbf{Q}(\sqrt{d})$ and let M be an integral module belonging to $\mathfrak{O}$ such that $\gcd(N(M), \mathscr{f}) = 1$.
    (a) Prove that $M \cdot I_d = M_0$ is an integral module belonging to I_d such that $N(M_0) = N(M)$.
    (b) Prove that $M_0 \cap \mathfrak{O} = M$.
    (c) Prove that if M^* is another integral module belonging to $\mathfrak{O}$ and such that $\gcd(N(M^*), \mathscr{f}) = 1$, then $(MM^*)_0 = M_0(M^*)_0$.

11. (Exercise 10, continued) Let M_0 be any ideal of I_d such that $\gcd(N(M_0), \mathscr{f}) = 1$.
    (a) Show that $M_0 \cap \mathfrak{O} = M$ is an integral module belonging to $\mathfrak{O}$ such that $N(M) = N(M_0)$.
    (b) Show that if $P_0 = M_0$ is prime, then so is $P = P_0 \cap \mathfrak{O}$.
    (c) Show that if $M = M_0 \cap \mathfrak{O}$, then $M \cdot I_d = M_0$.
    (d) Prove that the procedure of passing from M_0 to M preserves multiplication of modules.

12. Use Exercises 10 and 11 to show that the prime modules P of $\mathfrak{O} = \{1, \mathscr{f}\omega_d\}$ such that $\gcd(N(P), \mathscr{f}) = 1$ are precisely the modules $P_0 \cap \mathfrak{O}$, where P_0 is a prime ideal of I_d satisfying $\gcd(N(P_0), \mathscr{f}) = 1$.

13. Use Exercise 12 to give an alternative proof of Theorem 4.

9.5 The Prime Modules Belonging to $\mathfrak{O}$

Let us fix a coefficient ring $\mathfrak{O} = \{1, \mathscr{f}\omega_d\}$ of $\mathbf{Q}(\sqrt{d})$. In this section, we shall give an explicit description of all prime modules P which belong to $\mathfrak{O}$ and which satisfy $\gcd(N(P), \mathscr{f}) = 1$.

Let us begin by proving a property of prime modules which should be familiar from the reader's course in abstract algebra.

Proposition 1: Let P be a prime module belonging to $\mathfrak{O}$ such that $\gcd(N(P), \mathscr{f}) = 1$, and let α, β belong to $\mathfrak{O}$. If $\alpha\beta$ belongs to P, then either α is in P or β is in P.

Proof: If $\alpha\beta$ belongs to P, then $(\alpha\beta)\gamma$ belongs to P for any γ in $\mathfrak{O}$. Thus, $\alpha\beta\mathfrak{O} \subseteq P$, so that $P \mid \alpha\beta\mathfrak{O}$ by Theorem 4.6. But $\alpha\beta\mathfrak{O} = (\alpha\mathfrak{O})(\beta\mathfrak{O})$, so that by Corollary 4.10, either $P \mid \alpha\mathfrak{O}$ or $P \mid \beta\mathfrak{O}$. Suppose, for example, that $P \mid \alpha\mathfrak{O}$. Then $\alpha\mathfrak{O} \subseteq P$ by Theorem 4.6. Since 1 belongs to $\mathfrak{O}$, we see that $\alpha \cdot 1$ is contained in P. ■

Caution: Proposition 1 shows that a prime module P which belongs to $\mathfrak{O}$ and for which $\gcd(N(P), \mathcal{f}) = 1$ is a prime ideal of the ring $\mathfrak{O}$. However, not every prime ideal of $\mathfrak{O}$ is a prime module belonging to $\mathfrak{O}$.*

Let P be a prime module belonging to $\mathfrak{O}$ such that $\gcd(N(P), \mathcal{f}) = 1$. If α is any nonzero element of P, then $\alpha\alpha' = N(\alpha)$ belongs to P and is a nonzero rational integer. Thus, P contains nonzero rational integers. If n belongs to P, so does $-n$, and so P contains positive rational integers. Let p be the smallest positive rational integer in P. Then p is a rational prime, for if $p = ab$, $1 < a, b < p$, a, b rational integers, then ab belongs to P. By Proposition 1, either a or b belongs to P, contradicting the manner in which p was chosen. Thus, p is a rational prime and $p\mathfrak{O} \subseteq P$ since P belongs to $\mathfrak{O}$. Thus, $P \mid p\mathfrak{O}$ by Theorem 4.6. Moreover, $N(P) \mid N(p\mathfrak{O})$ and $N(p\mathfrak{O}) = p^2$, so that $p \nmid \mathcal{f}$. *Thus, we see that $P \mid p\mathfrak{O}$ for some rational prime p, $p \nmid \mathcal{f}$.*

Conversely, let p be any rational prime such that $p \nmid \mathcal{f}$. By our fundamental theorem, we may write

$$p\mathfrak{O} = P_1 \cdots P_t,$$

where $P_1, \ldots, P_t$ are prime modules belonging to $\mathfrak{O}$. Then

$$p^2 = N(p\mathfrak{O}) = N(P_1) \cdots N(P_t).$$

Since $N(P_i) > 1$ $(1 \leq i \leq t)$ (Proposition 2.13), we see that $t \leq 2$. In fact, there are precisely three cases:

Case 1: $t = 1$, $p\mathfrak{O} = P_1$, $N(P_1) = p^2$. In this case, we say that p is an *inert prime* (with respect to $\mathfrak{O}$).

Case 2: $t = 2$, $p\mathfrak{O} = P_1 P_2$, $P_1 \neq P_2$, $N(P_1) = N(P_2) = p$. In this case, we say that p is a *decomposed prime* (with respect to $\mathfrak{O}$).

Case 3: $t = 2$, $p\mathfrak{O} = P_1 P_1 = P_1^2$, $N(P_1) = p$. In this case, we say that p is a *ramified prime* (with respect to $\mathfrak{O}$).

Note that we are defining the concepts of inert, decomposed, and ramified primes only for p such that $p \nmid \mathcal{f}$, for otherwise, our fundamental theorem tells us nothing about the factorization of $p\mathfrak{O}$ into prime modules.

Suppose that p is a decomposed prime. Then $p\mathfrak{O} = P_1 P_2$ for prime modules P_1, P_2, $P_1 \neq P_2$, $N(P_1) = N(P_2) = p$. By Theorem 3.8, we have $P_1 P_1' = N(P_1)\mathfrak{O} = p\mathfrak{O}$. Therefore, since $P_1 P_1' = P_1 P_2$, our fundamental theorem implies that $P_1' = P_2$. Thus, $p\mathfrak{O} = P_1 P_1'$. If p is a ramified prime, then $p\mathfrak{O} = P_1^2$ and $P_1 P_1' = N(P_1)\mathfrak{O} = p\mathfrak{O}$, so that $P_1 = P_1'$. Thus, we have the following result:

*Note, however, that when $\mathfrak{O} = I_d$, then the two concepts are the same.

Theorem 2: Let P be a prime module belonging to $\mathcal{O}$ such that $\gcd(N(P),\mathcal{f}) = 1$. Then there is a rational prime p, $p \nmid \mathcal{f}$ such that $P \mid p\mathcal{O}$. Conversely, if p is a rational prime such that $p \nmid \mathcal{f}$, then there are three possibilities:

(i) p is inert. Then $p\mathcal{O} = P$ is a prime module and $N(P) = p^2$.

(ii) p is decomposed. Then $p\mathcal{O} = PP'$, where P and P' are distinct prime modules and $N(P) = N(P') = p$.

(iii) p is ramified. Then $p\mathcal{O} = P^2$, where P is a prime module such that $P = P'$ and $N(P) = p$.

Let us explicitly compute the prime module P in parts (i)–(iii) of Theorem 2. This will, by Theorem 2, allow us to explicitly determine all prime modules having norm relatively prime to $\mathcal{f}$.

Let Δ (respectively, Δ_0) denote the discriminant of $\mathcal{O}$ (respectively, I_d). Then a trivial calculation shows that

$$\Delta = \mathcal{f}^2\Delta_0$$

and

$$\Delta_0 = \begin{cases} d & \text{if } d \equiv 1 (\text{mod } 4), \\ 4d & \text{if } d \equiv 2 \text{ or } 3 (\text{mod } 4). \end{cases}$$

Note that $\Delta_0 \equiv 0$ or $1 (\text{mod } 4)$, so that $\Delta \equiv 0$ or $1 (\text{mod } 4)$. Next, note that

$$\frac{\Delta + \sqrt{\Delta}}{2} = \begin{cases} \mathcal{f}\omega_d + \dfrac{\mathcal{f}^2(d-1)}{2} + \dfrac{\mathcal{f}(\mathcal{f}-1)}{2} & \text{if } d \equiv 1 (\text{mod } 4), \\ \mathcal{f}\omega_d + 2\mathcal{f}^2 d & \text{if } d \equiv 2 \text{ or } 3 (\text{mod } 4). \end{cases}$$

Since $\mathcal{f}^2(d-1)/2 + \mathcal{f}(\mathcal{f}-1)/2$ (resp. $2d\mathcal{f}^2$) is a rational integer if $d \equiv 1 (\text{mod } 4)$ (resp. $d \equiv 2$ or $3 (\text{mod } 4)$), we see that

$$\mathcal{O} = \left\{1, \frac{\Delta + \sqrt{\Delta}}{2}\right\}.$$

The choice of this basis for $\mathcal{O}$ allows us to avoid having to make too many case distinctions in our calculations.

Let p be a rational prime such that $p \nmid \mathcal{f}$. We shall now explicitly factor $p\mathcal{O}$ into prime modules belonging to $\mathcal{O}$.

First, suppose that p is decomposed or ramified, so that $p\mathcal{O} = PP'$, $N(P) = p$. Since $N(P) = p$, a complete residue system modulo P contains p elements. We assert that such a complete residue system is given by $0, 1, \ldots, p-1$. Indeed, if two of these integers, say a, b $(a > b)$, are such that $a \equiv b (\text{mod } P)$, then $0 < a - b < p$, and $a - b$ belongs to P. Thus, $(a-b)\mathcal{O} \subseteq P$. By Theorem 4.6, we see that $P \mid (a-b)\mathcal{O}$. Therefore, $N(P) \mid N((a-b)\mathcal{O})$, so that $p \mid (a-b)^2$, which contradicts the fact that p is prime and $0 < a - b < p$. Thus, $0, 1, \ldots, p-1$ is a complete residue system modulo P.

Therefore, there exists a rational integer r such that

$$\frac{\Delta + \sqrt{\Delta}}{2} \equiv r(\text{mod } P).$$

Thus, $(\Delta + \sqrt{\Delta})/2 - r$ belongs to P, so that $(2r - \Delta) - \sqrt{\Delta}$ belongs to $2P$. Thus, $(2r - \Delta - \sqrt{\Delta})\mathfrak{O} \subseteq 2P$, so that $2P | (2r - \Delta - \sqrt{\Delta})\mathfrak{O}$ by Theorem 4.6. But then $N(2P) | N((2r - \Delta - \sqrt{\Delta})\mathfrak{O})$, so that $4p | (2r - \Delta)^2 - \Delta$. In other words,

$$(2r - \Delta)^2 \equiv \Delta(\text{mod } 4p).$$

Thus, if $p\mathfrak{O} = PP'$, $N(P) = p$, *then the congruence* $x^2 \equiv \Delta(\text{mod } 4p)$ *has the solution* $x = 2r - \Delta$ (recall that $p \nmid \mathfrak{f}$ has been assumed).

Conversely, suppose that the congruence $x^2 \equiv \Delta(\text{mod } 4p)$ is solvable. Let us consider two cases:

Case 1: $p \nmid \Delta$. Note that x and Δ have the same parity, so that $x + \Delta$ is even, say $x + \Delta = 2r$. Let P denote the module

$$P = \left\{p, r - \frac{\Delta + \sqrt{\Delta}}{2}\right\}.$$

Let us first use Lemma 8.6.13 to show that $\mathfrak{O}_P = \mathfrak{O}$. In the notation of Lemma 8.6.13, we have $\gamma = (r - (\Delta + \sqrt{\Delta})/2)/p$, so that $\text{Tr}(\gamma) = (2r - \Delta)/p = x/p$ and $\text{N}(\gamma) = ((2r - \Delta)/2p)^2 - \Delta/4p^2 = (1/p)((x^2 - \Delta)/4p)$. Since $x^2 \equiv \Delta(\text{mod } 4p)$ and $p \nmid \Delta$, we have $p \nmid x$ and $(x^2 - \Delta)/4p$ is a rational integer. Thus, the a in Lemma 8.6.13 is just p and

$$\mathfrak{O}_P = \{1, a\gamma\} = \left\{1, r - \frac{\Delta + \sqrt{\Delta}}{2}\right\} = \left\{1, \frac{\Delta + \sqrt{\Delta}}{2}\right\} = \mathfrak{O}.$$

From Theorem 2.6, it is clear that $\text{N}(P) = p$, so that by Theorem 3.8, we have $PP' = \text{N}(P)\mathfrak{O}_P = p\mathfrak{O}$. However, $P' = \{p, r - (\Delta - \sqrt{\Delta})/2\}$ and $P \neq P'$. (For example, $r - (\Delta + \sqrt{\Delta})/2$ belongs to P and not to P' (exercise).) Thus, we conclude that *if* $p \nmid \Delta$ *and* $x^2 \equiv \Delta(\text{mod } 4p)$ *is solvable, then* p *is decomposed.*

Case 2: $p | \Delta, p \nmid \mathfrak{f}$. The congruence $x^2 \equiv \Delta(\text{mod } 4p)$ is always solvable in this case. Our argument will proceed as in Case 1, except we shall set $r = 0$. Unfortunately we must consider the cases p odd and $p = 2$ separately.

(i) p odd. Set

$$P = \left\{p, \frac{\Delta + \sqrt{\Delta}}{2}\right\} = \left\{p, \frac{\Delta + \sqrt{\Delta}}{2} - \Delta\right\} = \left\{p, \frac{\Delta - \sqrt{\Delta}}{2}\right\} = P'.$$

Again using Lemma 8.6.13, we show $\mathfrak{O}_P = \mathfrak{O}$. Here $\gamma = (\Delta + \sqrt{\Delta})/2p$, $\text{Tr}(\gamma) = \Delta/p$, $\text{N}(\gamma) = \Delta(\Delta - 1)/4p^2$. Since $\Delta \equiv 0, 1(\text{mod } 4)$, we see $\Delta(\Delta - 1)/4p$ is a rational integer. Since $p | \Delta = \mathfrak{f}^2\Delta_0$ and $p \nmid \mathfrak{f}$, we see

$p \nmid \Delta(\Delta - 1)/4p$. Thus the a in Lemma 8.6.13 is just p and $\mathfrak{O} = \{1, a\gamma\} = \{1, (\Delta + \sqrt{\Delta})/2\} = \mathfrak{O}$. Moreover, Theorem 2.6 implies $\mathrm{N}(P) = p$. Thus from Theorem 3.8 we see $P^2 = p\mathfrak{O}$. That is, *if $p \mid \Delta$, $p \nmid \mathcal{f}$ and p odd, then p is ramified.*

(ii) $p = 2$. Since $p \nmid \mathcal{f}$, $\mathcal{f}$ must be odd and $\Delta_0 \equiv 0 \pmod 4$. Then $\Delta_0 = 4d$, where $d \equiv 2$ or $3 \pmod 4$, d square-free. We leave it to the reader to show that $p\mathfrak{O} = P^2$, where

$$P = \begin{cases} \left\{2, \sqrt{\dfrac{\Delta}{4}}\right\} & \text{if } d \equiv 2 \pmod 4, \\ \left\{2, 1 + \sqrt{\dfrac{\Delta}{4}}\right\} & \text{if } d \equiv 3 \pmod 4. \end{cases}$$

Again, P is ramified.

We may summarize our analysis in the following theorem:

Theorem 3: Let p be a rational prime such that $p \nmid \mathcal{f}$.

(i) Assume that $p \nmid \Delta$. Then p is decomposed if and only if the congruence $x^2 \equiv \Delta \pmod{4p}$ is solvable. Otherwise, p is inert. In the case p is decomposed, say $p\mathfrak{O} = PP'$, then

$$P = \left\{p, r - \frac{\Delta + \sqrt{\Delta}}{2}\right\},$$

where $x = 2r - \Delta$ is a solution of $x^2 \equiv \Delta \pmod{4p}$.

(ii) Assume that $p \mid \Delta$. Then p is ramified. If $p\mathfrak{O} = P^2$, then

$$P = \begin{cases} \left\{p, \dfrac{\Delta + \sqrt{\Delta}}{2}\right\} & \text{if } p \text{ is odd}, \\ \left\{2, \sqrt{\dfrac{\Delta}{4}}\right\} & \text{if } p = 2,\ d \equiv 2 \pmod 4, \\ \left\{2, 1 + \sqrt{\dfrac{\Delta}{4}}\right\} & \text{if } p = 2,\ d \equiv 3 \pmod 4. \end{cases}$$

We can state the results of Theorem 3, part (i), as follows: Assume that p is odd. Note that x satisfies the congruence $x^2 \equiv \Delta \pmod{4p}$ if and only if $x^2 \equiv \Delta \pmod p$ and $x^2 \equiv \Delta \pmod 4$. Assume that we are given an x satisfying $x^2 \equiv \Delta \pmod p$. Then $p - x$ satisfies the same congruence. Since Δ is congruent to 0 or $1 \pmod 4$ and since p is odd, we see that either $x^2 \equiv \Delta \pmod 4$ or $(p - x)^2 \equiv \Delta \pmod 4$. Thus, $x^2 \equiv \Delta \pmod{4p}$ is solvable if and only if $x^2 \equiv \Delta \pmod p$ is solvable, that is, if and only if $\left(\dfrac{\Delta}{p}\right) = +1$. On the other hand, assume that $p = 2$. Since $p \nmid \Delta$, we must have $\Delta \equiv 1 \pmod 4$, and so $\Delta \equiv 1$ or $5 \pmod 8$. By Theorem 3, p is decomposed if and only if $x^2 \equiv \Delta \pmod 8$ is solvable. But since a perfect square is congruent to 0, 1, or

4(mod 8), we see that this last condition is equivalent to $\Delta \equiv 1 \pmod{8}$. Thus, we can restate Theorem 3 as follows:

Theorem 4: Let $\mathfrak{O}$ be a ring of coefficients of discriminant Δ, p a rational prime such that $p \nmid \Delta$. Then p is not ramified in $\mathfrak{O}$.

(i) If p is odd, then p is decomposed if and only if $\left(\frac{\Delta}{p}\right) = +1$.

(ii) If $p = 2$, then p is decomposed if $\Delta \equiv 1 \pmod{8}$ and inert if $\Delta \equiv 5 \pmod{8}$.

Corollary 5: Let $\mathfrak{O}$ be a ring of coefficients of discriminant Δ, p a rational prime, $p \nmid \mathscr{f}$. If p ramifies, then $p \mid \Delta$, and, conversely, every prime dividing Δ and not $\mathscr{f}$ ramifies in $\mathfrak{O}$.

By using Theorems 3 and 4, we have completely settled the question of determining how a rational prime p, $p \nmid \mathscr{f}$, factors into prime modules in $\mathfrak{O}$. We may use the law of quadratic reciprocity to give a very elegant formulation of our law of factorization.

Theorem 6: Let p and q be rational primes not dividing $\mathscr{f}$, and assume that $p \equiv q \pmod{\Delta_0}$. Then p and q factor in the same way in $\mathfrak{O}$. That is, p is ramified (resp. inert, decomposed) in $\mathfrak{O}$ if and only if q is.

Proof: Without loss of generality, assume $p \neq q$. Thus, at most one of p, q can be even. Let us consider separately the two cases p, q both odd and $p = 2$, q odd.

Case 1: p, q odd. Recall that $\Delta = \mathscr{f}^2\Delta_0$. Since $p \equiv q \pmod{\Delta_0}$, we see that $p \mid \Delta_0$ implies $p = q$. Thus, we may assume that $p \nmid \Delta_0$, $q \nmid \Delta_0$. By Theorem 4, it suffices to show that

$$\left(\frac{\Delta}{p}\right) = \left(\frac{\Delta}{q}\right).$$

But

$$\left(\frac{\Delta}{p}\right) = \left(\frac{\mathscr{f}^2}{p}\right)\left(\frac{\Delta_0}{p}\right) = \left(\frac{\Delta_0}{p}\right)$$

$$\left(\frac{\Delta}{q}\right) = \left(\frac{\Delta_0}{q}\right),$$

so that it suffices to show that

$$\left(\frac{\Delta_0}{p}\right) = \left(\frac{\Delta_0}{q}\right).$$

The simplest approach to proving this is to make use of the generalized Legendre-Jacobi symbol, Exercises 9–18 in Section 4.4: Let n an odd rational integer > 1, and let a be a rational integer such that $\gcd(a,n) = 1$. We define the symbol

$$\left(\frac{a}{n}\right) = \left(\frac{a}{p_1}\right)^{e_1} \cdots \left(\frac{a}{p_t}\right)^{e_t},$$

where $n = p_1^{e_1} \cdots p_t^{e_t}$ is the decomposition of n into powers of distinct primes. Then this symbol has the following properties for rational integers a, a' with $\gcd(a,n) = \gcd(a',n) = 1$ and n odd:

$$\left(\frac{a}{n}\right) = \left(\frac{a'}{n}\right) \qquad \text{if } a \equiv a' (\text{mod } n) \tag{1}$$

$$\left(\frac{a}{n}\right)\left(\frac{a'}{n}\right) = \left(\frac{aa'}{n}\right) \tag{2}$$

$$\left(\frac{a^2}{n}\right) = 1 \tag{3}$$

$$\left(\frac{n}{m}\right)\left(\frac{m}{n}\right) = (-1)^{((n-1)/2)((m-1)/2)} \qquad \text{if } m \text{ is positive and odd, } \gcd(m, n) = 1 \tag{4}$$

$$\left(\frac{-1}{n}\right) = (-1)^{(n-1)/2} \tag{5}$$

$$\left(\frac{2}{n}\right) = (-1)^{(n^2-1)/8}. \tag{6}$$

Let us break Case 1 into two cases.

Case 1′: $\Delta_0 \equiv 0 (\text{mod } 4)$. Then $\Delta_0 = 4d$, where $d \equiv 2$ or $3 (\text{mod } 4)$, d square-free. In this case,

$$\left(\frac{\Delta_0}{p}\right) = \left(\frac{2}{p}\right)^2\left(\frac{d}{p}\right) = \left(\frac{d}{p}\right)$$

$$\left(\frac{\Delta_0}{q}\right) = \left(\frac{d}{q}\right).$$

Then, since $p \equiv q (\text{mod } 4d)$, the first form of the law of quadratic reciprocity (Theorem 4.4.1) implies that if $d > 0$, then

$$\left(\frac{\Delta_0}{p}\right) = \left(\frac{\Delta_0}{q}\right).$$

If $d < 0$, then we have $p \equiv q$ (mod $4|d|$) and, in particular, $p \equiv q$ (mod 4), so that $\left(\frac{-1}{p}\right) = \left(\frac{-1}{q}\right)$. Hence

$$\left(\frac{\Delta_0}{p}\right) = \left(\frac{d}{p}\right) = \left(\frac{-1}{p}\right)\left(\frac{|d|}{p}\right) = \left(\frac{-1}{q}\right)\left(\frac{|d|}{q}\right) = \left(\frac{d}{q}\right) = \left(\frac{\Delta_0}{q}\right).$$

Case 1″: $\Delta_0 \equiv 1 (\text{mod } 4)$. Then $\Delta_0 = d$, d square-free. Suppose that $\Delta_0 = \epsilon|\Delta_0|$, where $\epsilon = \pm 1$. Since $\Delta_0 \equiv 1 (\text{mod } 4)$, we see that $|\Delta_0| \equiv 1 (\text{mod } 4)$ if $\epsilon = +1$ and $|\Delta_0| \equiv 3 (\text{mod } 4)$ if $\epsilon = -1$.

$$\left(\frac{\Delta_0}{p}\right) = \left(\frac{\epsilon}{p}\right)\left(\frac{|\Delta_0|}{p}\right)$$

$$= (-1)^{((\epsilon-1)/2)((p-1)/2)}\left(\frac{|\Delta_0|}{p}\right) \qquad \text{(by Eq. (5))}$$

$$= (-1)^{((\epsilon-1)/2)((p-1)/2)+((|\Delta_0|-1)/2)((p-1)/2)}\left(\frac{p}{|\Delta_0|}\right) \qquad \text{(by Eq. (4))}$$

$$= (-1)^{((p-1)/2)((\epsilon-1)/2)+((|\Delta_0|-1)/2)}\left(\frac{p}{|\Delta_0|}\right)$$

$$= \left(\frac{p}{|\Delta_0|}\right) \qquad \left(\text{since } \frac{\epsilon-1}{2}+\frac{|\Delta_0|-1}{2} \text{ is even.}\right)$$

Similarly,

$$\left(\frac{\Delta_0}{q}\right) = \left(\frac{q}{|\Delta_0|}\right).$$

And since $p \equiv q(\text{mod } \Delta_0)$, we see that

$$\left(\frac{p}{|\Delta_0|}\right) = \left(\frac{q}{|\Delta_0|}\right)$$

by Eq. (1). Therefore,

$$\left(\frac{\Delta_0}{p}\right) = \left(\frac{\Delta_0}{q}\right).$$

Case 2: $p = 2$, q odd. Since $p \equiv q(\text{mod } \Delta_0)$, we see that Δ_0 is odd, so that $\Delta_0 = d$, $d \equiv 1(\text{mod } 4)$ and square-free. Since $p \nmid \mathcal{f}$, we see that $\mathcal{f}$ is odd, so that $\Delta \equiv \Delta_0(\text{mod } 8)$. By Theorem 4, p is decomposed if $\Delta_0 \equiv 1(\text{mod } 8)$ and inert if $\Delta_0 \equiv 5(\text{mod } 8)$. On the other hand, $q \equiv 2(\text{mod } \Delta_0)$. Set $\Delta_0 = \epsilon|\Delta_0|$, $\epsilon = \pm 1$. Then, since q is odd,

$$\left(\frac{\Delta_0}{q}\right) = \left(\frac{\epsilon|\Delta_0|}{q}\right)$$

$$= (-1)^{((\epsilon-1)/2)((q-1)/2)}\left(\frac{|\Delta_0|}{q}\right) \qquad \text{(by Eq. (5))}$$

$$= (-1)^{((\epsilon-1)/2)((q-1)/2)+((|\Delta_0|-1)/2)((q-1)/2)}\left(\frac{q}{|\Delta_0|}\right) \qquad \text{(by Eq. (4))}$$

$$= (-1)^{(((\epsilon-1)/2)+((|\Delta_0|-1)/2))((q-1)/2)}\left(\frac{q}{|\Delta_0|}\right)$$

$$= (-1)^{(((\epsilon-1)/2)+((|\Delta_0|-1)/2))(q-1)/2)}\left(\frac{2}{|\Delta_0|}\right) \qquad \text{since } q \equiv 2(\text{mod } \Delta_0))$$

$$= (-1)^{(|\Delta_0|^2-1)/8} \qquad \left(\text{since } \frac{\epsilon-1}{2}+\frac{|\Delta_0|-1}{2} \text{ is even and by Eq. (6)}\right)$$

$$= \begin{cases} +1 & \text{if } \Delta_0 \equiv 1(\text{mod } 8), \\ -1 & \text{if } \Delta_0 \equiv 5(\text{mod } 8). \end{cases}$$

Thus, q is decomposed if $\Delta_0 \equiv 1 \pmod 8$ and inert if $\Delta_0 \equiv 5 \pmod 8$ by Theorem 4. ■

Example 7: Let $\mathfrak{O} = \{1, \omega_5\}$. Here $\Delta = \Delta_0 = 5$, $\mathcal{f} = 1$. The only ramified prime is 5, and the factorization of primes is determined by their respective residue classes modulo $\Delta_0 = 5$. By Theorem 4, since $\Delta_0 \equiv 5 \pmod 8$, we see that 2 is inert. Now suppose that p is odd and that $p \neq 5$. To determine whether p is inert or decomposed, we must determine $\left(\frac{5}{p}\right)$. But from our discussion in Chapter 4,

$$\left(\frac{5}{p}\right) = \left(\frac{p}{5}\right) = \begin{cases} +1 & \text{if } p \equiv 1, 4 \pmod 5, \\ -1 & \text{if } p \equiv 2, 3 \pmod 5). \end{cases}$$

Thus, an odd prime p is decomposed if $p \equiv 1$ or $4 \pmod 5$ and inert if $p \equiv 2$ or $3 \pmod 5$. Thus, for example, 17 is inert, but 101 is decomposed.

9.5 Exercises

1. Factor the ideals $2I_3$, $5I_{-5}$, $3I_{-7}$, $11I_{-22}$ into products of prime ideals.
2. Find all prime ideals of norm ≤ 20 in I_{-5}.
3. Determine whether the prime p ramifies, is inert, or is decomposed in I_7, where
   (a) $p = 5$. (b) $p = 11$.
   (c) $p = 7$. (d) $p = 41$.
4. Determine which primes are ramified, inert, or decomposed in I_{11} and I_{-17}.

*5. Give an example of a prime $p \mid \Delta$ such that $p\mathfrak{O}$ is not the square of a prime module belonging to $\mathfrak{O}$.

*6. Give an example to show that unique factorization into prime modules does not necessarily hold for modules whose norm is not relatively prime to $\mathcal{f}$.

7. Determine whether $p = 2, 3, 5, 7$ are inert, ramified, or decomposed in $\mathfrak{O} = \{1, \mathcal{f}\omega_3\}$, where $\mathcal{f} = 1, 2, 6, 12, 21$.

9.6 Finiteness of the Class Number

As an application of our factorization theory we shall now prove one of the fundamental theorems in this theory, namely the finiteness of the class number of a coefficient ring. This result will be extremely valuable in our treatment of binary quadratic forms in Chapter 11 and in our attack on the Bachet equation in Chapter 10.

Let us fix a quadratic field $\mathbf{Q}(\sqrt{d})$ and a ring of coefficients $\mathcal{O} = \{1, f\omega_d\}$ contained in $\mathbf{Q}(\sqrt{d})$. Let us begin by dividing the modules belonging to $\mathcal{O}$ into classes.

Definition 1: Let M_1, M_2 be modules. We say that M_1 and M_2 are *similar* provided that there is a nonzero γ in $\mathbf{Q}(\sqrt{d})$ such that $M_1 = \gamma M_2$. Write $M_1 \sim M_2$.

Proposition 2: Similarity of modules is an equivalence relation.

Proof: Exercise. ∎

Let $[M]$ denote the set of all modules similar to M. If M belongs to $\mathcal{O}$, note that every module in $[M]$ also belongs to $\mathcal{O}$ (Lemma 8.6.12). The collection of modules $[M]$ is called the *similarity class of* $\mathcal{O}$ *determined by* M.

Definition 3: The number of distinct similarity classes of $\mathcal{O}$ is called the *class number of* $\mathcal{O}$, denoted $h_{\mathcal{O}}$. In the case $\mathcal{O} = I_d$, the ring of all integers of $\mathbf{Q}(\sqrt{d})$, the class number of $\mathcal{O}$ is also called the *class number of* $\mathbf{Q}(\sqrt{d})$ and is denoted h_d.

A priori, the class number of $\mathcal{O}$ may be infinite. However, we have the following fundamental theorem:

Theorem 4: $h_{\mathcal{O}}$ is finite.

To prove Theorem 4, we shall require several lemmas.

Lemma 5: Let M be any module belonging to $\mathcal{O}$. There exists a rational integer $k \neq 0$ such that $kM \subseteq \mathcal{O}$. Thus, any module M belonging to $\mathcal{O}$ is similar to an integral module.

Proof: Let g be the positive integer of Lemma 8.8.5, so that $gM \subseteq I_d$. Then we may set $k = fg$. ∎

Lemma 5 shows us that, insofar as similarity is concerned, we may restrict ourselves to integral modules belonging to $\mathcal{O}$.

Lemma 6: Let $M \subseteq \mathcal{O}$ be an integral module belonging to $\mathcal{O}$. Then there exists a nonzero γ in M such that

$$|N(\gamma)| \leq N(M)\,|\Delta_{\mathcal{O}}|.$$

Proof: Since $\mathcal{O} = \{1, f\omega_d\}$, we easily see that $\Delta_{\mathcal{O}} = f^2 d$ or $4f^2 d$ according to whether $d \equiv 1 \pmod 4$ or $d \equiv 2$ or $3 \pmod 4$. Consider the following set of elements of $\mathcal{O}$:

$$a + bf\omega_d \qquad (0 \leq a, b \leq [N(M)^{1/2}],\ a, b \text{ rational integers}).$$

There are $([N(M)^{1/2}] + 1)^2$ such elements, and they are all different. But since

$$([N(M)^{1/2}] + 1)^2 > N(M),$$

there are more than $N(M)$ elements in the set. However, since there are only $N(M)$ residue classes modulo M (see Definition 2.7), two of the elements must be congruent modulo M. In other words, there exist rational integers a, a', b, b' such that

$$0 \leq a, b, a', b' \leq [N(M)^{1/2}]$$

and either $a \neq a'$ or $b \neq b'$, satisfying

$$a + b\mathcal{f}\omega_d \equiv a' + b'\mathcal{f}\omega_d (\text{mod } M).$$

Set $\gamma = (a - a') + (b - b')\mathcal{f}\omega_d$. Then $\gamma \neq 0$ and γ is in M. If $\gamma = r + s\mathcal{f}\omega_d$, then $|r| = |a - a'| \leq N(M)^{1/2}$, $|s| = |b - b'| \leq N(M)^{1/2}$. Thus, in the case $d \equiv 2$ or $3 (\text{mod } 4)$, we have

$$|N(\gamma)| = |r^2 - \mathcal{f}^2 ds^2| \leq N(M)(1 + \mathcal{f}^2 |d|) < N(M) \cdot 4\mathcal{f}^2 d = N(M)|\Delta_\mathfrak{O}|.$$

In the case $d \equiv 1 (\text{mod } 4)$, we have

$$\begin{aligned} |N(\gamma)| &= \left| r^2 + rs\mathcal{f} + \mathcal{f}^2 \frac{d-1}{4} s^2 \right| \\ &\leq N(M)\left(1 + \mathcal{f} + \mathcal{f}^2 \frac{|d|}{4}\right) < N(M)\mathcal{f}^2 |d| = N(M)|\Delta_\mathfrak{O}|. \end{aligned}$$

■

Proposition 7: Every module belonging to $\mathfrak{O}$ is similar to an integral module M such that $N(M) \leq |\Delta_\mathfrak{O}|$.

Proof: As remarked above, we may restrict ourselves to consideration of integral modules M_1 belonging to $\mathfrak{O}$. By Theorem 3.8, we have

$$M_1 M_1' = N(M_1)\mathfrak{O}, \tag{1}$$

where M_1' is the conjugate module of M_1. Since $M_1' \subseteq \mathfrak{O}$ and $N(M_1') = N(M_1)$ (exercise), Lemma 6 shows that there exists a nonzero γ in M_1' such that

$$|N(\gamma)| \leq N(M_1)|\Delta_\mathfrak{O}|. \tag{2}$$

Since γ is in M_1' and since $\mathfrak{O}$ is the coefficient ring of M_1', we see that $\gamma\mathfrak{O} \subseteq M_1'$. Thus, since $\gamma\mathfrak{O}$ has $\mathfrak{O}$ as its coefficient ring, Theorem 4.6 implies that there exists an integral module M_2 belonging to $\mathfrak{O}$ such that $M_1' M_2 = \gamma\mathfrak{O}$. Using Eq. (1), we see that

$$(M_1 M_1')M_2 = N(M_1)M_2 = M_1(M_1' M_2) = M_1 \gamma\mathfrak{O} = \gamma M_1.$$

Thus, $M_2 = (\gamma / N(M_1))M_1$ is similar to M_1. Moreover, using (2), we obtain

$$\begin{aligned} N(M_2) &= N\left(\frac{\gamma}{N(M_1)} M_1\right) = \frac{|N(\gamma)|}{N(M_1)^2} N(M_1) \\ &= \frac{|N(\gamma)|}{N(M_1)} \leq |\Delta_\mathfrak{O}|. \end{aligned}$$

■

From Proposition 7, we see that in order to prove the finiteness of $h_{\mathcal{O}}$, it suffices to establish the following result:

Proposition 8: Let C be a fixed positive number. Then there exist only finitely many integral modules M belonging to $\mathcal{O}$ and such that $N(M) \leq C$.

Proof: Let M be any integral module belonging to $\mathcal{O}$. We showed in Corollary 2.5 that M has a basis of the following very special type: $M = \{a, b + c\mathcal{f}\omega_d\}$, where $a > 0$ is the smallest rational integer in M and $b + c\mathcal{f}\omega_d$ is the number in M with $c > 0$ least and $0 \leq b < a$. Moreover, we showed in Theorem 2.6 that $N(M) = ac$. Thus, $N(M) \leq C$ implies $ac \leq C$, and so there are only a finite number of possible values of a and c. Since $0 \leq b < a$, there are only finitely many possible values for b. Thus, there are only finitely many different modules $M = \{a, b + c\mathcal{f}\omega_d\}$ such that $N(M) \leq C$. ■

Theorem 4 is now completely proved.

Note that Propositions 7 and 8 actually provide us with a method for computing the class number of $\mathcal{O} = \{1, \mathcal{f}\omega_d\}$, for by Proposition 7, every module belonging to $\mathcal{O}$ is similar to an integral module with

$$N(M) \leq |\Delta_{\mathcal{O}}|. \tag{3}$$

If we specialize to the case where $\mathcal{f} = 1$, then such an M may be factored into a product of prime modules, say $M = P_1 \cdots P_t$. However, from Theorem 5.2, we know that $N(P_i) = p_i$ or p_i^2, where p_i is a rational prime. Condition (3) is equivalent to

$$N(M) = N(P_1) \cdots N(P_t) \leq |\Delta_{\mathcal{O}}|.$$

Thus, we need only consider products of prime modules for which

$$p_1 \cdots p_t \leq |\Delta_{\mathcal{O}}|. \tag{4}$$

There are only finitely many such prime modules, and we even described them in Theorem 5.3. Thus, we can find all integral modules satisfying (3). From these we can then determine which are similar and eventually arrive at a set of integral modules belonging to $\mathcal{O}$ such that every module belonging to $\mathcal{O}$ is similar to one and only one module of the set. This is for $\mathcal{f} = 1$. Modifications will be given in the exercises for $\mathcal{f} > 1$.

The above procedure is quite viable computationally. However, Proposition 7 is very clumsy because $|\Delta_{\mathcal{O}}|$ is usually quite large. But $|\Delta_{\mathcal{O}}|$ can be replaced by a much smaller number. Namely, we have

Proposition 7′ (Minkowski): Every module belonging to $\mathcal{O}$ is similar to an integral module M such that

$$N(M) \leq \frac{\sqrt{3}}{3} |\Delta_{\mathcal{O}}|^{1/2}.$$

The main advantage of Proposition 7′ is that $|\Delta_{\mathfrak{O}}|$ is replaced by $|\Delta_{\mathfrak{O}}|^{1/2}$, which is much smaller. We shall outline a proof of Proposition 7′ in the exercises. However, let us use Proposition 7′ to calculate a few examples.

Example 9: Let $d = 29$, $\mathfrak{f} = 1$. Let $\mathfrak{O} = I_{29} = \{1, (1 + \sqrt{29})/2\}$. We shall compute the class number h_{29} of $\mathfrak{O}$. From Proposition 7′ we see that every module is similar to an integral module (= ideal) M, where

$$N(M) \leq \frac{\sqrt{3}}{3}|\Delta_{\mathfrak{O}}|^{1/2} = \sqrt{3}\frac{\sqrt{29}}{3} < 4,$$

since $\Delta_{\mathfrak{O}} = 29$ (because $29 \equiv 1 \pmod 4$). Thus, M has norm 1, 2, or 3. If M has norm 1, then $M = I_{29}$. Moreover, since $2 \nmid \Delta_{\mathfrak{O}}$ and $29 \equiv 5 \pmod 8$, Theorem 5.4 implies that 2 is inert. Also by Theorem 5.4 we see that 3 is inert since $\left(\frac{3}{29}\right) = \left(\frac{29}{3}\right) = \left(\frac{2}{3}\right) = -1$. Since $2\mathfrak{O}$ has norm 4 and $3\mathfrak{O}$ has norm 9, we see there are no modules of norm 2 or 3. Thus, every module belonging to $\mathfrak{O}$ is similar to $\mathfrak{O}$. That is, $h_{\mathfrak{O}} = 1$.

Let us do another example.

Example 10: Let $d = -6$, $\mathfrak{f} = 1$. Since $-6 \equiv 2 \pmod 4$, we see that $\Delta_{\mathfrak{O}} = -24$. From Proposition 7′ we see that every module (= ideal) of $\mathfrak{O} = I_{-6}$ is similar to a module M such that

$$N(M) \leq \frac{\sqrt{3}}{3}|\Delta_{-6}|^{1/2} < 3.$$

Therefore, M has norm 1 or 2. If $N(M) = 1$, then $M = I_{-6}$. Since $2 | \Delta_{\mathfrak{O}}$, we see that 2 is ramified:

$$2\mathfrak{O} = P_2^2.$$

Finally we must determine whether P_2 and I_{-6} are similar. If $P_2 = \gamma I_{-6}$, then writing $\gamma = a + b\sqrt{-6}$ for a, b rational integers, we must have $a^2 + 6b^2 = N(\gamma) = N(\gamma I_{-6}) = N(P_2) = 2$. However, it is trivially checked that $a^2 + 6b^2 = 2$ cannot be solved. Therefore, P_2 is not principal. Thus, every module is similar to I_{-6} or P_2 and they are not similar, and so we conclude that $h_{-6} = 2$.

The following easily proved theorem shows the close connection between the class number of $\mathfrak{O} = I_d$ and the presence (or absence) of unique factorization in I_d.

Theorem 11: Let h_d denote the class number of I_d. Then $h_d = 1$ if and only if I_d is a unique factorization domain.

Proof: Exercise. ■

We shall close this section by observing that the set of similarity classes of modules belonging to $\mathfrak{O}$ forms a group in a natural way.

First let $\mathfrak{M}$ denote the collection of all modules belonging to $\mathcal{O}$. Among the modules of $\mathfrak{M}$, we have defined the operation of multiplication of modules. By Corollary 3.11 the product of two modules in $\mathfrak{M}$ belongs to $\mathfrak{M}$. From Proposition 3.6 we know that this operation is commutative and associative and that $\mathcal{O}$ is the identity. Finally, from Theorem 3.8, we see that if M is in $\mathfrak{M}$ and $M_1 = (1/(N(M))M'$, then $MM_1 = \mathcal{O}$, and so M_1 is an inverse for M. Thus, $\mathfrak{M}$ is an abelian group with respect to multiplication of modules.

Now let $\mathcal{P}$ denote the subgroup of $\mathfrak{M}$ consisting of all $\gamma\mathcal{O}$ such that $\gamma \neq 0$ belongs to $\mathbf{Q}(\sqrt{d})$. We observe that the quotient group $\mathfrak{M}/\mathcal{P}$ is just the set of all similarity classes of modules belonging to $\mathcal{O}$ (exercise). Denote $\mathcal{J}_{\mathcal{O}} = \mathfrak{M}/\mathcal{P}$.

Definition 12: $\mathcal{J}_{\mathcal{O}}$ is called the *ideal class group of* $\mathcal{O}$.

Theorem 13: $\mathcal{J}_{\mathcal{O}}$ is a finite abelian group of order $h_{\mathcal{O}}$ with respect to the operation $[M_1][M_2] = [M_1M_2]$. Note that the identity is $[\mathcal{O}]$ and that

$$[M]^{-1} = [M'].$$

The ideal class group is of major importance in number theory. We shall make use of it in Sections 10.1 and 11.5.

9.6 Exercises

1. Let M be any module of $\mathbf{Q}(\sqrt{d})$. Show that $N(M) = N(M')$.
2. Prove Proposition 2.
3. Determine whether the following modules are similar:
   (a) $M_1 = \{1, \sqrt{3}\}$, $M_2 = \{3 + \sqrt{3}, 3 + 3\sqrt{3}\}$.
   (b) $M_1 = \{1, \sqrt{10}\}$, $M_2 = \{2, \sqrt{10}\}$.
   (c) $M_1 = \{2 + \sqrt{3}, 1 + \sqrt{3}\}$, $M_2 = \{(7 + 5\sqrt{3})/2, 4 + \sqrt{3}\}$.
   (d) $M_1 = \{2, 3\sqrt{2}\}$, $M_2 = \{5, 3\sqrt{2}\}$.
4. Verify that $h_{-1} = 1$ using the Minkowski bound.
5. Compute the following class numbers: $h_{-2}, h_3, h_5, h_{10}, h_{-5}$.
6. Verify that $h_{-23} = 3$.
7. Compute the class numbers of the coefficient rings $\mathcal{O}_f$ in $\mathbf{Q}(\sqrt{-1})$ for $f = 2, 3, 5$.
8. Determine a γ satisfying Lemma 6 for $M = \{100, 50 + \sqrt{-2}\}$. For $M = \{100, 49 + \sqrt{-1}\}$.
9. Give an example of an element in I_{10} which does not have unique factorization. Do the same for I_{-10}.

**10. Prove the following: Let $n > 1$ be a rational integer. Let I_d be the integers in $\mathbf{Q}(\sqrt{d})$. Then for any ideal M there is an ideal M_1 similar to

M such that $\gcd(n, N(M_1)) = 1$. (This allows us to adapt the method for computing $h_{\mathfrak{O}}$ for $\mathcal{f} = 1$ given in the text for $\mathcal{f} > 1$.)

11. Prove Proposition 8 for I_d using unique factorization.

12. Prove that for an arbitrary coefficient ring $\mathfrak{O}$ there can only be a finite number of distinct factorizations for any given module M associated with $\mathfrak{O}$.

Exercises 13–19 constitute a proof of the Minkowski theorem (Proposition 7′).

Definition: Denote the Euclidean plane by $\mathbf{R}^2$. Let A_0, B_0 be in $\mathbf{R}^2$ and assume that they are linearly independent. By a *lattice* Λ we mean the collection of points in $\mathbf{R}^2$ of the form $aA_0 + bB_0$ such that a, b are rational integers. Call A_0, B_0 a *basis* of Λ.

13. Let A'_0, B'_0 belong to Λ. Show that A'_0, B'_0 is also a basis of Λ if and only if there are rational integers r, s, t, u such that $ru - st = \pm 1$ and

$$A'_0 = rA_0 + sB_0$$
$$B'_0 = tA_0 + uB_0.$$

14. Define the *determinant* of Λ to be the absolute value of the determinant of the 2×2 matrix (A_0, B_0) (denoted $det\ \Lambda$).
    (a) Show that the number $det\ \Lambda$ is the same for any possible choice of bases.
    (b) Show that $det\ \Lambda = \|A_0\| \cdot \|B_0\| \sin\theta$, where θ is the angle between A_0 and B_0, $0 < \theta < 180°$, and where $\|A_0\|$ is the length of A_0.

    Now choose two points X_1, X_2 in Λ as follows: Let X_1 be the smallest nonzero vector in Λ. Let X_2 be the smallest vector in Λ linearly independent of X_1 (i.e., $X_2 \neq tX_1$ for all real t).

15. (a) Show that X_1, X_2 is a basis of Λ.
    (b) Show that for all rational integers a, b with $b \neq 0$ we have

$$\|X_2\| \leq \|aX_1 + bX_2\|.$$

    (c) Let θ be the angle between X_1 and X_2, where $0 < \theta < 180°$. Use part (b) with $a = 1, b = \pm 1$ to show that, in fact, $60° \leq \theta \leq 120°$.

16. (Minkowski's theorem) Use Exercises 15(a) and (c) and 14(b) to show that

$$\|X_1\|^2 \leq \frac{2\sqrt{3}}{3}\ det\ \Lambda.$$

    That is, for any lattice Λ in $\mathbf{R}^2$ there is an $X \neq 0$ in Λ such that

$$\|X\|^2 \leq \frac{2\sqrt{3}}{3}\ det\ \Lambda.$$

17. Let M be a module in $\mathbf{Q}(\sqrt{d})$ for $d < 0$. Thus, M is a subset of $\mathbf{C}$, the complex numbers. View $\mathbf{C}$, as usual, as $\mathbf{R}^2$ via the real and imaginary parts of a complex number.
    (a) Show that M is a lattice.
    (b) Show that $\mathit{det}\ M = \frac{1}{2}|\Delta_M|^{1/2}$.
    (c) Show that there is a $\gamma \neq 0$ in M such that

$$|N(\gamma)| \leq \frac{\sqrt{3}}{3}|\Delta_M|^{1/2}.$$

    (d) Note that part (c) is the best possible for $M = I_{-3}$ (i.e., the constant $\sqrt{3}/3$ cannot be any smaller).

18. Let M be a module in $\mathbf{Q}(\sqrt{d})$ for $d > 0$. Let M^* be the subset of $\mathbf{R}^2$ consisting of all pairs (α, α') such that α belongs to M.
    (a) Show that M^* is a lattice.
    (b) Show that $\mathit{det}\ M^* = |\Delta_M|^{1/2}$.
    (c) Show that there is a $\gamma \neq 0$ in M such that

$$|N(\gamma)| \leq \frac{\sqrt{3}}{3}|\Delta_M|^{1/2}.$$

    (*Hint:* Exercise 16 guarantees that there is a $\gamma \neq 0$ in M such that $\gamma^2 + \gamma'^2 \leq (2\sqrt{3}/3)|\Delta_M|^{1/2}$.)

19. Let $\mathfrak{O}$ be a coefficient ring. Show that every integral module belonging to $\mathfrak{O}$ is similar to an integral module M belonging to $\mathfrak{O}$ such that

$$N(M) \leq \frac{\sqrt{3}}{3}|\Delta_{\mathfrak{O}}|^{1/2}.$$

20. Prove Theorem 11.

10

Applications of the Factorization Theory to Diophantine Equations

10.1 The Diophantine Equation $y^2 = x^3 + k$

In this section, we shall use the methods of quadratic fields to study the Bachet equation $y^2 = x^3 + k$, where k is a given rational integer. We described the history of this Diophantine equation in Chapter 3. The important fact to recall is that Mordell proved that the Bachet equation always has a finite number of solutions (x, y). Explicit bounds for solutions (x, y) in terms of k have very recently been obtained by Alan Baker and Harold Stark. These results are far beyond the scope of the present book, but we can verify them for special classes of k, using the machinery of Chapters 8 and 9. Our most striking result will be that the only solutions to $y^2 = x^3 - 74$ are given by $x = 99$, $y = \pm 985$. The methods of this section provide the reader with a fairly typical example of the way in which the theory of quadratic fields can be applied to solve Diophantine equations. The reader should compare the results of this section with the results on the Bachet equation obtained using more elementary ideas in Section 3.3 ($k = 23$), Section 4.5 ($k = 45$), and Exercise 12 in Section 3.3.

Let us write the Bachet equation in the form

$$(y + \sqrt{k})(y - \sqrt{k}) = x^3 \tag{1}$$

and interpret this equation as an equation involving elements of $\mathbf{Q}(\sqrt{k})$. For simplicity let us make the following assumption throughout this section: *k is square-free.* Equation (1) implies the following relation among ideals of I_k:

$$(y + \sqrt{k})I_k \cdot (y - \sqrt{k})I_k = (xI_k)^3. \tag{2}$$

We shall show that for certain values of k, the ideals $(y + \sqrt{k})I_k$ and $(y - \sqrt{k})I_k$ have no prime factors in common, so that, by the unique factorization theorem (Corollary 9.4.11), we may conclude that both $(y + \sqrt{k})I_k$ and $(y - \sqrt{k})I_k$ are cubes of ideals. But it is very difficult for these ideals to be cubes because of the very special form of $y + \sqrt{k}$, namely that the coefficient of $\sqrt{k}$ is 1. An analysis of the various possibilities leads to a determination of all solutions of the Bachet equation. Unfortunately it is not known (as of this writing) how to carry out such an analysis for general k. However, we shall show how various restrictive assumptions about k suffice to make the analysis possible.

The basic tool to be used in our study of the Bachet equation is the ideal class group of I_k, defined in Definition 9.6.12. It allows us to prove the following basic lemma:

Lemma 1: Let A be an ideal of I_k, and h_k the class number of $\mathbf{Q}(\sqrt{k})$.

(i) A^{h_k} is a principal ideal.
(ii) Let m be a positive rational integer such that $\gcd(m, h_k) = 1$. Then, if A^m is a principal ideal, A is also a principal ideal.

Proof: Let $\mathcal{I}_k$ denote the ideal class group of I_k (See Definition 9.6.12). We observe that an ideal B is principal if and only if $[B] = [I_k]$. Since the order of $\mathcal{I}_k$ is h_k and the identity of $\mathcal{I}_k$ is $[I_k]$, we know from elementary group theory that $[A]^{h_k} = [I_k]$, so that $[A]^{h_k} = [A^{h_k}]$ immediately gives part (i). To show part (ii) we need to show that $[A]^m = [I_k]$ implies that $[A] = [I_k]$. This again is an elementary fact from group theory. Indeed, $\gcd(m, h_k) = 1$ implies that there are rational integers x, y such that $mx + h_k y = 1$, and so

$$[A] = [A]^{mx+h_ky} = ([A]^m)^x([A]^{h_k})^y$$
$$= [I_k]^x[I_k]^y = [I_k]. \qquad \blacksquare$$

Let us consider three possible restrictive conditions on k:

Case 1: $k < -1$, $k \equiv 2$ or $3 \pmod 4$, $3 \nmid h_k$. We immediately conclude from Theorem 8.3.2 that $I_k = \{1, \sqrt{k}\}$. Moreover, since $k \equiv 2$ or $3 \pmod 4$, the relation $y^2 = x^3 + k$ implies that $\gcd(x, 2k) = 1$. For if $2 | x$, then $y^2 \equiv k \pmod 8$; but $y^2 \equiv 0, 1$, or $4 \pmod 8$ and $k \equiv 2, 3, 6$, or $7 \pmod 8$, which gives a contradiction. Further, if p is an odd prime such that $p | x$ and $p | k$, then $p | y$ and $p^2 | y^2 - x^3$, so that $p^2 | k$, a contradiction to the assumption that k is square-free. Thus, $\gcd(x, 2k) = 1$.

Let us now prove that the ideals $(y + \sqrt{k})I_k$, $(y - \sqrt{k})I_k$ have no prime factors in common. Let P be a prime ideal of I_k such that

$$P | (y + \sqrt{k})I_k \quad \text{and} \quad P | (y - \sqrt{k})I_k.$$

By Theorem 9.4.6, $y + \sqrt{k}$ and $y - \sqrt{k}$ belong to P, and thus

$$2\sqrt{k} = (y + \sqrt{k}) - (y - \sqrt{k})$$

is in P, and also

$$x^3 = (y + \sqrt{k})(y - \sqrt{k})$$

is in P. But then $P \mid 2\sqrt{k} I_k$ and $P \mid x^3 I_k$. But $x^3 I_k = (xI_k)^3$ and P is a prime ideal, so that $P \mid xI_k$. Therefore,

$$N(P) \mid N(2\sqrt{k} I_k) = 4|k|, \qquad N(P) \mid N(xI_k) = x^2,$$

which implies that $\gcd(x^2, 4k) > 1$, contradicting the fact that $\gcd(x, 2k) = 1$. Thus, we have proved that $(y + \sqrt{k})I_k$ and $(y - \sqrt{k})I_k$ have no common prime factors.

From Eq. (2) we can now deduce that each of the ideals $(y + \sqrt{k})I_k$, $(y - \sqrt{k})I_k$ is the cube of an ideal (by unique factorization of ideals). In particular, there exists an ideal A of I_k such that

$$(y + \sqrt{k})I_k = A^3. \tag{3}$$

Let us now use the hypothesis $3 \nmid h_k$. From Eq. (3), we have

$$[A]^3 = [I_k],$$

so that from Lemma 1, $[A] = [I_k]$ and A is a principal ideal, say $A = (a + b\sqrt{k})I_k$ for rational integers a, b. Then (3) can be rewritten

$$(y + \sqrt{k})I_k = (a + b\sqrt{k})^3 I_k.$$

By Lemma 9.1.10, we can deduce that

$$y + \sqrt{k} = \epsilon(a + b\sqrt{k})^3 \tag{4}$$

for some unit ϵ of I_k. However, since $k < -1$, $k \equiv 2, 3 \pmod 4$, we see that $k = -2$ or $k < -3$, so the only units of I_k are ± 1 (Theorem 8.7.5).

Thus,

$$y + \sqrt{k} = \pm(a + b\sqrt{k})^3.$$

By multiplying this last equation out, we see that

$$y = \pm a(a^2 + 3b^2k)$$

$$1 = \pm b(3a^2 + b^2k).$$

The last equation immediately implies that $b = \pm 1$ and $3a^2 + k = \pm 1$. In particular, k must be of the form

$$k = \pm 1 - 3a^2 \tag{5}$$

or there are no solutions. If Eq. (5) holds, then

$$y = \pm a(a^2 + 3k).$$

We may obtain x from

$$x = N(a + b\sqrt{k}) = a^2 - k.$$

Thus, we have proved the following result:

Theorem 2: Let k be a negative, square-free rational integer, $k \equiv 2$ or $3 \pmod 4$, $k \neq -1$. Assume that $3 \nmid h_k$. Then the Diophantine equation

$y^2 = x^3 + k$ can be solved if and only if there is a rational integer a such that $k = \pm 1 - 3a^2$. In this case, there are two solutions $(x, y) = (a^2 - k, \pm a(a^2 + 3k))$.

Let us find some specific values of k to which Theorem 2 applies. If $|k| \leq 100$, $k = \pm 1 - 3a^2$, then a is one of 1, 2, 3, 4, 5. Moreover, if $k \equiv 2$ or 3(mod 4) and is square-free, the only possibilities are $k = -2, -13, -26, -74$. Referring to Table 3, we see that $h_{-26} = 6$, so that -26 must be eliminated. All of $k = -2, -13, -74$ have corresponding class numbers not divisible by 3. Thus, we obtain

Corollary 3: The solutions of

(i) $y^2 = x^3 - 2$ are given by $(x, y) = (3, \pm 5)$.
(ii) $y^2 = x^3 - 13$ are given by $(x, y) = (17, \pm 70)$.
(iii) $y^2 = x^3 - 74$ are given by $(x, y) = (99, \pm 985)$.

If $k = -5, -6, -10, -14, -17, -21, -22, -30, -33, -34, -37, -41, -42, -46, -57, -58, -62, -65, -66, -69, -70, -73, -77, -78, -82, -85, -86, -89, -93, -94$, or -97, the equation $y^2 = x^3 + k$ has no solutions.

Proof: The list of k given includes all square-free, negative k, $|k| \leq 100$, $k \equiv 2$ or 3(mod 4), $3 \nmid h_k$, k not of the form $\pm 1 - 3a^2$. (Again, consult Table 3.) ■

Let us now give an example of the above sort of analysis for k's which are positive.

Case 2: $k > 0$, $k \equiv 2$ or 3(mod 4), $3 \nmid h_k$. The main difficulty in this case is caused by the presence of nontrivial units in I_k. Let ϵ_0 be the fundamental unit of I_k. Then any unit ϵ of I_k is of the form $\pm\epsilon_0^n$. The reasoning of the previous case works up to and including Eq. (4). Suppose that $\epsilon = \pm\epsilon_0^n$. If $N(\epsilon_0) = 1$, write $n = 3m + r$, $r = -1, 0, 1$; if $N(\epsilon_0) = -1$, write $n = 3m + r$, $r = -2, 0, 2$ (possible since $-2, 0, 2$ is a complete residue system modulo 3). Then, in either case,

$$y + \sqrt{k} = \epsilon_0^r(\pm\epsilon_0^m(a + b\sqrt{k}))^3. \tag{6}$$

Let $\eta_0 = \epsilon_0$ if $N(\epsilon_0) = +1$, $\eta_0 = \epsilon_0^2$ if $N(\epsilon_0) = -1$. Since $\pm\epsilon_0^m(a + b\sqrt{k})$ belongs to I_k, ϵ in Eq. (4) may be replaced by one of $\eta_0^{-1}, 1, \eta_0$. That is, (4) may be rewritten in the form

$$y + \sqrt{k} = \eta(a + b\sqrt{k})^3, \tag{7}$$

where $\eta = \eta_0, 1$, or η_0^{-1}. Suppose that $\eta_0 = u + v\sqrt{k}$. Then $u^2 - kv^2 = +1$, and $u > 0$, $v > 0$ by Lemma 6.6.6.

If $\eta = 1$, then from Eq. (7), we deduce that $1 = b(3a^2 + kb^2)$, so that $b = \pm 1$, $k = \pm 1 - 3a^2$. If $|a| \geq 1$, there are no solutions since $k > 0$. If

$a = 0$, there are no solutions since $k \equiv 2, 3 \pmod 4$. Thus, since $\eta_0^{-1} = u - v\sqrt{k}$, we see that $\eta = u \pm v\sqrt{k}$. Multiplying out Eq. (7) and equating the coefficients of $\sqrt{k}$, we obtain

$$u(3a^2 + kb^2)b \pm v(a^2 + 3kb^2)a = 1 \tag{8}$$

or

$$\pm va^3 + 3ua^2b \pm 3vkab^2 + ukb^3 = 1. \tag{9}$$

Recall that u, v are known. (We can compute them in a finite number of steps for given k.) We must examine Eq. (9) for solutions (a, b). In general, this is a quite delicate task. However, in some cases, congruence considerations suffice.

For example, let $k = 7$. Then $h_7 = 1$, $u = 8$, $v = 3$ (see Table 2 at the end of the book). Then (9) becomes

$$\pm 3a^3 + 24a^2b \pm 63ab^2 + 56b^3 = 1. \tag{10}$$

By examining congruences modulo 3, Eq. (10) reads $56b^3 \equiv 1 \pmod 3$ or $b \equiv -1 \pmod 3$. However, by (10), read modulo 9, we obtain that

$$\pm 3a^3 + 6a^2b + 2b^3 \equiv 1 \pmod 9.$$

Since $b \equiv -1 \pmod 3$ implies that $b^3 \equiv -1 \pmod 9$, we obtain

$$\pm 3a^3 + 6a^2b \equiv 3 \pmod 9,$$

so that $\pm a^3 + 2a^2b \equiv 1 \pmod 3$, which implies that $\pm a^3 - 2a^2 \equiv 1 \pmod 3$ (since $b \equiv -1 \pmod 3$). However, checking the three cases, $a \equiv 1, 0, -1 \pmod 3$, we see that the last congruence is always impossible to satisfy. *Thus, the Diophantine equation $y^2 = x^3 + 7$ has no solutions.*

Case 3: $k = -31$. In this case, we avoid the difficulty $k > 0$, but we have the additional difficulty that $3 \mid h_{-31}$, since from Table 3 at the end of the book, $h_{-31} = 3$. Let us show how even an example of this sort can be handled by the methods of the theory of quadratic fields. Let us assume that

$$y^2 = x^3 - 31. \tag{11}$$

First assume that x is odd. If $x \equiv 1 \pmod 4$, then (11) implies that $y^2 \equiv 2 \pmod 4$, which is impossible. If $x \equiv -1 \pmod 4$, we may write (11) in the form

$$y^2 + 4 = (x - 3)(x^2 + 3x + 9),$$

and $x^2 + 3x + 9 \equiv -1 \pmod 4$, so that $x^2 + 3x + 9$ has a prime divisor p such that $p \equiv -1 \pmod 4$. But $y^2 \equiv -4 \pmod p$, and so $\left(\frac{-4}{p}\right) = 1$. However, $\left(\frac{-4}{p}\right) = \left(\frac{-1}{p}\right) = -1$ since $p \equiv -1 \pmod 4$. Thus, $x \not\equiv -1 \pmod 4$ and x must be even, say $x = 2x_1$. It then follows from (11) that y is odd. We may write (11) in the form

$$\frac{y + \sqrt{-31}}{2}\,\frac{y - \sqrt{-31}}{2} = 2x_1^3. \tag{12}$$

Since $-31 \equiv 1 \pmod 4$, we see that $I_{-31} = \{1, (1 + \sqrt{-31})/2\}$. Thus, since y is odd, we have that $(y \pm \sqrt{-31})/2$ belongs to I_{-31}. Moreover, by using the same argument as in Case 1, we can show that the two ideals

$$\frac{y + \sqrt{-31}}{2} I_{-31}, \qquad \frac{y - \sqrt{-31}}{2} I_{-31}$$

are relatively prime. We leave the details as an exercise.

Since $-31 \equiv 1 \pmod 8$, we see that 2 is decomposed in I_{-31} (Theorem 9.5.4). In fact, we proved that

$$2I_{-31} = P_2 P_2', \qquad P_2 = \{2, 1 + \omega\},$$

where $\omega = (1 + \sqrt{-31})/2$. We assert that P_2 is not principal, for if $P_2 = \gamma I_{-31}$, then $N(\gamma) = N(P_2) = 2$, but if $\gamma = a + b\omega$ (a, b rational integers), then $2 = a^2 + ab + 8b^2$, which is impossible (exercise).

Writing Eq. (12) as an equation involving ideals, we see that the unique factorization theorem implies that (note that $2 \nmid (y \pm \sqrt{-31})/2$)

$$P_2 A^3 = \gamma I_{-31}, \tag{13}$$

where A is some ideal of I_{-31} and γ is one of $(y \pm \sqrt{-31})/2$. But, since $h_{-31} = 3$, Corollary 3 implies that A^3 is principal. Thus, by (13), P_2 is principal, which contradicts what we just proved. *Therefore, the Diophantine equation* $y^2 = x^3 - 31$ *has no solutions.*

10.1 Exercises

1. Show that the Diophantine equation $x^2 + xy + 8y^2 = 2$ has no solutions.
2. Show that the Diophantine equation $y^2 + 4 = x^3$ has as its only solutions $(x, y) = (2, \pm 2), (5, \pm 11)$ (Fermat).
3. Find all solutions of the Diophantine equation $y^2 + 3 = x^5$.
4. Let k be a square-free negative integer such that $k = 2, 3 \pmod 4$ and $5 \nmid h_k$. Find all solutions of $y^2 + k = x^5$.
5. Find all solutions of $y^2 = x^3 - 1$.

10.2 Proof of Fermat's Last Theorem for $n = 3$

In the first half of this book, we stated Fermat's Last Theorem, which asserts that for a given integer $n \geq 3$, there are no solutions of the Diophantine equation $x^n + y^n = z^n$ in rational integers x, y, z with $xyz \neq 0$. In this section we shall give a proof of this assertion in the special case $n = 3$. Our proof will rest on the arithmetic properties of I_{-3}, the ring of integers in $\mathbf{Q}(\sqrt{-3})$.

From Theorem 8.3.2, noting that $\omega_{-3} = (1 + \sqrt{-3})/2$, we have

$$I_{-3} = \left\{1, \frac{1 + \sqrt{-3}}{2}\right\} = \left\{1, \frac{-1 + \sqrt{-3}}{2}\right\}. \tag{1}$$

Let $\zeta = (-1 + \sqrt{-3})/2$. Then it is easy to verify that $\zeta^3 = 1$. It will be more convenient to deal with ζ than with ω_{-3}. Since $\zeta^3 = 1$, we see that $(\zeta - 1)(\zeta^2 + \zeta + 1) = 0$. Therefore, since $\zeta \neq 1$, we have

$$\zeta^2 + \zeta + 1 = 0. \tag{2}$$

Equation (2) is very helpful in computations. We use the above relations involving ζ to factor $x^3 + y^3$ in I_{-3} as

$$x^3 + y^3 = (x + y)(x + \zeta y)(x + \zeta^2 y).$$

Now let us prove the basic fact about the arithmetic of I_{-3}.

Theorem 1: I_{-3} is a unique factorization domain (UFD).

Proof: By Theorem 9.6.11, it suffices to show that $h_{-3} = 1$. By using Minkowski's bound (Proposition 9.6.7′) we see that every class of ideals of I_{-3} contains an integral ideal A such that $N(A) \leq (\sqrt{3}/3)\sqrt{3} = 1$ (since the discriminant of I_{-3} is -3). That is, $N(A) = 1$, and so $A = I_{-3}$, which is the principal ideal $1 \cdot I_{-3}$. Thus, every ideal is principal and $h_{-3} = 1$. ■

A direct proof of this result, which exactly parallels the proof that I_{-1} is a UFD, can easily be given (exercise).

It is necessary to know the units of I_{-3}. We recall the following from Theorem 8.7.5

Lemma 2: The units of I_{-3} are precisely the numbers ± 1, $\pm(1 - \sqrt{-3})/2$, $\pm(1 + \sqrt{-3})/2$, that is, ± 1, $\pm\zeta$, $\pm\zeta^2$.

The result we are going to prove is

Theorem 3: The Diophantine equation

$$x^3 + y^3 = z^3$$

has no solutions in rational integers x, y, z, where $xyz \neq 0$.

We note that since $x^3 + y^3 = z^3$ is solvable if and only if $x^3 + y^3 + (-z)^3 = 0$ is solvable, Theorem 3 is equivalent to the insolubility of $x^3 + y^3 + z^3 = 0$.

Our proof of Theorem 3 will involve a descent argument (see the discussion in Chapter 6 concerning $x^4 + y^4 = z^2$). However, it is necessary to use the factorization of $x^3 + y^3$ in I_{-3} given above, and consequently our descent argument will give us new solutions to our equation with x, y, z lying in I_{-3}. Thus we are forced to prove the more general

Theorem 4: The equation

$$\xi^3 + \eta^3 + \theta^3 = 0 \tag{3}$$

has no solutions for ξ, η, θ in I_{-3} with $\xi\eta\theta \neq 0$.

Actually the statement made immediately above the statement of Theorem 4 is not quite accurate. The descent argument is based upon how divisible the three numbers ξ, η, θ are by $\lambda = 1 - \zeta$. We shall eventually wind up proving a more general assertion than Theorem 4 (Theorem 7). The point we want to emphasize is that we do not prove Theorem 7 because it is more general: We are forced to prove Theorem 7 just to be able to prove Theorem 3.

Lemma 5: Let $\lambda = 1 - \zeta = (3 - \sqrt{-3})/2$. Then $N(\lambda) = 3$, so that λ is a prime in I_{-3}. Moreover $\lambda \mid 3$.

Proof: It is trivial to compute that $N(\lambda) = \lambda\lambda' = 3$. Thus, also, $\lambda \mid 3$ since λ' is in I_{-3}. ■

We recall the concept of a congruence given at the beginning of Section 9.2. We change the notation slightly. Let γ belong to I_{-3}. We say α *is congruent to* β *mod* γ if

$$\alpha \equiv \beta (\text{mod } \gamma I_{-3}),$$

and write $\alpha \equiv \beta (\text{mod } \gamma)$. That is, $\alpha \equiv \beta (\text{mod } \gamma)$ if and only if $\gamma \mid \alpha - \beta$. We assume the properties of congruences given in Section 9.2.

Lemma 6: For any α in I_{-3} we have $\alpha \equiv -1, 0, 1 (\text{mod } \lambda)$.

Proof: Since $N(\lambda) = 3$ and $N(\lambda I_{-3}) = N(\lambda)$, we know from Theorem 9.2.11 that any complete residue system in I_{-3} mod λI_{-3} contains three elements. Thus, it suffices to show that no pair of the three numbers $-1, 0, 1$ are congruent mod λ. That is, we need to prove that λ does not divide 1, 2. But this is clear since $\lambda \mid a$ (for some rational integer a) implies that $N(\lambda) \mid a^2$, and $3 \nmid 1$ and $3 \nmid 4$. ■

If $\lambda \nmid \alpha$, then $\alpha \equiv \pm 1 (\text{mod } \lambda)$. Thus, $\alpha = \pm 1 + \kappa\lambda$ for some κ in I_{-3}. Thus,

$$\begin{aligned} \alpha^3 &= \pm 1 + 3\kappa\lambda \pm 3\kappa^2\lambda^2 + \kappa^3\lambda^3 \\ &= \pm 1 - \zeta^2\kappa\lambda^3 \mp \zeta^2\kappa^2\lambda^4 + \kappa^3\lambda^3 \qquad (\text{since } 3 = -\zeta^2\lambda^2). \end{aligned}$$

Thus, $\lambda \nmid \alpha$ *implies that*

$$\alpha^3 \equiv \pm 1 (\text{mod } \lambda^3). \tag{4}$$

Let us divide the proof of Theorem 4 into two cases.

Case 1: Equation (3) has no solutions in integers ξ, η, θ in I_{-3} such that $\xi\eta\theta \neq 0$ and $\lambda \nmid \xi\eta\theta$.

Case 2: Equation (3) has no solutions in integers ξ, η, θ in I_{-3} such that $\xi\eta\theta \neq 0$ and $\lambda \mid \xi\eta\theta$.

The first case turns out to be very easy. We proved the similar assertion for ξ, η, θ rational integers at the end of Section 3.2.

Proof of Case 1: Assume that ξ, η, θ satisfies (3) and that $\lambda \nmid \xi\eta\theta$. Then none of ξ, η, θ are divisible by λ, and so by Eq. (4), we have

$$0 = \xi^3 + \eta^3 + \theta^3 \equiv \pm 1 \pm 1 \pm 1 (\text{mod } \lambda^3).$$

If all the signs are the same, we obtain $\pm 3 \equiv 0(\text{mod } \lambda^3)$, or $\lambda^3 \mid 3$. But $N(\lambda^3) = N(\lambda)^3 = 27$, $N(3) = 9$ and $27 \nmid 9$. If two of the signs are different, then two of the terms cancel, and we obtain $\pm 1 \equiv 0(\text{mod } \lambda^3)$. That is, $\lambda^3 \mid 1$. But λ is a prime, so this is also absurd. ∎

Proof of Case 2: Now we assume that ξ, η, θ is a solution to (3) with $\xi\eta\theta \neq 0$ and $\lambda \mid \xi\eta\theta$. Then, since λ is a prime, λ divides one of ξ, η, θ. If λ divides two of ξ, η, θ, it must divide the third since $\xi^3 + \eta^3 + \theta^3 = 0$. But we may assume that ξ, η, θ have no common factors since if γ is a common factor of ξ, η, θ, then ξ/γ, η/γ, θ/γ is another solution to (3). Henceforth, we shall assume that ξ, η, θ have no common factors. Then we see that λ divides precisely one of ξ, η, θ. Without loss of generality we may assume that $\lambda \mid \theta$, $\lambda \nmid \xi$, $\lambda \nmid \eta$. Write $\theta = \lambda^s \phi$ for some rational integer $s \geq 1$, where $\lambda \nmid \phi$. Now we have

$$\xi^3 + \eta^3 + \lambda^{3s}\phi^3 = 0, \qquad \xi\eta\phi \neq 0, \lambda \nmid \xi\eta\phi. \tag{5}$$

Our descent argument goes as follows. If Eq. (5) has solutions, then it has solutions for a least value of $s \geq 1$. But given an s such that (5) has a solution ξ, η, ϕ we shall find another triple ξ, η, ϕ satisfying (5) with s replaced by $s - 1$. This contradiction shows that Eq. (5) has no solutions, and thus Theorem 4 is true.

Actually we do not quite manage to replace a given ξ, η, ϕ with another triple satisfying (5). Instead, given ξ, η, ϕ, we may replace them with another triple satisfying

$$\xi^3 + \eta^3 + \epsilon\lambda^{3(s-1)}\phi^3 = 0,$$

where ϵ is some unit. This further complication forces us to prove the even more general

Theorem 7: Let ϵ be a unit of I_{-3} and let $s \geq 1$ be a rational integer. Then the equation

$$\xi^3 + \eta^3 + \epsilon\lambda^{3s}\phi^3 = 0, \tag{6}$$

has no solution for ξ, η, ϕ in I_{-3} with $\xi\eta\phi \neq 0$ and $\lambda \nmid \xi\eta\phi$.

Proof: If Eq. (6) does have solutions, then let $s \geq 1$ be the least integer such that $\xi, \eta, \phi, \epsilon$ exist satisfying (6). Without loss of generality we may assume that ξ, η, ϕ have no common factor except units. Then, since $s \geq 1$,

$$\xi^3 + \eta^3 \equiv \xi^3 + \eta^3 + \epsilon\lambda^{3s}\phi^3 \equiv 0 (\text{mod } \lambda).$$

Therefore, one of ξ, η is congruent to 1 and the other to $-1(\text{mod } \lambda)$ (see Lemma 6). Without loss of generality assume that $\xi \equiv 1(\text{mod } \lambda)$ and $\eta \equiv -1(\text{mod } \lambda)$.

We assert that $s \geq 2$. Let $\xi = 1 + \lambda\alpha$ and $\eta = -1 + \lambda\beta$ for α, β in I_{-3}. Then

$$\begin{aligned}\xi^3 + \eta^3 &= 3\lambda(\alpha + \beta) + 3\lambda^2(\alpha^2 - \beta^2) + \lambda^3(\alpha^3 + \beta^3)\\ &= -\zeta^2\lambda^3(\alpha + \beta) - \zeta^2\lambda^4(\alpha^2 - \beta^2) + \lambda^3(\alpha^3 + \beta^3) \qquad (\text{since } 3 = -\zeta^2\lambda^2)\\ &\equiv \lambda^3(-\zeta^2(\alpha + \beta) + \alpha^3 + \beta^3)(\text{mod } \lambda^4). \qquad (7)\end{aligned}$$

Moreover, since $\zeta \equiv 1(\text{mod } \lambda)$,

$$-\zeta^2(\alpha + \beta) + \alpha^3 + \beta^3 \equiv (\alpha^3 - \alpha) + (\beta^3 - \beta)(\text{mod } \lambda). \qquad (8)$$

Now we note that in general we have (see Lemma 6)

$$\gamma^3 \equiv \gamma(\text{mod } \lambda). \qquad (9)$$

It follows from (7), (8), and (9) that

$$\xi^3 + \eta^3 \equiv 0(\text{mod } \lambda^4).$$

But $\xi^3 + \eta^3 = -\epsilon\lambda^{3s}\phi^3$, so that $\lambda \nmid \phi$ implies that $\lambda^4 \mid \lambda^{3s}$. Thus, $s \geq 2$ as desired.

Let

$$\xi_1 = \frac{\xi + \zeta\eta}{\lambda}, \qquad \eta_1 = \frac{\zeta(\xi + \zeta^2\eta)}{\lambda}, \qquad \theta_1 = \frac{\zeta^2(\xi + \eta)}{\lambda}. \qquad (10)$$

Since $\xi \equiv 1(\text{mod } \lambda)$, $\eta \equiv -1(\text{mod } \lambda)$, $\zeta \equiv 1(\text{mod } \lambda)$, $\zeta^2 \equiv 1(\text{mod } \lambda)$, it is easily seen that ξ_1, η_1, θ_1 are in I_{-3}. Moreover, ξ_1, η_1, θ_1 are pairwise relatively prime. For example, if $\pi \mid \xi_1$ and $\pi \mid \eta_1$, then

$$\pi \left| \frac{\xi + \zeta\eta}{\lambda} - \frac{\xi + \zeta^2\eta}{\lambda} = \zeta\eta \right.$$

$$\pi \left| \frac{\zeta(\xi + \zeta\eta)}{\lambda} - \frac{\xi + \zeta^2\eta}{\lambda} = -\xi, \right.$$

and so $\pi \mid \eta$ and $\pi \mid \xi$. Since $\lambda \nmid \eta$ and $\lambda \nmid \xi$, we see that $\gcd(\lambda, \pi) = 1$ (λ is a prime). Thus, by (6), $\pi^3 \mid \epsilon\lambda^{3s}\phi^3$ implies that $\pi^3 \mid \phi^3$, which implies that $\pi \mid \phi$ (since I_{-3} is a UFD). But then π is a common factor of ξ, η, ϕ whose only common factors are units. Thus, the only common factors of ξ_1 and η_1 are units. Similarly, $\gcd(\xi_1, \theta_1) = \gcd(\eta_1, \theta_1) = 1$.

We now factor $\xi^3 + \eta^3$. (The whole point of working in I_{-3} is that this may be done.) We have

$$\xi^3 + \eta^3 = (\xi + \zeta\eta)(\xi + \zeta^2\eta)(\xi + \eta) = -\epsilon\lambda^{3s}\phi^3,$$

so that

$$\xi_1\eta_1\theta_1 = -\epsilon\lambda^{3(s-1)}\phi^3. \tag{11}$$

Furthermore, recalling from (2) that $1 + \zeta + \zeta^2 = 0$, we see that

$$\xi_1 + \eta_1 + \theta_1 = 0. \tag{12}$$

From (11), the fact that I_{-3} is a UFD, and the fact that ξ_1, η_1, θ_1 are pairwise relatively prime, we see that there exist elements α, β, γ and units $\epsilon_1, \epsilon_2, \epsilon_3$ such that

$$\xi_1 = \epsilon_1\alpha^3, \qquad \eta_1 = \epsilon_2\beta^3, \qquad \theta_1 = \epsilon_3\gamma^3. \tag{13}$$

Since $s \geq 2$ and since ξ_1, η_1 and θ_1 are pairwise relatively prime, Eq. (11) implies that λ divides precisely one of ξ_1, η_1, θ_1. Therefore, λ divides precisely one of the α, β, γ, say $\lambda \mid \gamma$. Then (11) and (13) imply that

$$\gamma = \lambda^{s-1}\mu, \qquad \lambda \nmid \mu. \tag{14}$$

Moreover, by (11), (12), (13), and (14), we see that

$$\alpha^3 + \epsilon_4\beta^3 + \epsilon_5\lambda^{3(s-1)}\mu^3 = 0, \tag{15}$$

where ϵ_4 and ϵ_5 are units of I_{-3}. Since $s \geq 2$,

$$\alpha^3 + \epsilon_4\beta^3 \equiv 0 \pmod{\lambda^3}. \tag{16}$$

But by (4) and the fact that $\lambda \nmid \alpha$, $\lambda \nmid \beta$, we see that

$$\pm 1 \pm \epsilon_4 \equiv 0 \pmod{\lambda^3}. \tag{17}$$

But $\epsilon_4 = \pm 1, \pm\zeta, \pm\zeta^2$. A simple verification shows that if (17) holds, then $\epsilon_4 = +1$ or -1. In the former case, (15) implies that

$$\alpha^3 + \beta^3 + \epsilon_5\lambda^{3(s-1)}\mu^3 = 0, \tag{18}$$

while in the latter case

$$\alpha^3 + (-\beta)^3 + \epsilon_5\lambda^{3(s-1)}\mu^3 = 0. \tag{19}$$

In either case, (18) and (19) contradict the original choice of s. Thus, a contradiction is established, and Theorem 7 is proved. ■

10.2 Exercise

1. Show that the Diophantine equation $x^3 + y^3 + az^3 = 0$ has no solutions, where a is a prime $\equiv 2$ or $5 \pmod 8$.

10.3 Norm Form Equations

In Chapter 8 we considered the theory of quadratic Diophantine equations in two variables as an outgrowth of the theory of quadratic fields. That entire discussion may be generalized to include other Diophantine equations called *norm form equations*. To develop this theory, it is first necessary to study general *algebraic number fields*, which are generalizations of the quadratic fields studied in Chapters 8 and 9. In this section, we shall give an outline of some of the basic results concerning norm form equations and algebraic number fields. We shall include no proofs since we intend this section to be a point of departure for further study rather than a complete exposition.

Let us begin by describing the object which generalizes a quadratic field. Note that every quadratic field is of the form $\mathbf{Q}(\delta)$, where $\delta = \sqrt{d}$ and d is a rational integer. Moreover, δ satisfies the equation $x^2 - d = 0$. Let us generalize this as follows: Let δ be a complex number which satisfies an equation of the form

$$P(x) = x^n + d_{n-1}x^{n-1} + \cdots + d_0 = 0, \tag{1}$$

where $d_0, \ldots, d_{n-1}$ are rational integers. Note that if we can factor $P(x)$ into $Q(x)R(x)$, where $Q(x)$ and $R(x)$ are polynomials with rational coefficients, then

$$0 = P(\delta) = Q(\delta)R(\delta),$$

so that either $Q(\delta) = 0$ or $R(\delta) = 0$. Thus, by successively factoring, we may suppose that $P(x)$ is *irreducible over the rationals;* i.e., $P(x)$ cannot be written as a product of two polynomials with rational coefficients and positive degrees.

Let us define the *algebraic number field* $\mathbf{Q}(\delta)$ to be the set of all complex numbers of the form

$$a_0 + a_1\delta + \cdots + a_{n-1}\delta^{n-1}, \qquad a_0, \ldots, a_{n-1} \text{ rational.}$$

The following properties of $\mathbf{Q}(\delta)$ are not too hard to check:

Lemma 1:

(i) Let α belong to $\mathbf{Q}(\delta)$. Then α can be written in the form

$$\alpha = a_0 + a_1\delta + \cdots + a_{n-1}\delta^{n-1}, \qquad a_0, \ldots, a_{n-1} \text{ rational,}$$

where $a_0, \ldots, a_{n-1}$ are uniquely determined by α.

(ii) Let α, β belong to $\mathbf{Q}(\delta)$. Then $\alpha \pm \beta, \alpha\beta, \alpha/\beta$ $(\beta \neq 0)$ all belong to $\mathbf{Q}(\delta)$.

Let us now generalize the notion of the norm. To do this, we require the following result about the complex numbers:

The Fundamental Theorem of Algebra: Let $f(x)$ be a nonconstant polynomial with complex coefficients. Then $f(x)$ can be written in the form $f(x) = a_0(x - \alpha_1) \cdots (x - \alpha_n)$, where $a_0, \alpha_1, \ldots, \alpha_n$ are complex numbers.

From the above result, we see that

$$P(x) = (x - \delta^{(1)})(x - \delta^{(2)}) \cdots (x - \delta^{(n)}).$$

Since $P(\delta) = 0$, it is clear that $\delta = \delta^{(i)}$ for some i $(1 \le i \le n)$. Suppose that $\delta = \delta^{(1)}$. Then, if $\alpha = a_0 + a_1\delta + \cdots + a_{n-1}\delta^{n-1}$ belongs to $\mathbf{Q}(\delta)$, set

$$\alpha^{(i)} = a_0 + a_1\delta^{(i)} + \cdots + a_{n-1}(\delta^{(i)})^{n-1}. \tag{2}$$

Then $\alpha^{(1)} = \alpha, \alpha^{(2)}, \ldots, \alpha^{(n)}$ are called the *conjugates* of α. Let us define the *norm* of α, denoted $N(\alpha)$, by

$$N(\alpha) = \alpha^{(1)} \cdots \alpha^{(n)}. \tag{3}$$

The following properties of the norm may be proved:

Lemma 2: Let α, β be in $\mathbf{Q}(\delta)$. Then

(i) $N(\alpha)$ is a rational number.

(ii) If $\alpha = a_0 + a_1\delta + \cdots + a_{n-1}\delta^{n-1}$ with all a_i rational integers, then $N(\alpha)$ is a rational integer.

(iii) $N(\alpha) = 0$ if and only if $\alpha = 0$.

(iv) $N(\alpha\beta) = N(\alpha)N(\beta)$.

Now let us turn to the generalization of the notion of a module. Let us say that $\alpha_1, \ldots, \alpha_m$ are *linearly independent* if and only if whenever

$$a_1\alpha_1 + a_2\alpha_2 + \cdots + a_m\alpha_m = 0,$$

for rational numbers $a_1, a_2, \ldots, a_m$, we have

$$a_1 = a_2 = \cdots = a_m = 0.$$

If $\alpha_1, \ldots, \alpha_m$ are not linearly independent, then we say that $\alpha_1, \ldots, \alpha_m$ are *linearly dependent*. Call $\alpha_1, \ldots, \alpha_m$ a *basis* of $\mathbf{Q}(\delta)$ if and only if every α in $\mathbf{Q}(\delta)$ can be written uniquely in the form

$$\alpha = a_1\alpha_1 + a_2\alpha_2 + \cdots + a_m\alpha_m,$$

where $a_1, \ldots, a_m$ are rational numbers. It is easily shown that $\alpha_1, \ldots, \alpha_m$ in $\mathbf{Q}(\delta)$ form a basis of $\mathbf{Q}(\delta)$ if and only if $m = n$ and $\alpha_1, \ldots, \alpha_n$ are linearly independent. (This is just elementary linear algebra.) For example, $1, \delta, \ldots, \delta^{n-1}$ is a basis of $\mathbf{Q}(\delta)$.

A *module* in $\mathbf{Q}(\delta)$ is a subset M of $\mathbf{Q}(\delta)$ consisting of all numbers of the form

$$\alpha = x_1\alpha_1 + x_2\alpha_2 + \cdots + x_n\alpha_n, \tag{4}$$

where $x_1, x_2, \ldots, x_n$ range through all rational integers and $\alpha_1, \alpha_2, \ldots, \alpha_n$ is a fixed basis of $\mathbf{Q}(\delta)$. In the above circumstances, we write

$$M = \{\alpha_1, \alpha_2, \ldots, \alpha_n\}.$$

It is easily seen that the conjugates $\alpha^{(1)}, \ldots, \alpha^{(n)}$ of α in (4) are given by

$$\alpha^{(i)} = x_1\alpha_1^{(i)} + x_2\alpha_2^{(i)} + \cdots + x_n\alpha_n^{(i)} \qquad (1 \le i \le n).$$

Thus, for all α in M,

$$N(\alpha) = (x_1\alpha_1^{(1)} + \cdots + x_n\alpha_n^{(1)})(x_1\alpha_1^{(2)} + \cdots + x_n\alpha_n^{(2)})\cdots (x_1\alpha_1^{(n)} + \cdots + x_n\alpha_n^{(n)})$$

is a rational number. $N(\alpha)$ is also a polynomial in $x_1, \ldots, x_n$ of degree n. Set

$$f(x_1, \ldots, x_n) = N(\alpha).$$

Since for arbitrary rational integer values of $x_1, \ldots, x_n$, $f(x_1, \ldots, x_n)$ is a rational number, we can see that $f(x_1, \ldots, x_n)$ must have rational coefficients. We call $f(x_1, \ldots, x_n)$ a *norm form.* If $n = 2$, $f(x_1, x_2)$ is simply a binary quadratic form.

Example 3: Let $\delta = \sqrt[3]{2}$, which satisfies the polynomial $x^3 - 2 = 0$. Consider

$$M = \{1, \delta, \delta^2\}.$$

Then M is a module of $\mathbf{Q}(\delta)$. We shall consider the norm form

$$f(x_1, x_2, x_3) = N(x_1 + x_2\delta + x_3\delta^2).$$

We first determine the conjugates of δ. Set

$$\zeta = \frac{-1 + \sqrt{-3}}{2}.$$

Then $\zeta^3 = 1$. Thus,

$$\delta^3 = 2, \quad (\zeta\delta)^3 = 2, \quad \text{and} \quad (\zeta^2\delta)^3 = 2.$$

Since δ, $\zeta\delta$, $\zeta^2\delta$ are easily seen to be different, we must have

$$x^3 - 2 = (x - \delta)(x - \zeta\delta)(x - \zeta^2\delta).$$

We may set $\delta^{(1)} = \delta$, $\delta^{(2)} = \zeta\delta$, $\delta^{(3)} = \zeta^2\delta$. Thus,

$$f(x_1, x_2, x_3)$$
$$= (x_1 + x_2\delta + x_3\delta^2)(x_1 + x_2\zeta\delta + x_3\zeta^2\delta^2)(x_1 + x_2\zeta^2\delta + x_3\zeta\delta^2).$$

Using the relations $\zeta + \zeta^2 = -1$ and $\zeta^3 = 1$ and $\delta = \sqrt[3]{2}$ we can easily show that

$$f(x_1, x_2, x_3) = x_1^3 + 2x_2^3 + 4x_3^3 - 6x_1x_2x_3.$$

Assume that we are given a norm form $f(x_1, \ldots, x_n)$ obtained from a module M in a field $\mathbf{Q}(\delta)$. Let m be a rational number. We say that $f(x_1, \ldots, x_n)$ *represents* m if and only if there exist rational integers $x_1, \ldots, x_n$ such that

$$f(x_1, \ldots, x_n) = m, \tag{5}$$

that is, provided the Diophantine equation (5) can be solved. We may ask the same questions here that we asked concerning binary quadratic forms in Section 8.4. Namely, (i) give a procedure for solving (5) or (ii) determine all

m such that (5) is solvable. Problem (i) may be solved by imitating the methods of Section 8.8, which sufficed to solve problem (i) in the case $n = 2$. Problem (ii) has not been solved in general, and there are many open problems concerning it.

We shall now outline some of the results concerning problem (i). We let

$$\mathfrak{O} = \{\gamma \text{ in } \mathbf{Q}(\delta) \mid \gamma M \subseteq M\}.$$

Then $\mathfrak{O}$ is called the *ring of coefficients* of M. Just as in the case $n = 2$ we have

Lemma 4:

(i) For any numbers γ_1, γ_2 in $\mathfrak{O}$, $\gamma_1 \pm \gamma_2$ and $\gamma_1\gamma_2$ belong to $\mathfrak{O}$.
(ii) $\mathfrak{O}$ is a module in $\mathbf{Q}(\delta)$.
(iii) There is a unique coefficient ring I_δ in $\mathbf{Q}(\delta)$ such that $\mathfrak{O} \subseteq I_\delta$ for every coefficient ring $\mathfrak{O}$.

The coefficient ring I_δ is called the *ring of integers* of $\mathbf{Q}(\delta)$. It is characterized as the set of all γ in $\mathbf{Q}(\delta)$ satisfying a polynomial equation of the form

$$\gamma^n + c_{n-1}\gamma^{n-1} + \cdots + c_0 = 0,$$

where $c_{n-1}, \ldots, c_0$ are rational integers. (See the quadratic case.)

We call ϵ in $\mathfrak{O}$ a *unit* if and only if there is an ϵ_1 in $\mathfrak{O}$ such that $\epsilon\epsilon_1 = 1$. It can be shown that ϵ is a unit if and only if $N(\epsilon) = \pm 1$. Suppose that $N(\epsilon) = 1$ and that ϵ is in $\mathfrak{O}$. Let $x_1, \ldots, x_n$ satisfy (5); that is, for α given in (4), we have $N(\alpha) = m$. Then

$$N(\epsilon\alpha) = N(\epsilon)N(\alpha) = N(\alpha) = m,$$

so that $\epsilon\alpha = y_1\alpha_1 + \cdots + y_n\alpha_n$ gives a new solution to (5) with $x_1 = y_1, \ldots, x_n = y_n$. Also, it may be shown that all these solutions are different. We call two solutions $(x_1, \ldots, x_n)$ and $(y_1, \ldots, y_n)$ *associates* if and only if they correspond to $\alpha = x_1\alpha_1 + \cdots + x_n\alpha_n$ and $\beta = y_1\alpha_1 + \cdots + y_n\alpha_n$, respectively, with $\alpha = \epsilon\beta$ for some unit ϵ of $\mathfrak{O}$. Then it can be shown that

Theorem 5: There are only a finite number of nonassociate solutions of

$$f(x_1, \ldots, x_n) = m. \tag{6}$$

For completeness we shall describe the units of $\mathfrak{O}$. Once this is done then the solution of problem (i) is given by first finding the units of $\mathfrak{O}$ and then finding the finite number of nonassociate solutions of (6).

Let us number $\delta = \delta^{(1)}, \delta^{(2)}, \ldots, \delta^{(n)}$ in such a way that $\delta^{(1)}, \ldots, \delta^{(s)}$ are real numbers ($s \geq 0$) and $\delta^{(s+1)}, \ldots, \delta^{(n)}$ are complex numbers but not real numbers. Since δ satisfies (1) with rational integers $d_0, \ldots, d_{n-1}$ it is easy to see that for any root $\delta^{(i)}$ of $P(x)$, $\overline{\delta^{(i)}}$ is a root of $P(x)$ ($\overline{\delta^{(i)}}$ denotes the complex conjugate of $\delta^{(i)}$). Thus, there is an even number of the complex numbers $\delta^{(s+1)}, \ldots, \delta^{(n)}$; say there are $2t$ of them. Set $r = s + t - 1$.

Lemma 6: There are numbers $\zeta, \epsilon_1, \ldots, \epsilon_r$ in $\mathcal{O}$ such that $\zeta^k = 1$ for some integer $k > 0$ (we may suppose that $k > 0$ is least) and such that for every unit ϵ in $\mathcal{O}$ there exist unique integers $\ell, n_1, \ldots, n_r$ such that $0 \leq \ell < k$ and

$$\epsilon = \zeta^{\ell}\epsilon_1^{n_1} \cdot \epsilon_2^{n_2} \cdots \epsilon_r^{n_r}. \tag{7}$$

Example 7: If we specialize to the case $n = 2$, $\delta = \sqrt{d}$, then there are two cases (see Theorem 8.7.10 and 8.7.5):

(i) $d > 0$. Here $s = 2$, $t = 0$, and so $r = 1$. Also $\zeta = -1$. Thus, any unit ϵ has the form $\epsilon = \pm \epsilon_1^{n_1}$.

(ii) $d < 0$. Here $s = 0$, $t = 1$, and so $r = 0$. We showed before that there is one case ($d = -3$) where $\zeta = (-1 + \sqrt{-3})/2$ and one case ($d = -1$) where $\zeta = \sqrt{-1}$; in all other cases $\zeta = -1$.

Example 8: Let us continue Example 3, where $\delta = \sqrt[3]{2}$. The conjugates of δ are $\delta, \zeta\delta, \zeta^2\delta$. Now $\overline{\zeta\delta} = \zeta^2\delta$, and $\zeta\delta$ and $\zeta^2\delta$ are complex. Thus, $t = s = 1$ and $r = s + t - 1 = 1$. Also $\zeta = -1$, since δ is real. Thus, just as in the real quadratic case, the units of any coefficient ring $\mathcal{O}$ are all of the form $\epsilon = \pm\epsilon_1^{n_1}$ for some fundamental unit ϵ_1.

It is an easy matter to determine which units ϵ in (7) have $N(\epsilon) = +1$ by taking into account the values of $N(\zeta), N(\epsilon_1), \ldots, N(\epsilon_r)$. Thus, we may conclude that

Theorem 9: Let $f(x_1, \ldots, x_n) = N(x_1\alpha_1 + \cdots + x_n\alpha_n)$ be given as above. Let $M = \{\alpha_1, \ldots, \alpha_n\}$ and $\mathcal{O}$ be its associated ring of coefficients. Let m be a given rational integer. Let $\gamma_1, \ldots, \gamma_u$ be a complete set of nonassociate numbers in M such that $N(\gamma_i) = m$. Then for rational integers $x_1, \ldots, x_n$, we have

$$f(x_1, \ldots, x_n) = m$$

if and only if

$$x_1\alpha_1 + \cdots + x_n\alpha_n = \epsilon\gamma_i$$

for some i, $1 \leq i \leq u$, and some unit ϵ as given in (7) with $N(\epsilon) = 1$.

Corollary 10: Assume that $n > 2$ or $n = 2$ and that δ is real. Then if

$$f(x_1, \ldots, x_n) = m$$

has a solution, it has an infinite number of solutions.

Now let us remark on the equivalence classes of modules. If M_1, M_2 are modules in $\mathbf{Q}(\delta)$, we say that M_1 is *similar* to M_2 if and only if there is a γ in $\mathbf{Q}(\delta)$ such that $\gamma M_1 = M_2$.

Theorem 11: Let $\mathcal{O}$ be a coefficient ring of $\mathbf{Q}(\delta)$. Then there are only a finite number of nonsimilar modules whose coefficient ring is $\mathcal{O}$.

The number h of nonsimilar modules whose coefficient ring is $\mathcal{O}$ is called the class number of $\mathcal{O}$. If $\mathcal{O} = I_\delta$, then h is called the *class number* of $\mathbf{Q}(\delta)$.

The study of $\mathbf{Q}(\delta)$ and the arithmetic of its integers may be pushed much further than we have suggested above. The resulting subject is called *algebraic number theory*. We have given some motivation for studying it, and we hope that many of you will. Some suggested readings are *The Theory of Algebraic Numbers* by H. Pollard (Wiley & Sons, New York, 1950), *Number Theory* by Z. I. Borevich and I. R. Shafarevich (Academic Press, New York, 1965) and *Algebraic Number Theory* by S. Lang (Addison-Wesley, Reading, Mass., 1971).

11

The Representation of Integers by Binary Quadratic Forms

11.1 Equivalence of Forms

Let us now turn to the second problem about binary quadratic forms which was stated in Section 8.4.

Definition 1: Let $f(x, y) = ax^2 + bxy + cy^2$ be a binary quadratic form and let m be given. If the Diophantine equation $f(x, y) = m$ is solvable (in rational integers x, y), then we say that $f(x, y)$ *represents m*.

We may phrase the problem we shall consider in this chapter as follows:

Problem: Determine all numbers which $f(x, y)$ represents.

Often, we can solve this problem by clever observation and known results. For example, let $f(x, y) = 2x^2 - 10xy + 13y^2$. At first, it seems very difficult to determine which integers $f(x, y)$ represents. However, note that

$$f(x, y) = (x - 2y)^2 + (-x + 3y)^2 = X^2 + Y^2,$$

where

$$X = x - 2y, \qquad Y = -x + 3y. \tag{1}$$

Thus, we see that for any rational integers x and y, $f(x, y)$ is a sum of two squares, X^2 and Y^2. Moreover, if X and Y are any two given rational integers, then we can find x and y such that $f(x, y) = X^2 + Y^2$, since from Eq. (1) we may set $y = X + Y, x = 3X + 2Y$. Thus, $f(x, y)$ represents the same integers as the form $g(X, Y) = X^2 + Y^2$. And the integers represented by this form are described in Theorem 6.4.5. Thus, we have solved our problem for the form $f(x, y) = 2x^2 - 10xy + 13y^2$.

Let us consider further the phenomenon exhibited in the above example. There are two facts which explain the example: (i) The form $f(x, y)$ was obtained from the form $g(X, Y) = X^2 + Y^2$ by making the linear substitution (1). (ii) The linear substitution (1) has the following property: As x, y run over all rational integers, so do X, Y.

Let us consider linear substitutions of the form

$$X = rx + sy, \qquad Y = tx + uy, \tag{2}$$

where r, s, t, u are given rational integers. By solving Eqs. (2) for x, y, it is easy to see that fact (ii) holds for Eq. (2) if and only if $ru - st = \pm 1$. A substitution with this property is called a unimodular substitution. For various technical reasons, we shall restrict ourselves to those unimodular substitutions whose determinant $ru - st = +1$. Henceforth, the term *unimodular substitution* will be used only to refer to substitutions (2) for which $ru - st = +1$.

We may now generalize our above example. Suppose that $f(x, y)$ and $g(x, y)$ are binary quadratic forms and suppose that there exists a unimodular substitution of the form (2) such that

$$f(x, y) = g(rx + sy, tx + uy). \tag{3}$$

Then it is clear that the numbers represented by f are the same as the numbers represented by g. Equation (3) is useful in transforming the problem of representation of numbers by f into the corresponding problem for g. This is useful if we know something about g, as we did in the above example, where $g(x, y) = x^2 + y^2$. This idea will be the principal theme in this section. Equation (3) suggests the following definition:

Definition 2: Let f and g be binary quadratic forms. We say that f is *equivalent** to g, denoted $f \approx g$, provided that there exists a unimodular substitution

$$X = rx + sy, \qquad Y = tx + uy$$

such that

$$f(x, y) = g(rx + sy, tx + uy).$$

From the above discussion, we have the following result:

Proposition 3: Let f and g be equivalent binary quadratic forms. Then f and g represent the same numbers.

---

*Note that some authors define equivalence of forms to mean that Eq. (3) holds for r, s, t, u satisfying $ru - st = \pm 1$. In such circumstances, our notion of equivalence is called *strict equivalence* or *proper equivalence*.

Assume that f and g are equivalent with respect to the unimodular substitution $X = rx + sy$, $Y = tx + uy$ (as in Eq. (3)). Suppose that $g(x, y) = ax^2 + bxy + cy^2$, $f(x, y) = Ax^2 + Bxy + Cy^2$. Then, a simple computation shows that

$$\begin{aligned} A &= g(r, t) \\ C &= g(s, u) \\ B &= 2ars + b(ru + st) + 2ctu. \end{aligned} \tag{4}$$

Moreover, a further calculation shows that

$$\begin{aligned} B^2 - 4AC &= (ru - st)^2(b^2 - 4ac) \\ &= b^2 - 4ac \end{aligned}$$

since $ru - st = 1$. Thus, we have proved

Proposition 4: Let f and g be equivalent binary quadratic forms. Then f and g have the same discriminant.

The notion of equivalence of forms will be basic to everything that follows in this chapter. Therefore, let us state and prove a few more elementary properties of equivalence.

Proposition 5: Equivalence of forms is an equivalence relation.

Proof:

(i) Reflexivity. $f \approx f$ with respect to the substitution $X = 1 \cdot x + 0 \cdot y$, $Y = 0 \cdot x + 1 \cdot y$.

(ii) Symmetry. Suppose that $f(x, y) = g(rx + sy, tx + uy)$, where r, s, t, u are rational integers such that $ru - st = 1$. Then

$$\begin{aligned} g(x, y) &= g(r(ux - sy) + s(-tx + ry), t(ux - sy) + u(-tx + ry)) \\ &= f(ux - sy, -tx + ry). \end{aligned}$$

Thus, $g \approx f$.

(iii) Transitivity. Suppose that $f(x, y) = g(rx + sy, tx + uy)$, $g(x, y) = h(r'x + s'y, t'x + u'y)$. Then

$$f(x, y) = h(Rx + Sy, Tx + Uy),$$

where $R = r'r + s't$, $S = r's + s'u$, $T = t'r + u't$, $U = t's + u'u$. Moreover,

$$RU - TS = (ru - ts)(r'u' - t's') = 1.$$

Thus, $f \approx h$. ∎

Definition 6: Let f be a binary quadratic form. Denote by $\{f\}$ the equivalence class of f with respect to $\approx$. Then $\{f\}$ is called the *class* of forms determined by f. Thus, $\{f\}$ is the set of all forms equivalent to f and consists of all the forms

$f(rx + sy, tx + uy)$, where r, s, t, u are rational integers such that $ru - st = 1$. By Proposition 3, the numbers represented by all forms in a class are the same. Let us call these *the numbers represented by the class*. Also, by Proposition 4, all forms in a class have the same discriminant, which we call the *discriminant of the class*.

If we are interested only in the problem of representation of numbers by forms, we may study the problem only for one form selected from each class of forms. Is this any simpler than our original problem? Indeed it is, by virtue of the following amazing and deep theorem of Gauss.

Theorem 7 (Gauss): The number of classes of forms having a given discriminant is finite.

According to Gauss' theorem, we may list the classes of forms of discriminant D in a finite list, say $\{f_1\}, \ldots, \{f_t\}$. In terms of quadratic Diophantine equations, this means that in order to determine the representability of a number by forms of discriminant D, it suffices to study its representability by $f_1, \ldots, f_t$. Thus, our original problem, although not completely settled, is reduced to a finite series of questions. To give the reader some idea how this can help, consider the following example. The form $x^2 + y^2$ has discriminant $D = -4$. We shall prove that there is only one class of forms of discriminant $D = -4$. Thus, the form $2x^2 - 10xy + 13y^2$, which has discriminant $D = 100 - 4 \cdot 2 \cdot 13 = -4$, is equivalent to $x^2 + y^2$, and the rational integers represented by $2x^2 - 10xy + 13y^2$ are given by the two-square theorem. Note that we did this example once before. But with the aid of some information about the classes, we are able to establish the equivalence of $2x^2 - 10xy + 13y^2$ with $x^2 + y^2$ without exhibiting the linear substitution which effects the equivalence. Thus, we can replace some of the accident and luck present in our preceding analysis with what seems to be the initial step of a general method. This is precisely Gauss' plan.

But Gauss' method is even more powerful than the above might suggest. Let us illustrate its power by giving yet another proof of the two-square theorem (Theorem 6.4.5). The essential part of the two-square theorem is the statement which asserts that if p is a prime and $p \equiv 1 \pmod 4$, then

$$x^2 + y^2 = p$$

is solvable in integers x, y. Let us prove this statement using the fact that every form of discriminant -4 is equivalent to $x^2 + y^2$. If p is a prime such that $p \equiv 1 \pmod 4$, Theorem 3.3.5 implies that there exists a rational integer u such that

$$u^2 \equiv -1 \pmod p.$$

Set $b = 2u$. Then

$$b^2 \equiv -4 \pmod{4p},$$

and thus $b^2 = -4 + 4pc$ for some c. Set

$$g(x, y) = px^2 + bxy + cy^2.$$

Then the discriminant of g is $b^2 - 4pc = -4$, and thus $g(x, y)$ is equivalent to $x^2 + y^2$, so that $x^2 + y^2$ represents the same rational integers as $g(x, y)$. In particular, since $g(1, 0) = p$, $x^2 + y^2$ represents p, which is what we desired to prove.

Let $f(x, y) = ax^2 + bxy + cy^2$ be a binary quadratic form, a, b, c rational numbers. It is an easy exercise to show that there is a unique positive rational number e such that if $a = ea_1$, $b = eb_1$, $c = ec_1$, then a_1, b_1, c_1 are rational integers with no common factor > 1. Then $f(x, y) = ef_1(x, y)$, where $f_1(x, y) = a_1x^2 + b_1xy + c_1y^2$. Thus, we see that m is represented by $f(x, y)$ if and only if $m_1 = m/e$ is represented by $f_1(x, y)$. Thus, insofar as determining which numbers are represented by a form are concerned, we may always restrict ourselves to forms with rational integer coefficients having no common factor greater than 1. Such forms are called *primitive forms*. For example, $f(x, y) = 2x^2 + 3xy + y^2$ is primitive, but $f(x, y) = 2x^2 + 6xy + 8y^2$ is not primitive. The connection between arbitrary binary quadratic forms and primitive forms is summarized in the following lemma:

Lemma 8: Let $f(x, y) = ax^2 + bxy + cy^2$ be a binary quadratic form, a, b, c rational numbers. Then $f(x, y)$ can be uniquely written in the form

$$f(x, y) = ef_1(x, y),$$

where e is a positive rational number and $f_1(x, y)$ is a primitive form.

We shall see that it is more convenient to work with primitive forms. Note that if one form in a class is primitive, then all are primitive. Thus, we may speak of *primitive classes*. We may state Gauss' theorem equivalently in the following form:

Theorem 7′: There are only finitely many classes of primitive forms of discriminant D.

We leave it as an exercise for the reader to prove that Theorem 7′ implies Theorem 7. We shall prove Theorem 7′ in Section 3. One distinction which will be important in our proof of Gauss' theorem is given in the following definition:

Definition 9: Let $f(x, y) = ax^2 + bxy + cy^2$ be a binary quadratic form of discriminant D. If $D > 0$, we say that f is *indefinite*. If $D < 0$, we say that $f(x, y)$ is *positive-definite* if $a > 0$ and *negative-definite* if $a < 0$.

It is clear that if $f(x, y)$ is negative-definite, then $-f(x, y)$ is positive-definite and conversely. Moreover, $f(x, y)$ represents m if and only if

$-f(x, y)$ represents $-m$. Therefore, insofar as our problem of representability goes, we may safely ignore the negative-definite forms. This will be convenient in what follows.

The reason for the terminology above is as follows:

Proposition 10: Let $f(x, y)$ be a binary quadratic form of discriminant D.

(i) f is positive-definite if and only if $f(x, y) \geq 0$ for all x, y.

(ii) If f is indefinite, then $f(x, y)$ assumes both positive and negative values.

Proof: Exercise. ■

11.1 Exercises

1. Determine whether $f \approx g$ if
   (a) $f = 2x^2 + 3xy + 3y^2$, $g = x^2 + y^2$.
   (b) $f = 2x^2 + 3y^2$, $g = x^2 + 6y^2$.
   (c) $f = x^2 + 8y^2$, $g = 2x^2 + 4y^2$.
   (d) $f = x^2 + xy + 5y^2$, $g = x^2 + 5xy + 11y^2$.
   (*Hint:* Before beginning a search for a unimodular substitution, check discriminants and the integers obviously represented by each form.)
2. (a) Show that the two forms $f = x^2 + 2y^2$, $g = 17x^2 + 20xy + 6y^2$ are equivalent.
   (b) Note that $f(1, 3) = 19$. Find rational integers x, y such that $g(x, y) = 19$.
3. (a) Make a table of all integers ≤ 100 represented by $f = x^2 + 2y^2$.
   (b) Can you make a conjecture concerning the integers so represented? (*Hint:* Factor the numbers represented and look at the statement of the two-square theorem.)
   (c) By imitating the proof of the two-square theorem, prove your conjecture.
4. Describe $\{f\}$ for $f = x^2 + 5y^2$.
5. Prove that equivalent forms have the same discriminant.
6. Prove Proposition 10.
7. Prove that Theorem 7′ implies Theorem 7.
8. Let $f = x^2 + 6y^2$.
   (a) Does f represent 415?
   (b) Does f represent 31?
9. Verify directly that $\{x^2 + 2y^2\} = \{17x^2 + 20xy + 6y^2\}$ (see Exercise 2(a)).

Exercises 10–13 will outline a proof of Gauss' theorem (Theorem 7′) in the case where the discriminant D is negative. The method outlined is called *reduction theory* and can also be developed for positive D, although the results are much more complicated. The reduction theory for binary quadratic forms was discovered by Gauss. Let us write (a, b, c) instead of $ax^2 + bxy + cy^2$.

10. Let a, b, c be rational integers.
    (a) Show that if (a, b, c) is positive-definite, then $a > 0$ and $c > 0$.
    (b) Show that $(a, b, c) \approx (c, -b, a)$.
    (c) Show that (a, b, c) is equivalent to some form (a, b', c'), where $|b'| \leq a$. (Apply a substitution of the form $X = x + ky$, $Y = y$.)
    (d) Let (a, b, c) be positive-definite. Show that by alternately using equivalences (*b*) and (*c*), we can find a form (a', b', c') equivalent to (a, b, c) such that either (i) $c > a$ and $-a < b \leq a$ or (ii) $c = a$ and $0 \leq b \leq a$. A form satisfying one of the conditions (i) or (ii) is called a *reduced form*. (e) Show that every positive-definite form is equivalent to *one and only one* reduced form.
11. Use operations (b) and (c) of Exercise 10 to find the reduced form associated to the following positive-definite quadratic forms:
    (a) $(2, 3, 8)$. (b) $(1, 0, 3)$. (c) $(2, 4, 17)$.
12. Let (a, b, c) be a positive-definite reduced form of discriminant D.
    (a) Show that $3ac \leq |D|$.
    (b) Show by using part (a) that the number h_D of classes of primitive forms of discriminant D is finite.
    (c) Show that $|b| \leq \sqrt{|D|/3}$.
    (d) Show that one may enumerate all positive-definite reduced forms of discriminant D by considering for each b, such that $|b| \leq \sqrt{|D|/3}$, the pairs of positive integers (a, c) such that $ac = (D + b^2)/4$. Then determine those forms (a, b, c) which are reduced.
13. Use the results of Exercise 12 to compute a complete set of reduced forms of discriminant D for $D = -3, -4, -7, -8, -11, -12, -15, -16, -19, -20$. In each case, compute the number of classes of positive-definite forms of discriminant D.

11.2 Strict Similarity of Modules

We plan to discuss our problem of the representation of integers by a binary form by translating the problem into an equivalent problem about modules. In the next section, we shall describe the means for translating the problem from forms to modules and back to forms. There, we shall construct a one-to-one correspondence between certain classes of forms and certain classes of modules. This correspondence will allow us to effect the translation. However, the correct classes to use in this correspondence are, unfortunately, not the similarity classes introduced in Section 9.6. We need a new notion of class, that of strict similarity class. This section is devoted to discussing this concept.

Definition 1: Let M_1 and M_2 be modules of $\mathbf{Q}(\sqrt{d})$. We say that M_1 and M_2 are *strictly similar** if there exists a nonzero γ in $\mathbf{Q}(\sqrt{d})$ such that $N(\gamma) > 0$ and $M_1 = \gamma M_2$. If M_1 and M_2 are strictly similar, we write $M_1 \approx M_2$.

Clearly, if $M_1 \approx M_2$, then $M_1 \sim M_2$. But the converse is not necessarily true. There are modules which are similar but not strictly similar. (See the proof of Theorem 3 for an example.) Just as was the case with similarity, the notion of strict similarity can be used to divide all modules into *strict similarity classes.* If M is any module, then $\{M\}$ will denote the strict similarity class containing M and equals the collection of all modules strictly similar to M. As was the case with similarity classes, every module belongs to one and only one strict similarity class. Moreover, since strictly similar modules are similar, all modules belonging to a single strict similarity class have the same ring of coefficients.

Definition 2: Let $\mathfrak{O}$ be a ring of coefficients of $\mathbf{Q}(\sqrt{d})$. The number of strict similarity classes of modules belonging to $\mathfrak{O}$ is called the *strict class number of* $\mathfrak{O}$ and is denoted $h_{\mathfrak{O}}^+$.

We do not have to give a long complicated proof that the strict class number of $\mathfrak{O}$ is finite, since it is simple to compute $h_{\mathfrak{O}}^+$ in terms of $h_{\mathfrak{O}}$, the class number of $\mathfrak{O}$.

Theorem 3: Let $\mathfrak{O} = \{1, f\omega_d\}$ be a ring of coefficients of $\mathbf{Q}(\sqrt{d})$. Then the strict class number of $\mathfrak{O}$ is given as follows:

$$h_{\mathfrak{O}}^+ = \begin{cases} h_{\mathfrak{O}} & \text{if } d < 0, \\ h_{\mathfrak{O}} & \text{if } d > 0 \text{ and } N(\epsilon_f) = -1, \\ 2h_{\mathfrak{O}} & \text{if } d > 0 \text{ and } N(\epsilon_f) = 1, \end{cases}$$

where $\epsilon_f =$ the fundamental unit of $\mathfrak{O}$.

Proof:

Case 1: $d < 0$. Let $\gamma = r + s\sqrt{d}$ be an element of $\mathbf{Q}(\sqrt{d})$, where r, s are rational. Then $N(\gamma) = r^2 - s^2 d \geq 0$ since $d < 0$. Therefore, all nonzero elements of $\mathbf{Q}(\sqrt{d})$ have positive norm, so there is no difference between similarity and strict similarity of modules. Thus, $h_{\mathfrak{O}}^+ = h_{\mathfrak{O}}$.

*Some authors use *similar in the narrow sense.*

Case 2: $d > 0$. Suppose that $N(\epsilon_{\mathscr{f}}) = -1$. If M_1 and M_2 are modules belonging to $\mathfrak{O}$ and if M_1 and M_2 are similar, there exists γ in $\mathbf{Q}(\sqrt{d})$, $\gamma \neq 0$, such that $M_1 = \gamma M_2$. If $N(\gamma) > 0$, M_1 and M_2 are strictly similar. If $N(\gamma) < 0$, then $N(\epsilon_{\mathscr{f}}\gamma) = N(\epsilon_{\mathscr{f}})N(\gamma) > 0$ since $N(\epsilon_{\mathscr{f}}) = -1$. Moreover, since $\epsilon_{\mathscr{f}}$ is a unit of $\mathfrak{O}$, we have $\epsilon_{\mathscr{f}} M_2 = M_2$, so that

$$(\epsilon_{\mathscr{f}}\gamma)M_2 = \gamma(\epsilon_{\mathscr{f}} M_2) = \gamma M_2 = M_1.$$

Thus, M_1 and M_2 are also strictly similar, and so $h_{\mathfrak{O}}^+ = h_{\mathfrak{O}}$.

Next, suppose that $N(\epsilon_{\mathscr{f}}) = +1$. First note that $M \approx M_1$ implies that $M \sim M_1$, and so $\{M\} \subseteq [M]$. Let α_0 be any element of $\mathbf{Q}(\sqrt{d})$ of negative norm (e.g., $\alpha_0 = \sqrt{d}$). Then we shall show that for each module M belonging to $\mathfrak{O}$ the similarity class, $[M]$, breaks up into precisely two strict similarity classes: $[M] = \{M\} \cup \{\alpha_0 M\}$ with $\{M\} \neq \{\alpha_0 M\}$. It then follows immediately that $h_{\mathfrak{O}}^+ = 2h_{\mathfrak{O}}$. It is obvious that $\{M\} \cup \{\alpha_0 M\} \subseteq [M]$, since any module in $\{M\}$ (respectively, $\{\alpha_0 M\}$) is strictly similar to M (respectively, $\alpha_0 M$) and is thus similar to M. Conversely, let M_1 belong to $[M]$. Then $M_1 \sim M$, and so there is a $\gamma \neq 0$ in $\mathbf{Q}(\sqrt{d})$ such that $M_1 = \gamma M$. If $N(\gamma) > 0$, then M_1 is in $\{M\}$. If $N(\gamma) < 0$, then $N(\alpha_0^{-1}\gamma) = N(\alpha_0)^{-1}N(\gamma) > 0$ and $M_1 = (\alpha_0^{-1}\gamma)(\alpha_0 M)$, so that M_1 is in $\{\alpha_0 M\}$. Thus, $[M] \subseteq \{M\} \cup \{\alpha_0 M\}$, and hence $[M] = \{M\} \cup \{\alpha_0 M\}$. Finally, we see that $\{M\} \neq \{\alpha_0 M\}$, for if $\{M\} = \{\alpha_0 M\}$, then there exists a γ in $\mathbf{Q}(\sqrt{d})$ such that $N(\gamma) > 0$ and $M = \gamma(\alpha_0 M)$. From this last equality, we have $\alpha_0\gamma M = M$ and $(\alpha_0\gamma)^{-1}M = M$. Therefore, both $\alpha_0\gamma$ and $(\alpha_0\gamma)^{-1}$ belong to $\mathfrak{O}$, so that $\alpha_0\gamma$ is a unit of $\mathfrak{O}$. Also $N(\alpha_0\gamma) = N(\alpha_0)N(\gamma) < 0$. But by Theorem 8.7.10 all the units of $\mathfrak{O}$ are of the form $\pm\epsilon_{\mathscr{f}}^n$ and so have positive norm. This contradiction establishes the result. ■

Recall that when we change bases in a module $M = \{\alpha, \beta\}$, the new basis α_1, β_1 is given by a substitution: $\alpha = r\alpha_1 + s\beta_1$, $\beta = t\alpha_1 + u\beta_1$, where $ru - st = \pm 1$. Conversely, every such substitution gives rise to a new basis of M via the above formula. On the other hand, in defining equivalence of forms, we restricted such substitutions to having the determinant $ru - st = +1$. Since we wish to set up a one-to-one correspondence between strict similarity classes of modules and equivalence classes of forms, we must somehow avoid shifting from one basis to another when the determinant of the transformation is -1. This can be done by an easy clever subterfuge. Let us agree to always arrange the basis elements α, β of M in a particular order, so that it is impossible to get from one such ordered basis to another by a unimodular substitution of determinant -1.

Definition 4: Let $M = \{\alpha, \beta\}$ be a module. We say that α, β is an *ordered basis* provided that the determinant

$$\delta = \begin{vmatrix} \alpha & \alpha' \\ \beta & \beta' \end{vmatrix}$$

satisfies $\delta/\sqrt{d} > 0$.

If $d > 0$, α, β are real, and so our definition certainly makes sense. If $d < 0$, $\delta/\sqrt{d}$ is a real number, because $\alpha' = \bar{\alpha}$, $\beta' = \bar{\beta}$, the complex conjugates of α, β respectively, so that

$$\bar{\delta} = \overline{\alpha\bar{\beta} - \bar{\alpha}\beta} = \bar{\alpha}\beta - \alpha\bar{\beta} = -\delta.$$

Thus, δ is purely imaginary, and $\delta/\sqrt{d} = \delta/\sqrt{-1}\sqrt{|d|}$ is real.

Note that if α, β is not an ordered basis of M, then β, α is an ordered basis. *Therefore, every module has an ordered basis.*

If M has ring of coefficients $\mathfrak{O}$, we have $N(M) = |\Delta_M/\Delta_{\mathfrak{O}}|^{1/2}$; moreover,

$$\Delta_{\mathfrak{O}} = \begin{cases} f^2 d & \text{if } d \equiv 1 \pmod 4, \\ 4f^2 d & \text{if } d \equiv 2, 3 \pmod 4, \end{cases}$$

and $\delta^2 = \Delta_M$. Thus, we have

$$N(M) = \begin{cases} \dfrac{1}{f}\left|\dfrac{\delta}{\sqrt{d}}\right| & \text{if } d \equiv 1 \pmod 4 \\ \dfrac{1}{2f}\left|\dfrac{\delta}{\sqrt{d}}\right| & \text{if } d \equiv 2, 3 \pmod 4. \end{cases}$$

Hence we see that α, β *is an ordered basis of* M *if and only if*

$$N(M) = \frac{\delta}{\sqrt{\Delta_{\mathfrak{O}}}} = \frac{\alpha\beta' - \alpha'\beta}{\sqrt{\Delta_{\mathfrak{O}}}}. \tag{1}$$

The most important property of ordered bases is given by the following:

Proposition 5: Let α, β and α_1, β_1 be two ordered bases for the module M. Then there exists a unimodular substitution

$$X = rx + ty$$
$$Y = sx + uy$$

such that $ru - st = +1$ and

$$\begin{aligned} \alpha_1 &= r\alpha + t\beta \\ \beta_1 &= s\alpha + u\beta. \end{aligned} \tag{2}$$

Conversely, if α, β is any ordered basis of M and if $ru - st = +1$, then α_1, β_1 defined by (2) is an ordered basis for M.

Proof: We already know that there exists a unimodular substitution such that (2) holds. Note that

$$\delta_1 = \begin{vmatrix} \alpha_1 & \alpha_1' \\ \beta_1 & \beta_1' \end{vmatrix} = \begin{vmatrix} r\alpha + t\beta & r\alpha' + t\beta' \\ s\alpha + u\beta & s\alpha' + u\beta' \end{vmatrix}$$
$$= (ru - st)(\alpha\beta' - \beta\alpha') = (ru - st)\delta.$$

Therefore,

$$\frac{\delta_1}{\sqrt{d}} = (ru - st)\frac{\delta}{\sqrt{d}}.$$

Since α, β and α_1, β_1 are both ordered bases, $\delta_1/\sqrt{d}$ and $\delta/\sqrt{d}$ are both positive, so that $ru - st = +1$. We leave the second statement as an exercise. ■

Proposition 6: Let $M = \{\alpha, \beta\}$ be a module with α, β an ordered basis, and let γ in $\mathbf{Q}(\sqrt{d})$ satisfy $N(\gamma) > 0$. Then $\gamma\alpha, \gamma\beta$ is an ordered basis of γM.

Proof: Indeed, if

$$\delta_1 = \begin{vmatrix} \gamma\alpha & (\gamma\alpha)' \\ \gamma\beta & (\gamma\beta)' \end{vmatrix}$$

then

$$\delta_1 = (\gamma\gamma')(\alpha\beta' - \beta\alpha') = N(\gamma)(\alpha\beta' - \beta\alpha'),$$

so that $\delta_1/\sqrt{d} = N(\gamma)(\delta/\sqrt{d}) > 0$, since α, β is an ordered basis of M and $N(\gamma) > 0$. Thus, $\gamma\alpha, \gamma\beta$ is an ordered basis for γM. ■

11.2 Exercises

1. Show that $1, \mathscr{f}\omega_d$ is an ordered basis of the coefficient ring $\mathfrak{O}_{\mathscr{f}} = \{1, \mathscr{f}\omega_d\}$.
2. Determine ordered bases for the following modules:
   (a) $M = \{2 + \sqrt{3}, 1 - \sqrt{3}\}$.
   (b) $M = \{1 - \sqrt{-5}, 3 + \sqrt{-5}\}$.
3. Let α, β be an ordered basis of the module M. Let r, s, t, u be rational integers such that $ru - st = 1$. Show that $\alpha_1 = r\alpha + t\beta, \beta_1 = s\alpha + u\beta$ is an ordered basis of M.
4. Let $f(x, y)$ be a binary quadratic form of nonsquare discriminant Δ. Show that $\Delta = \mathscr{f}^2\Delta_0$, where $\mathscr{f}$ is a rational integer and Δ_0 is the discriminant of the integers in some quadratic field. Show that $\mathscr{f}$ and Δ_0 are unique.
5. Show that strict similarity of modules is an equivalence relation.
6. Let $\{\alpha, \beta\}$ be a module such that the basis α, β is not ordered. Show that the basis α_1, β_1, of (2) with $ru - st = -1$ is ordered.
7. Determine the strict class number of I_2, I_3, I_6, I_{10}, and I_{15}. Determine the strict class numbers of the coefficient rings $\mathfrak{O}_{\mathscr{f}}$ contained in $\mathbf{Q}(\sqrt{2})$ for $\mathscr{f} = 2, 3, 4, 5, 6, 7$.
8. (a) Give an example of two modules which are similar but not strictly similar.
   (b) In your example in part (a), show how the classes are broken up into strict classes.
9. Give an example of a module M such that $[M] \neq [M']$.

10. Determine whether the following modules are strictly similar:
    (a) $\{1, \sqrt{3}\}, \{2, \sqrt{3}\}$.
    (b) $\{\sqrt{3} + 1, 3 - \sqrt{3}\}, \{4 + 2\sqrt{3}, 2\sqrt{3}\}$.
    (c) $\{5, \sqrt{3}\}, \{8 + 3\sqrt{3}, 5 - 2\sqrt{3}\}$.

11. Here is an algorithm for determining whether two modules M_1 and M_2 are similar in a finite number of steps. Verify that it works and that all steps can be carried out in a finite number of steps.
    (a) Determine $\mathfrak{O}_{M_1}$ and $\mathfrak{O}_{M_2}$. If $\mathfrak{O}_{M_1} \neq \mathfrak{O}_{M_2}$, then M_1 and M_2 are not similar.
    (b) Assume that $\mathfrak{O}_{M_1} = \mathfrak{O}_{M_2} = \mathfrak{O}$. By replacing M_1 by a module similar to M_1, we may assume that $M_1 \subset M_2$.
    (c) Let $n = (M_2 : M_1)$. Show that n belongs to $\mathfrak{O}$. Show that if $\gamma M_2 = M_1$, then $|N(\gamma)| = n$ and γ is in $\mathfrak{O}$.
    (d) Let $\gamma_1, \ldots, \gamma_r$ be a complete set of nonassociated elements of $\mathfrak{O}$ such that $N(\gamma_i) = n$ $(1 \leq i \leq r)$. Then M_1 is similar to M_2 if and only if M_1 is one of $\gamma_1 M_2, \gamma_2 M_2, \ldots, \gamma_r M_2$.

12. Use the algorithm of Exercise 11 to determine whether $\{2 - 3\sqrt{5}, 5 - 8\sqrt{5}\}$ and $\{3, 4 + \sqrt{5}\}$ are similar.

13. Modify the algorithm of Exercise 11 to determine when two modules M_1 and M_2 are strictly similar.

11.3 The Correspondence Between Modules and Forms

In this section, we shall set up our correspondence between strict similarity classes of modules and equivalence classes of forms. *Throughout this section, all forms will be assumed to be primitive with discriminant not equal to a perfect square.*

Let us begin with a module $M = \{\alpha, \beta\}$ of $\mathbf{Q}(\sqrt{d})$, where the basis α, β is ordered. To this module, let us associate the class of forms $C_M = \{f_{\alpha,\beta}\}$, where

$$f_{\alpha,\beta}(x, y) = \frac{1}{N(M)} N(\alpha x + \beta y).$$

Lemma 1: C_M depends only on M and not on the particular ordered basis of M.

Proof: If α_1, β_1 is another ordered basis of M, then there exist rational integers r, s, t, u such that $\alpha_1 = r\alpha + s\beta$, $\beta_1 = t\alpha + u\beta$, $ru - st = 1$. But

$$\begin{aligned} f_{\alpha_1,\beta_1}(x, y) &= \frac{1}{N(M)} N(\alpha_1 x + \beta_1 y) \\ &= \frac{1}{N(M)} N(\alpha(rx + ty) + \beta(sx + uy)) \\ &= f_{\alpha,\beta}(rx + ty, sx + uy). \end{aligned} \tag{1}$$

Since $ru - st = 1$, we see that $f_{\alpha_1,\beta_1} \approx f_{\alpha,\beta}$, so that $\{f_{\alpha_1,\beta_1}\} = \{f_{\alpha,\beta}\}$. ■

Theorem 2: Let α, β be a basis of $\mathbf{Q}(\sqrt{d})$, $M = \{\alpha, \beta\}$. Then the binary quadratic form

$$f_{\alpha,\beta}(x, y) = \frac{1}{N(M)} N(\alpha x + \beta y)$$

has rational integral coefficients. Moreover, $f_{\alpha,\beta}(x, y)$ is primitive, and the discriminant of $f_{\alpha,\beta}$ equals the discriminant of $\mathfrak{O}_M$, the coefficient ring of M.

Proof: Let $\gamma = -\beta/\alpha$. Then γ satisfies a quadratic equation with rational coefficients. Therefore, γ satisfies an equation of the form $a\gamma^2 + b\gamma + c = 0$, where a, b, c are rational integers, having no common factor greater than 1, $a > 0$. Now

$$\begin{aligned} f_{\alpha,\beta}(x, y) &= \frac{1}{N(M)} N(\alpha x + \beta y) = \frac{N(\alpha)}{N(M)} N\left(x + \frac{\beta}{\alpha} y\right) \\ &= \frac{N(\alpha)}{N(M)} N(x - \gamma y) = \frac{N(\alpha)}{N(M)a}(ax^2 + bxy + cy^2), \end{aligned}$$

since $Tr(\gamma) = -b/a$, $N(\gamma) = c/a$. By Corollary 9.3.10, the module $\{1, -\gamma\} = \{1, \gamma\}$ has norm $1/a$, so that

$$N(M) = N(\alpha\{1, \gamma\}) = |N(\alpha)| N(\{1, \gamma\}) = \frac{|N(\alpha)|}{a}.$$

Thus, $f_{\alpha,\beta}(x, y) = \pm(ax^2 + bxy + cy^2)$. In particular $f_{\alpha,\beta}$ has integer coefficients and is primitive. By Lemma 8.6.13, the coefficient ring $\mathfrak{O}_M$ is equal to $\{1, a\gamma\}$, so that the discriminant of $\mathfrak{O}_M$ equals

$$\begin{aligned} \begin{vmatrix} 1 & 1 \\ a\gamma & a\gamma' \end{vmatrix}^2 &= a^2(\gamma' - \gamma)^2 \\ &= a^2((\gamma + \gamma')^2 - 4\gamma\gamma') \\ &= a^2(Tr(\gamma)^2 - 4N(\gamma)) \\ &= a^2\left(\frac{b^2}{a^2} - 4\frac{c}{a}\right) \\ &= b^2 - 4ac \\ &= \text{the discriminant of } f_{\alpha,\beta}. \end{aligned}$$

■

Corollary 3: Let M be a module. Then C_M is a primitive class of forms of discriminant $\Delta_{\mathfrak{O}_M}$.

The following result summarizes some important facts about discriminants of forms and coefficient rings:

Proposition 4:

(i) Let $\mathfrak{O} = \{1, \mathscr{f}\omega_d\}$. Then $\Delta_{\mathfrak{O}} = \mathscr{f}^2\Delta_0$, where $\Delta_0 =$ the discriminant of I_d; that is, $\Delta_0 = d$ if $d \equiv 1 (\text{mod } 4)$, and $\Delta_0 = 4d$ if $d \equiv 2, 3 (\text{mod } 4)$.

(ii) $\mathfrak{O}$ is uniquely determined by its discriminant.

(iii) Let D be the discriminant of a binary quadratic form. Then D is the discriminant of a unique coefficient ring, denoted* $\mathfrak{O}_D$.

Proof:

(i) Exercise.

(ii) If $\Delta_{\mathfrak{O}}$ is given, then $\mathscr{f}$ and d may be computed from $\Delta_{\mathfrak{O}}$ since d is square-free. Thus, from $\Delta_{\mathfrak{O}}$, we may reconstruct $\mathfrak{O} = \{1, \mathscr{f}\omega_d\}$.

(iii) Let $f = ax^2 + bxy + cy^2$ have discriminant D. Since $D = b^2 - 4ac \equiv b^2 (\text{mod } 4)$, we see that $D \equiv 0$ or $1 (\text{mod } 4)$. Write $D = \mathscr{f}_1^2 d$, where d is not divisible by a perfect square > 1. If D is odd, so are $\mathscr{f}_1$ and d. Since $\mathscr{f}_1^2$ is then $\equiv 1 (\text{mod } 4)$, d is $\equiv 1 (\text{mod } 4)$, and we may set $\mathfrak{O}_D = \{1, \mathscr{f}_1\omega_d\}$. If D is even, it is divisible by 4. And since d is square-free, we see that $d \equiv 2$ or $3 (\text{mod } 4)$ and $\mathscr{f}_1 = 2\mathscr{f}$. In this case, we may set $\mathfrak{O}_D = \{1, \mathscr{f}\omega_d\}$. ■

Let us fix a discriminant D of a binary quadratic form and let $\mathfrak{O}_D$ be the coefficient ring having the same discriminant. By Proposition 4 and Corollary 3, we see that if C_M is a class of forms of discriminant D, then M belongs to $\mathfrak{O}_D$. Moreover, we have

Proposition 5: If $M \approx M_1$, then $C_M = C_{M_1}$.

Proof: Suppose that $M_1 = \gamma M$ with $N(\gamma) > 0$, $M = \{\alpha, \beta\}$, α, β an ordered basis for M. Then $\gamma\alpha, \gamma\beta$ is an ordered basis for M_1 (Proposition 2.6). Therefore,

$$C_{M_1} = \{f_{\gamma\alpha, \gamma\beta}\},$$

where

$$\begin{aligned} f_{\gamma\alpha,\gamma\beta}(x, y) &= \frac{1}{N(M_1)} N(\gamma\alpha x + \gamma\beta y) \\ &= \frac{N(\gamma)}{|N(\gamma)|} \frac{1}{N(M)} N(\alpha x + \beta y) = f_{\alpha,\beta}(x, y) \end{aligned}$$

*Note that this notation is at variance with the notation $\mathfrak{O}_{\mathscr{f}}$ introduced in Chapter 8.

since $N(\gamma) > 0$. Thus,

$$C_{M_1} = \{f_{\alpha,\beta}\} = C_M. \qquad \blacksquare$$

By Proposition 5 and our above discussion, let us set up the following correspondence between strict similarity classes of modules and classes of forms: Let D be the discriminant of a binary quadratic form, and $\mathfrak{O}_D$ the coefficient ring of the same discriminant.

> To each strict similarity class $\{M\}$ of modules belonging to $\mathfrak{O}_D$, let us associate the class of primitive forms C_M of discriminant D. Denote this correspondence
>
> $$\{M\} \longrightarrow C_M.$$

We have proved above that this correspondence is well defined. The reason that this correspondence is useful is because the following theorem holds:

Theorem 6: The function $\{M\} \to C_M$, from the set of all strict similarity classes of modules belonging to $\mathfrak{O}_D$ to the set of all classes of forms of discriminant D, is one to one. Its image consists of all classes of primitive positive-definite or indefinite forms of discriminant D.

Before we prove Theorem 6, let us note that Gauss' theorem follows immediately. In fact, we have

Corollary 7: The number of classes of primitive binary forms of discriminant D is finite. In fact, this number equals h_D^+ $(D > 0)$ or $2h_D^+$ $(D < 0)$, where h_D^+ denotes the strict class number of $\mathfrak{O}_D$.

Proof: In the case $D > 0$, the result follows immediately from Theorem 6. If $D < 0$, note that the classes of positive-definite forms can be put in one to one correspondence with the classes of negative-definite forms by mapping $\{f\}$ into $\{-f\}$. $\blacksquare$

Let us prove Theorem 6 in two parts. Let us first identify the image of the function $\{M\} \to C_M$. Since C_M is a class of forms of discriminant D, these forms are indefinite in the case $D > 0$. If $D < 0$, then $f_{\alpha,\beta}$ is positive-definite since the coefficient of x^2 is $N(\alpha)/N(M)$, and $N(\alpha) > 0$ for all α when $D < 0$. Thus, the image of the function consists only of indefinite and positive-definite forms.

Let $f = ax^2 + bxy + cy^2$ be a primitive form of discriminant D which is either positive-definite or indefinite. If $a > 0$, set

$$M = \left\{1, \frac{b - \sqrt{D}}{2a}\right\}.$$

Then by Corollary 9.3.10, we have $N(M) = 1/a$, and thus

$$\frac{1}{N(M)}N\left(x + \frac{b - \sqrt{D}}{2a}y\right) = ax^2 + bxy + cy^2.$$

If $a < 0$, then we must have $D > 0$ since f cannot be negative-definite. In this case, set

$$M = \sqrt{D}\left\{1, \frac{b - \sqrt{D}}{2a}\right\}.$$

Then

$$N(M) = |N(\sqrt{D})|\,N\left(\left\{1, \frac{b - \sqrt{D}}{2a}\right\}\right) = \frac{D}{|a|}.$$

Therefore,

$$\frac{1}{N(M)}N\left(\sqrt{D}x + \sqrt{D}\frac{b - \sqrt{D}}{2a}y\right) = -\frac{D}{N(M)}N\left(x + \frac{b - \sqrt{D}}{2a}\right)$$
$$= ax^2 + bxy + cy^2.$$

In each case let $M = \{\alpha, \beta\}$. Then we have

$$\begin{vmatrix} \alpha & \alpha' \\ \beta & \beta' \end{vmatrix} = \begin{vmatrix} 1 & 1 \\ \dfrac{b - \sqrt{D}}{2a} & \dfrac{b + \sqrt{D}}{2a} \end{vmatrix} = \frac{\sqrt{D}}{a} \qquad (a > 0)$$

$$= \begin{vmatrix} \sqrt{D} & -\sqrt{D} \\ \dfrac{-D + b\sqrt{D}}{2a} & \dfrac{-D - b\sqrt{D}}{2a} \end{vmatrix} = -\frac{D\sqrt{D}}{a} \qquad (a < 0,\ D > 0).$$

Therefore, we see that α, β is an ordered basis, so that our correspondence associates $\{f\}$ to the strict similarity class of modules $\{M\}$, where

$$M = \left\{1, \frac{b - \sqrt{D}}{2a}\right\} \qquad (a > 0)$$
$$= \left\{\sqrt{D}, \frac{-D + b\sqrt{D}}{2a}\right\} \qquad (a < 0,\ D > 0).$$

This completes the study of the image of the function $\{M\} \to C_M$.

Example 8: The form $f(x, y) = 2x^2 + 3xy - 4y^2$ of discriminant $D = 41$ determines the strict equivalence class $\{f\}$ consisting of the forms

$$2(rx + sy)^2 + 3(rx + sy)(tx + uy) - 4(tx + uy)^2, \quad ru - st = +1. \tag{2}$$

And this class of forms corresponds to the strict similarity class of modules $\{M\}$, where

$$M = \left\{1, \frac{3 - \sqrt{41}}{4}\right\}.$$

In other words, to the collection of modules

$$\gamma\left\{1, \frac{3 - \sqrt{41}}{4}\right\}, \qquad \gamma \text{ in } \mathbf{Q}(\sqrt{41}),\ N(\gamma) > 0,$$

we associate the collection of forms (2).

Now let us prove that under our correspondence distinct classes of modules give rise to distinct classes of forms. This will complete the proof of Theorem 6.

Proposition 9: Suppose that M_1 and M_2 are modules and that $C_{M_1} = C_{M_2}$. Then $\{M_1\} = \{M_2\}$.

Proof: We may rephrase the proposition. Suppose that $M_1 = \{\alpha_1, \beta_1\}$, $M_2 = \{\alpha_2, \beta_2\}$, where α_1, β_1 (respectively, α_2, β_2) is an ordered basis for M_1 (respectively, M_2). Then $C_{M_1} = C_{M_2}$ if and only if $f_{\alpha_1,\beta_1} \approx f_{\alpha_2,\beta_2}$. Therefore, suppose that

$$f_{\alpha_2,\beta_2}(x, y) = f_{\alpha_1,\beta_1}(rx + sy, tx + uy), \qquad ru - st = +1.$$

We shall prove that $M_1 \approx M_2$. By Eq. (1) we see that

$$f_{\alpha_1,\beta_1}(rx + sy, tx + uy) = f_{r\alpha_1+t\beta_1,\, s\alpha_1+u\beta_1}(x, y).$$

Moreover, since $ru - st = +1$, we know that $r\alpha_1 + s\beta_1$, $t\alpha_1 + u\beta_1$ is an ordered basis of M_1. Let us replace the basis α_1, β_1 by this new basis of M_1. Then we see that we could have initially assumed that

$$f_{\alpha_1,\beta_1}(x, y) = f_{\alpha_2,\beta_2}(x, y). \tag{3}$$

Let us henceforth make this assumption. Now

$$\begin{aligned} f_{\alpha_1,\beta_1}(x, y) &= \frac{1}{N(M_1)} N(\alpha_1 x + \beta_1 y) \\ &= \frac{N(\alpha_1)}{N(M_1)}\left(x + \frac{\beta_1}{\alpha_1}y\right)\left(x + \frac{\beta_1'}{\alpha_1'}y\right). \end{aligned} \tag{4}$$

Similarly,

$$f_{\alpha_2,\beta_2}(x, y) = \frac{N(\alpha_2)}{N(M_2)}\left(x + \frac{\beta_2}{\alpha_2}y\right)\left(x + \frac{\beta_2'}{\alpha_2'}y\right). \tag{5}$$

Therefore, the zeros of $f_{\alpha_1,\beta_1}(x, 1)$ are $-\beta_1/\alpha_1$ and $-\beta_1'/\alpha_1'$, and the zeros of $f_{\alpha_2,\beta_2}(x, 1)$ are $-\beta_2/\alpha_2$ and $-\beta_2'/\alpha_2'$. Thus, by Eq. (3), we conclude that β_1/α_1 equals either β_2/α_2 or β_2'/α_2'. Let us prove that the latter cannot be true, for if $\beta_1/\alpha_1 = \beta_2'/\alpha_2'$, then $\beta_1'/\alpha_1' = \beta_2/\alpha_2$, and

$$\begin{aligned} \frac{1}{\sqrt{d}}\begin{vmatrix} \alpha_1 & \alpha_1' \\ \beta_1 & \beta_1' \end{vmatrix} &= \frac{\alpha_1\alpha_1'}{\sqrt{d}}\begin{vmatrix} 1 & 1 \\ \frac{\beta_1}{\alpha_1} & \frac{\beta_1'}{\alpha_1'} \end{vmatrix} = \frac{N(\alpha_1)}{\sqrt{d}}\begin{vmatrix} 1 & 1 \\ \frac{\beta_2'}{\alpha_2'} & \frac{\beta_2}{\alpha_2} \end{vmatrix} \\ &= -\frac{N(\alpha_1)}{\sqrt{d}}\begin{vmatrix} 1 & 1 \\ \frac{\beta_2}{\alpha_2} & \frac{\beta_2'}{\alpha_2'} \end{vmatrix} \\ &= -\frac{N(\alpha_1)}{N(\alpha_2)}\frac{1}{\sqrt{d}}\begin{vmatrix} \alpha_2 & \alpha_2' \\ \beta_2 & \beta_2' \end{vmatrix} < 0, \end{aligned}$$

since α_2, β_2 is an ordered basis, and $N(\alpha_1)$, $N(\alpha_2)$ have the same sign by Eqs.

(3), (4), and (5). But then

$$\frac{1}{\sqrt{d}}\begin{vmatrix}\alpha_1 & \alpha'_1\\ \beta_1 & \beta'_1\end{vmatrix} < 0,$$

contradicting the fact that α_1, β_1 is an ordered basis. Thus, $\beta_1/\alpha_1 = \beta'_2/\alpha'_2$ is impossible, and we must have $\beta_1/\alpha_1 = \beta_2/\alpha_2$. Set $\gamma = \beta_1/\alpha_1$. Then, $\beta_1 = \alpha_1\gamma$ and $\beta_2 = \alpha_2\gamma$, so that

$$M_1 = \{\alpha_1, \beta_1\} = \{\alpha_1, \alpha_1\gamma\} = \alpha_1\{1, \gamma\}$$
$$M_2 = \{\alpha_2, \beta_2\} = \{\alpha_2, \alpha_2\gamma\} = \alpha_2\{1, \gamma\}.$$

Therefore,

$$M_2 = \frac{\alpha_2}{\alpha_1}M_1. \tag{6}$$

Again we observe by Eqs. (3), (4), and (5) that $N(\alpha_1)$ and $N(\alpha_2)$ are of the same sign. Therefore, $N(\alpha_2/\alpha_1) > 0$, and Eq. (6) shows that M_1 and M_2 are strictly similar. Theorem 6 is now completely proved. ■

Example 10: Let $d = -6$, $\mathscr{f} = 1$. In this case, $\mathfrak{O} = \{1, \omega_{-6}\} = I_{-6}$. Since $d < 0$, strict similarity and similarity are the same. In Example 9.6.10, we showed that every module M belonging to $\mathfrak{O}$ is similar to one of the two modules $P_2 = \{2, \sqrt{-6}\}$, $\mathfrak{O} = \{1, \sqrt{-6}\}$. Therefore, there are two strict similarity classes of modules belonging to $\mathfrak{O}$, namely $\{P_2\}$ and $\{\mathfrak{O}\}$. Therefore, there are two equivalence classes of forms of discriminant $-24 = \Delta_{\mathfrak{O}}$. These two classes are

$$\{x^2 + 6y^2\}, \qquad \{2x^2 + 3y^2\}.$$

Example 11: Let us suppose that $\mathfrak{O} = \{1, \mathscr{f}\omega_d\}$ and $h^+_{\mathfrak{O}} = 1$. Thus, every module belonging to $\mathfrak{O}$ is strictly similar to $\mathfrak{O}$, and thus every form of discriminant $\Delta_{\mathfrak{O}} = \mathscr{f}^2\Delta_0$ (Δ_0 = the discriminant of $\mathbf{Q}(\sqrt{d})$, that is, of I_d) is equivalent to the form associated with $\mathfrak{O}$. This form is

$$\frac{1}{N(\mathfrak{O})}N(x + y\mathscr{f}\omega_d) = \begin{cases} x^2 - \mathscr{f}^2dy^2 & \text{if } d \equiv 2 \text{ or } 3 \pmod 4,\\ x^2 + \mathscr{f}xy + \mathscr{f}^2\dfrac{1-d}{4}y^2 & \text{if } d \equiv 1 \pmod 4.\end{cases}$$

11.3 Exercises

1. Let $f(x, y) = 3x^2 + 7xy + y^2$. Determine the strict module class associated with f.
2. Use Exercises 7 and 13 in Section 2 to determine the equivalence classes of primitive forms of discriminant $D = 8, 12, 24, 40, 32$.
3. On the basis of Exercise 11 in Section 2, devise an algorithm for determining the equivalence classes of forms of discriminant D in a finite number of steps.

11.4 The Representation of Integers by Binary Quadratic Forms

Let us now make use of the correspondence between modules and forms constructed in Section 3 to reformulate the problem of representation of integers by a binary form in terms of modules. Throughout this section we shall assume the following notation (see Proposition 3.4). Let $f(x, y)$ be a primitive binary quadratic form of discriminant D. We factor $D = \mathscr{f}^2\Delta_0$, where Δ_0 is the discriminant of $\mathbf{Q}(\sqrt{d})$ and $\mathscr{f}$ is a rational integer. We further write $d = \Delta_0$ if $\Delta_0 \equiv 1 \pmod 4$ and $d = \Delta_0/4$ if $\Delta_0 \equiv 0 \pmod 4$ (in this case $d \equiv 2, 3 \pmod 4$). Thus, d is a square-free rational integer. Set $\mathfrak{O} = \{1, \mathscr{f}\omega_d\}$, the coefficient ring of $\mathbf{Q}(\sqrt{d})$ of discriminant D. Assume that f is positive-definite or indefinite.

From our correspondence of Section 3 we can find an ordered basis α, β of $\mathbf{Q}(\sqrt{d})$ such that, if we set $M = \{\alpha, \beta\}$, then $\mathfrak{O}_M = \mathfrak{O}$ and

$$f(x, y) = \frac{1}{N(M)} N(\alpha x + \beta y).$$

Let m be a rational integer. Then we seek necessary and sufficient conditions on m for the Diophantine equation

$$f(x, y) = m$$

to be solvable. As we have previously seen, this is equivalent to determining all elements $\xi = \alpha x + \beta y$ in M such that $N(\xi) = mN(M)$. Since we have already given a constructive procedure for determining the units of $\mathfrak{O}$, it suffices to determine a complete set of nonassociate ξ's such that $N(\xi) = mN(M)$. Let us show how to solve this problem, at least in principle, in terms of modules. If $D < 0$, then we may take $m \geq 0$ since $f(x, y)$ is positive-definite. If $D > 0$ and $m < 0$, we may replace f by $-f$ and m by $-m$, so that we may again assume that $m \geq 0$. Thus, in any case, assume that $m \geq 0$.

We know from the construction of our correspondence that M may be replaced by any strictly similar module. Moreover, we have shown that there exists a rational integer g such that $gM \subseteq \mathfrak{O}$ (Lemma 9.6.5). Since $N(g) = g^2 > 0$, we see that $gM \approx M$. Thus, without loss of generality, let us assume that $M \subseteq \mathfrak{O}$; *that is, we may assume that M is an integral module.*

Let $\xi = \alpha x + \beta y$ belong to M and suppose that $N(\xi) = mN(M)$. Then $\xi\mathfrak{O} \subseteq M$, so that $M \mid \xi\mathfrak{O}$ by Theorem 9.4.6. Thus, there exists an integral module M_1 such that $MM_1 = \xi\mathfrak{O}$. But then

$$N(M)N(M_1) = |N(\xi)| = mN(M) \qquad \text{(since } m > 0).$$

Therefore, $N(M_1) = m$. From Theorem 9.3.8, we have that $MM' = N(M)\mathfrak{O}$, and thus multiplying by M_1 and recalling that $MM_1 = \xi\mathfrak{O}$, we see that

$$MM'M_1 = \xi M' = N(M)M_1,$$

and since $N(\xi) = mN(M)$, we see that $N(N(M)/\xi) = N(M)^2/N(\xi) > 0$. Therefore, we see that $M' \approx M_1$. Thus, from the element ξ of M of norm $mN(M)$, we have constructed the module M_1 having the following properties: $N(M_1) = m$, M_1 is integral, and $M_1 \approx M'$; in fact, $M_1 = (\xi/N(M))M'$. Let us try to classify elements ξ of M having norm m in terms of the corresponding modules M_1.

Suppose that ξ_1 and ξ_2 are associate elements of M. Then there exists a unit ϵ of $\mathfrak{O}$ such that $\xi_1 = \epsilon\xi_2$. But then, since $\epsilon M' = M'$, we have

$$\frac{\xi_1}{N(M)}M' = \frac{\epsilon\xi_2}{N(M)}M' = \frac{\xi_2}{N(M)}(\epsilon M') = \frac{\xi_2}{N(M)}M'.$$

Therefore, associate elements of M give rise to the same module M_1.

On the other hand, if ξ_1, ξ_2 belong to M, both have norm $mN(M)$, and if they give rise to the same module, then

$$\frac{\xi_1}{N(M)}M' = \frac{\xi_2}{N(M)}M',$$

so that $(\xi_1\xi_2^{-1})M' = M'$. Therefore, $\xi_1\xi_2^{-1}$ is a unit ϵ of $\mathfrak{O}_{M'} = \mathfrak{O}$ and $\xi_1 = \epsilon\xi_2$. In other words ξ_1 and ξ_2 are associates. To put it another way, *nonassociate ξ's give rise to different modules M_1.*

Theorem 1: Let m be a positive rational integer, and M be an integral module of $\mathbf{Q}(\sqrt{d})$. To each element ξ of M such that $N(\xi) = mN(M)$, associate the module

$$M_1 = \frac{\xi}{N(M)}M'.$$

Then M_1 is an integral module such that $N(M_1) = m$ and is strictly similar to M'. Moreover, two different elements ξ_1 and ξ_2 of M correspond to the same module M_1 if and only if ξ_1 and ξ_2 are associate. Finally, every integral module of norm m which is strictly similar to M' arises from some ξ in M of norm $mN(M)$.

Proof: We have established everything but the last statement. Let M_1 be an integral module of norm m which is strictly similar to M'. Then there exists ξ_1 with $N(\xi_1) > 0$ and $M_1 = \xi_1 M'$. Set $\xi = \xi_1 N(M)$, so that

$$M_1 = \frac{\xi}{N(M)}M'.$$

Then M_1 is the module corresponding to ξ. Let us conclude the proof by showing that $N(\xi) = mN(M)$ and that ξ is in M. Since $M_1 = \xi_1 M'$ and $N(M_1) = m$, $N(M') = N(M)$, $N(\xi_1) > 0$, we see that

$$N(\xi_1) = \frac{N(M_1)}{N(M')} = \frac{m}{N(M)},$$

and thus

$$N(\xi) = N(\xi_1)N(N(M)) = \frac{m}{N(M)}N(M)^2 = mN(M).$$

Moreover, $N(M)M_1 = \xi M'$ implies that $N(M)M_1M = \xi M'M$, so that $MM' = N(M)\mathfrak{O}$ implies that $N(M)M_1M = N(M)\xi\mathfrak{O}$, or, on cancelling $N(M)$, $\xi\mathfrak{O} = MM_1$. Since M and M_1 are integral, we see that $M \mid \xi\mathfrak{O}$, so that $\xi\mathfrak{O} \subseteq M$ by Theorem 9.4.6. Thus, ξ contained in $\xi\mathfrak{O}$ implies that ξ is in M. ■

Corollary 2: The number of nonassociated elements ξ of M having norm $mN(M)$ equals the number of integral modules M_1 of norm m which are strictly similar to M'.

Proof: Immediate from Theorem 1. ■

By using Theorem 1, we can outline a procedure for determining a complete set of nonassociated solutions of $f(x, y) = m$, where $f(x, y)$ is any primitive binary quadratic form with rational integral coefficients and non-square discriminant and not negative-definite, and m is any positive rational integer. Suppose that $f(x, y) = ax^2 + bxy + cy^2$, $D = b^2 - 4ac$. Set

$$M = \begin{cases} \left\{1, \dfrac{b - \sqrt{D}}{2a}\right\} & (a > 0) \\ \left\{\sqrt{D}, \dfrac{-D + b\sqrt{D}}{2a}\right\} & (a < 0,\ D > 0). \end{cases}$$

Then

$$f(x, y) = \frac{1}{N(M)}N(\alpha x + \beta y),$$

where $\alpha = 1$, $\beta = (b - \sqrt{D})/2a$ if $a > 0$ and $\alpha = \sqrt{D}$, $\beta = (-D + b\sqrt{D})/2a$ if $a < 0$ and $D > 0$. We may proceed as follows:

1. Determine all integral modules M_1 strictly similar to M' and of norm m.
2. Determine ξ such that $M_1 = (\xi/N(M))M'$, $N(\xi) > 0$.
3. Write $\xi = \alpha x + \beta y$.

Then, as M_1 ranges over all the modules defined in step 1, (x, y) ranges over a complete set of nonassociated solutions of $f(x, y) = m$.

In practice, step 1 is difficult to carry out. To determine all integral modules M_1 belonging to $\mathfrak{O}$ and having norm m is quite easy, at least if $\gcd(m, \mathfrak{f}) = 1$. This can be explicitly carried out using our factorization theory for modules developed in Chapter 9. But then, after determining all these modules, we must determine which are strictly similar to M'. That is, in general, very difficult to do using a reasonably short calculation. We outlined a somewhat unreasonable procedure in the exercises of Section 2. However, it is very easy if $h_{\mathfrak{O}}^{+} = 1$, for in this case, every module belonging to $\mathfrak{O}$ is strictly similar to every other module. Thus, in the case $h_{\mathfrak{O}}^{+} = 1$, every integral module of norm m is automatically similar to M'.

Example 3: Let us completely determine the integers represented by the form $f(x, y) = x^2 + xy - 7y^2$. Here $D = 29$ and $M = \{1, \omega_{29}\} = I_{29}$ and* $\mathfrak{O} = I_{29}$. It is possible to calculate the fundamental unit ϵ of $\mathfrak{O}$ by using the algorithm of Section 8.7. The result is $\epsilon = 2 + \omega_{29}$. Since $N(2 + \omega_{29}) = -1$, we see that $h_{\mathfrak{O}}^{+} = 1$ by Theorem 2.3 and Example 9.6.9. Thus, m is representable by the form $f(x, y)$ if and only if there exists an integral module belonging to $\mathfrak{O}$ of norm m. In fact, the number of nonassociated solutions of $f(x, y) = m$ equals the number of integral modules M_1 belonging to $\mathfrak{O}$ of norm m.

Let $m = p_1^{a_1} \cdots p_t^{a_t}$. If M_1 is a module belonging to $\mathfrak{O}$ and having norm m, then we may factor M_1 into a product of powers of distinct prime modules (note that in this example $\mathfrak{f} = 1$):

$$M_1 = P_1^{e_1} P_2^{e_2} \cdots P_k^{e_k}.$$

And since

$$N(M_1) = N(P_1)^{e_1} \cdots N(P_k)^{e_k}$$

and $N(P_i)$ is either a prime or the square of a prime, we see that if M_1 exists, $P_1, \ldots, P_k, e_1, \ldots, e_k$ are constrained by the relation

$$p_1^{a_1} \cdots p_t^{a_t} = N(P_1)^{e_1} \cdots N(P_k)^{e_k}. \tag{1}$$

In particular, the only possible value for $N(P_i)$ is a power of one of the primes $p_1, \ldots, p_t$. Let p be a given rational prime. What prime modules P of norm a power of p are there? Our factorization theory asserts that such a module must necessarily be a factor of the module $p\mathfrak{O}$. And there are three possible ways for $p\mathfrak{O}$ to factor:

1. p ramified. Then $p\mathfrak{O} = P^2$ and $N(P) = p$. This case can occur only if p divides the discriminant of $\mathfrak{O}$, which in this case is 29. Thus, only $p = 29$ is ramified and $29\mathfrak{O} = P_{29}^2$, $N(P_{29}) = 29$.
2. p inert. Then $p\mathfrak{O} = P$ and $N(P) = p^2$. If p is odd, this case occurs if and only if

$$\left(\frac{29}{p}\right) = -1.$$

But by the law of quadratic reciprocity

$$\left(\frac{29}{p}\right) = \left(\frac{p}{29}\right),$$

so that $\left(\frac{29}{p}\right) = -1$ if and only if

$$p \equiv \pm 2, \pm 3, \pm 8, \pm 10, \pm 11, \pm 12, \pm 14 \pmod{29}. \tag{2}$$

*Note that $f(x, y) = \frac{1}{N(M)} N(x + y\omega_{29}') = \frac{1}{N(M)} N(x + y\omega_{29})$.

Thus, if p satisfies one of conditions (2), then $p\mathfrak{O} = P$ and $N(P) = p^2$. If $p = 2$, then p is also inert by Theorem 9.5.4.

3. p decomposed. Then $p\mathfrak{O} = PP'$, $P \neq P'$, $N(P) = N(P') = p$. Again by our previous work, we know that p is decomposed if and only if $\left(\frac{29}{p}\right) = +1$, which occurs if and only if

$$p \equiv \pm 1, \pm 4, \pm 5, \pm 6, \pm 7, \pm 9, \pm 13 \pmod{29}. \tag{3}$$

Thus, if p satisfies one of conditions (3), then $p\mathfrak{O} = PP'$, $P \neq P'$, $N(P) = N(P') = p$.

Using this threefold classification of rational primes, let us see what integers can occur on the right-hand side of Eq. (1). It is clear that only even powers of inert primes can occur. However, any power of a decomposed or ramified prime can occur. Thus, we have the following result:

The positive integer $m = p_1^{a_1} \cdots p_t^{a_t}$ is represented by the form $x^2 + xy - 7y^2$ if and only if all inert primes (*those in the list in Eq.* (2)) *among $p_1, \ldots, p_t$ occur with an even exponent.*

Thus, for example, $2^2 \cdot 3^2 \cdot 17^{11} \cdot 13^7$ is not represented since 17 is inert and appears to an odd exponent.

To explicitly find all representations of m by $f(x, y)$ it is simplest to consider the case $m = p$ a prime. The general case is more tedious but follows the same lines as the prime case. From what we have just said we have the following:

The form $x^2 + xy - 7y^2$ represents a prime p if and only if either $p = 29$ or $p \equiv \pm 1, \pm 4, \pm 5, \pm 6, \pm 7, \pm 9, \pm 13 \pmod{29}$.

If $p = 29$, there is only one module P of norm 29, and so there is a single solution of $x^2 + xy - 7y^2 = 29$, up to associates. Thus, if (x_0, y_0) is any one solution, all solutions (x, y) can be obtained by writing each number

$$\pm(x_0 + y_0\omega_{29})(2 + \omega_{29})^{2n} \qquad (n = 0, \pm 1, \pm 2, \ldots)$$

in the form $x + y\omega_{29}$.

If p is a decomposed prime, that is, a prime satisfying Eq. (3), then there are two integral modules of norm p. If $p\mathfrak{O} = PP'$, then P and P' are the only integral modules of norm p. Thus, there are two nonassociated solutions to $x^2 + xy - 7y^2 = p$ in this case. If $\xi_1 = x_1 + y_1\omega_{29}$ corresponds to P, then

$$P = \frac{\xi_1}{N(\mathfrak{O})}\mathfrak{O} = \xi_1\mathfrak{O}.$$

Therefore, we may let $\xi_1' = (x_1 + y_1) - y_1\omega_{29}$ correspond to P'. Then, all solutions (x, y) are obtained by writing each of the numbers

$$\begin{aligned} &\pm(2 + \omega_{29})^{2n}(x_1 + y_1\omega_{29}), \\ &\pm(2 + \omega_{29})^{2n}(x_1 + y_1 - y_1\omega_{29}) \end{aligned} \qquad (n = 0, \pm 1, \pm 2, \ldots)$$

in the form $x + y\omega_{29}$. For (x_1, y_1) we can take any particular solution of $x^2 + xy - 7y^2 = p$.

In the case of representations of a composite number m, there generally will be more than two modules of norm m. We shall leave this subject to the exercises. However, we may summarize the result to be proved there as follows:

Let $m = p_1^{a_1} \cdots p_t^{a_t}$ be representable by the form $x^2 + xy - 7y^2$. Let $T =$ the sum of those a_i for which p_i is decomposed. Then the equation $x^2 + xy - 7y^2 = m$ has 2^T nonassociate solutions.

Let us complete this long-winded example by showing how the above theoretical calculations can be handled numerically. Let us determine all solutions of the equation $x^2 + xy - 7y^2 = 35$. Since $35 = 5 \cdot 7$ and

$$\left(\frac{29}{5}\right) = \left(\frac{29}{7}\right) = 1,$$

we see that the equation may be solved. Moreover,

$$5\Theta = 5I_{29} = P_5P_5'$$
$$7\Theta = 7I_{29} = P_7P_7'.$$

There are four modules of norm 35, namely P_5P_7, P_5P_7', $P_5'P_7$, $P_5'P_7'$, and so there are four nonassociated solutions. We may explicitly calculate P_5 and P_7 using the calculations in the proof of Theorem 9.5.3. We easily derive that*

$$P_5 = \left\{5, \frac{3 - \sqrt{29}}{2}\right\} = \frac{3 - \sqrt{29}}{2} I_{29}$$

$$P_7 = \left\{7, \frac{1 - \sqrt{29}}{2}\right\} = \frac{1 - \sqrt{29}}{2} I_{29}.$$

Set $\xi_1 = (3 - \sqrt{29})/2$, $\xi_2 = (1 - \sqrt{29})/2$. Then

$$\xi_1 = 2 - \omega_{29}, \qquad \xi_1' = 2 - \omega_{29}' = 2 - (1 - \omega_{29}) = 1 + \omega_{29}$$
$$\xi_2 = 1 - \omega_{29}, \qquad \xi_2' = 1 - (1 - \omega_{29}) = \omega_{29}.$$

Thus, the four nonassociate solutions of $x^2 + xy - 7y^2 = 35$ correspond to

$$\xi_1\xi_2 = (2 - \omega_{29})(1 - \omega_{29}) = 9 - 2\omega_{29}$$
$$\xi_1\xi_2' = (2 - \omega_{29})\omega_{29} = -7 + \omega_{29}$$
$$\xi_1'\xi_2 = (1 + \omega_{29})(1 - \omega_{29}) = -6 - \omega_{29}$$
$$\xi_1'\xi_2' = (1 + \omega_{29})\omega_{29} = 7 + 2\omega_{29}.$$

Thus, four nonassociated solutions are $(x, y) = (9, -2), (-7, 1), (-6, -1), (7, 2)$. All the solutions may be obtained from the nonassociated solutions by using the formula derived above.

---

*In this case it is easier to use ad hoc reasoning since we know that all modules belonging to I_{29} are principal (because $h_{29}^+ = 1$). In general, however, Theorem 9.5.3 must be used.

The above example was carried out in such great detail because the phenomena which are exhibited are really quite general. The only very special feature of the example is the fact that the class number $h_{\mathfrak{O}}^{+}$ is 1. Because of this, it was not necessary for us to check which of the modules M_1 of norm m are strictly similar to M'. This is because $h_{\mathfrak{O}}^{+} = 1$ implies that all modules belonging to $\mathfrak{O}$ are strictly similar. There is another case in which we can get around the problem of determining the strict similarity. Suppose, that instead of considering the problem of representing m by a single form $f(x, y)$, we consider the problem of representing m by *some* form of discriminant D. As we have seen, the strict equivalence classes of primitive forms of discriminant D correspond in one to one fashion with the strict similarity classes of modules belonging to $\mathfrak{O} = \{1, \mathcal{f}\omega_d\}$. Therefore, if

$$f_1(x, y), \ldots, f_h(x, y)$$

are a set of primitive forms of discriminant D, one form from each equivalence class, then their corresponding modules

$$M_1, M_2, \ldots, M_h$$

represent all the strict similarity classes of modules belonging to $\mathfrak{O}$. Since equivalent forms represent the same integers, determining whether m is represented by some form of discriminant D is the same as determining whether m is represented by one of the forms $f_1, \ldots, f_h$. And we know that this is the case if and only if there is a module M belonging to $\mathfrak{O}$, of norm m and strictly similar to one of $M'_1, M'_2, \ldots, M'_h$. However, any module is strictly similar to one of the modules $M'_1, \ldots, M'_h$. Therefore, we have the following result.

Theorem 4: Let m be a positive rational integer. Then m is represented by some primitive form of discriminant $D = \mathcal{f}^2\Delta_0$ if and only if there exists an integral module M belonging to $\mathfrak{O} = \{1, \mathcal{f}\omega_d\}$ such that $N(M) = m$.

If M is a module belonging to $\mathfrak{O}$, and if $\gcd(N(M), \mathcal{f}) = 1$, then we may factor M into a product of prime modules belonging to $\mathfrak{O}$:

$$M = P_1^{e_1} \cdots P_k^{e_k}, \qquad P_i \text{ distinct.}$$

Assume that $\gcd(m, \mathcal{f}) = 1$. If $m = p_1^{a_1} \cdots p_t^{a_t}$ is the factorization of m into powers of distinct primes, then $N(M) = m$ if and only if

$$p_1^{a_1} \cdots p_t^{a_t} = N(P_1)^{e_1} \cdots N(P_k)^{e_k}.$$

And now the same reasoning as used in the above example may be applied to yield the following corollary:

Corollary 5: Let m be a positive rational integer and let D be the discriminant of a primitive binary quadratic form. Adopt the notation given at the beginning of this section. Assume that $\gcd(m, \mathcal{f}) = 1$. Then m can be represented by some primitive binary form of discriminant D if and only if every prime appearing to an odd exponent in the factorization of m is either ramified or decomposed in $\mathfrak{O} = \{1, \mathfrak{f}\omega_d\}$.

Let us keep the same notation as in the corollary and let us make the further assumption that p is a prime and that $p \nmid \mathfrak{f}$. We showed in Theorem 9.5.4 that p was ramified, inert, or decomposed in $\mathfrak{O} = \{1, \mathfrak{f}\omega_d\}$ according as $p \mid \Delta_0$, $p \nmid \Delta_0$ and $\left(\frac{D}{p}\right) = -1$, or $p \nmid \Delta_0$ and $\left(\frac{D}{p}\right) = +1$, respectively. Therefore, we may rephrase our corollary.

Corollary 6: Let us continue the notation of Corollary 5. Then m can be represented by some primitive binary quadratic form of discriminant D if and only if every prime p appearing to an odd exponent in the factorization of m satisfies $\left(\frac{D}{p}\right) = +1$ or $p \mid \Delta_0$.

Now we can tell the full story concerning the representability of primes by forms of a given discriminant.

Corollary 7: Let us again continue the notations introduced at the beginning of the section. Let p be a prime such that $p \nmid \mathfrak{f}$. Then

(i) p can be represented by a primitive form of discriminant D if and only if either $p \mid \Delta_0$ or $\left(\frac{D}{p}\right) = +1$.

(ii) If q is a prime such that $q \nmid \mathfrak{f}$ and $q \equiv p \pmod{\Delta_0}$, then either p and q are both representable by a form of discriminant D or both are not representable.

Proof: Part (i) follows immediately from Corollary 6. Part (ii) follows immediately from Theorem 9.5.6. ■

Example 8: Let $D = 149 = 1^2 \cdot 149$. Then the primes represented by some form of discriminant 149 are $p = 149$ and those primes p such that $\left(\frac{149}{p}\right) = +1$. Since $149 \equiv 1 \pmod 4$, we see that $\Delta_0 = 149$, so that the primes represented are 149 and those lying in certain reduced residue classes modulo 149.

From Corollary 7, we may finally deduce the following result about the representability of arbitrary integers prime to $\mathfrak{f}$:

Corollary 9: Again continue the notation of Corollary 5. Let m be a positive integer such that $\gcd(\mathfrak{f}, m) = 1$. Write $m = as^2$, a square-free. Then m

is representable by a primitive form of discriminant D if and only if the congruence $x^2 \equiv D(\text{mod } 4a)$ is solvable.

We shall outline two proofs of Corollary 9 in the exercises.

11.4 Exercises

1. Show that the form $x^2 + 2y^2$ represents a prime p if and only if $p \equiv 1$ or $3(\text{mod } 8)$ or $p = 2$.

2. Show that $x^2 + 3y^2$ represents a positive rational integer m if and only if m is of the form $m = k^2 p_1 \cdots p_t$, where $p_1, \ldots, p_t$ are distinct primes $\equiv 1(\text{mod } 3)$.

3. Show that $x^2 + 7y^2$ represents a positive rational integer m if and only if m is of the form $m = k^2 p_1 \cdots p_t$, where $p_1, \ldots, p_t$ are distinct primes $\equiv 1, 2$, or $4(\text{mod } 7)$.

4. (a) Show that a prime p such that $p \nmid 6$ is represented by some form of discriminant -6 if and only if $p \equiv 1, 5, 7$, or $11(\text{mod } 24)$.
   (b) Use part (a) and the facts proved in the text, as well as a congruence argument, to show that $x^2 + 6y^2$ represents a prime p if and only if $p \equiv 1, 7(\text{mod } 24)$, whereas $2x^2 + 3y^2$ represents a prime p if and only if $p \equiv 5, 11(\text{mod } 24)$.

5. Use the theory of the text to find all solutions to the Diophantine equation $x^2 + xy - 3y^2 = 39$.

6. Let n be an odd rational integer and let $\left(\frac{d}{n}\right)$ denote the Jacobi symbol (see Section 4.4 Exercises 9-20).
   (a) Show that the number of integral modules of $\mathbf{Q}(\sqrt{d})$ belonging to I_d and of norm n equals $\sum_{m|n}\left(\frac{d}{m}\right)$.
   (b) Let $f(x, y)$ be a form of discriminant D such that $\gcd(D, n) = 1$, and let $h_D = 1$. Show that the number of nonassociated solutions to $f(x, y) = n$ is just $\sum_{m|n}\left(\frac{D}{m}\right)$.

7. Use Exercise 6 to calculate the number of representations of m by $x^2 + 7y^2$.

8. Same question as Exercise 7 except for $x^2 + 3y^2$.

Exercises 9–13 will be aimed at deriving many of the results of this section by means of a completely elementary method.

9. A rational integer n is said to be *properly represented* by the primitive form f if there exist rational integers x, y such that $f(x,y)=n$ and $\gcd(x,y)=1$.
   (a) Show that n can be represented by f if and only if n can be written in the form $n = k^2 n_0$, where n_0 is properly represented by f.
   (b) Show that n is properly represented by f if and only if there exists a form $g \approx f$ such that $g = nx^2 + b_1xy + c_1y^2$.
10. Show that if a nonzero rational integer n is properly represented by the primitive form f of discriminant D, then the congruence $x^2 \equiv D(\text{mod } 4|n|)$ is solvable. (*Hint:* Use Exercise 9(b).)
11. Show that if the congruence $x^2 \equiv D(\text{mod } 4|n|)$ is solvable, then n is represented by some primitive form of discriminant D.
12. Suppose that D is a nonsquare integer such that $h_D = 1$. Let f be a primitive form of discriminant D. Show that f represents n if and only if the congruence $x^2 \equiv D(\text{mod } 4|n|)$ is solvable.
13. Deduce Corollary 9 from Exercises 10 and 11.
14. Prove Corollary 9 by using the methods of the text.
15. Which primes are represented by forms of discriminant 15?

11.5 Composition Theory for Binary Quadratic Forms

Let us continue our analysis of the preceding section and derive more precise results concerning multiplication of modules and the representability of integers by forms. Our change in viewpoint in this section will be as follows: Whereas we previously considered the possibility of representing m by some form of discriminant D, we now wish to focus on the representability of m by a given form f. For this purpose, we shall define a product of classes of forms. This product, called the *composition* of the classes, was first introduced by Gauss in his *Disquisitiones Arithmeticae* and will enable us to significantly extend our results of Section 4. Gauss worked directly with forms, whereas our approach, as usual, will be to work with modules and then derive the results on forms from the results on modules. In fact, the theory of modules was developed by Dirichlet in order to simplify Gauss' presentation of the theory of composition.

Let us first recall our approach to Diophantine equations of Chapter 6. There, we studied a number of Diophantine equations using a number of formal identities. At the time, it must have seemed that these identities were pulled from the air, and the fact that such identities existed at all must have seemed like a total accident. In this section, we shall explain these identities and shall show how to systematically manufacture further such identities in order to study the representability of integers by binary quadratic forms. This, simply stated, is the goal of composition theory.

To get a concrete feel for the problems we are considering, let us recall the identities used in connection with the two-square problem and Pell's equation. In the two-square problem, we considered the following identity:

$$(x_1^2 + y_1^2)(x_2^2 + y_2^2) = (x_1x_2 - y_1y_2)^2 + (x_1y_2 + x_2y_1)^2. \qquad (1)$$

(or, to be more precise, we used (1) with y_1 replaced by $-y_1$). From the point of view of quadratic fields, this identity merely states the fact that if $\alpha_1 = x_1 + y_1\sqrt{-1}$ and $\alpha_2 = x_2 + y_2\sqrt{-1}$, then

$$N(\alpha_1)N(\alpha_2) = N(\alpha_1\alpha_2).$$

Recall that we used identity (1) to derive representations of composite rational integers m as a sum of two squares. For example, since $1^2 + 3^2 = 10$ and $2^2 + 5^2 = 29$, we were able to conclude (by setting $x_1 = 1$, $y_1 = 3$, $x_2 = 2$, $y_2 = 5$) that

$$x_1x_2 - y_1y_2 = 1 \cdot 2 - 3 \cdot 5 = -13$$

$$x_1y_2 + x_2y_1 = 1 \cdot 5 + 3 \cdot 2 = 11,$$

so that $13^2 + 11^2 = 10 \cdot 29 = 290$. One primary theoretical use of identity (1) was to reduce the problem of representing a positive rational integer m by the form $x^2 + y^2$ to the case of prime m.

When we considered Pell's equation, we considered a very similar identity, namely

$$(x_1^2 - dy_1^2)(x_2^2 - dy_2^2) = (x_1x_2 + dy_1y_2)^2 - d(x_1y_2 + x_2y_1)^2. \qquad (2)$$

(or, to be more precise, we used (2) with y_1 replaced by $-y_1$). If we set $\alpha_1 = x_1 + \sqrt{d}\, y_1$, $\alpha_2 = x_2 + \sqrt{d}\, y_2$, then identity (2) is equivalent to

$$N(\alpha_1)N(\alpha_2) = N(\alpha_1\alpha_2).$$

Identity (2) can be used to derive solutions to $x^2 - dy^2 = m$. For example, since $3^2 - 3 \cdot 1^2 = 6$ and $5^2 - 3 \cdot 2^2 = 13$, upon setting $d = 3$, $x_1 = 3$, $y_1 = 1$, $x_2 = 5$, $y_2 = 2$, we derive

$$x_1x_2 + 3y_1y_2 = 21, \qquad x_1y_2 + x_2y_1 = 11,$$

so that $21^2 - 3 \cdot 11^2 = 6 \cdot 13 = 78$. Can we use identity (2) to reduce the problem of representing m in the form $x^2 - dy^2$ to the case where m is a prime? This is what happened in the two square-problem. But, unfortunately, the phenomenon does not repeat itself here. For example, set $d = -6$. Then $2^2 + 6 \cdot 1^2 = 10$, so that $x^2 + 6y^2$ represents $10 = 2 \cdot 5$, but $x^2 + 6y^2$ represents neither 2 nor 5. This seeming numerical accident is explained by the fact that, in addition to identity (2), which in this case reads

$$(x_1^2 + 6y_1^2)(x_2^2 + 6y_2^2) = (x_1x_2 - 6y_1y_2)^2 + 6(x_1y_2 + x_2y_1)^2, \qquad (2')$$

we have the further identity

$$(2x_1^2 + 3y_1^2)(2x_2^2 + 3y_2^2) = (2x_1x_2 - 3y_1y_2)^2 + 6(x_1y_2 + y_1x_2)^2. \quad (3)$$

Therefore, the product of two integers represented by the form $f_2(x, y) = 2x^2 + 3y^2$ is represented by $f_1(x, y) = x^2 + 6y^2$. And note that 2 and 5 are represented by $f_2(x, y)$, since $f_2(1, 0) = 2, f_2(1, 1) = 5$. Thus, by identity (3), $f_1(x, y)$ represents 10, as we observed above. Thus, there are two ways to build up solutions of $x^2 + 6y^2 = m$. We may use identity (2′) or identity (3). Are there any solutions which cannot be built up in this way from primes? Surprisingly, the answer is no. Indeed, we may prove* that *every* rational integer m represented by $x^2 + 6y^2$ is of the form $s^2p_1 \cdots p_k$, where s is a rational integer and $p_1, \ldots, p_k$ are primes which are represented by either $x^2 + 6y^2$ or $2x^2 + 3y^2$ and where the number of p_i represented by $2x^2 + 3y^2$ is even. Thus, to reduce the problem of representing integers by the form $x^2 + 6y^2$ to the case of representing primes, we must necessarily consider two problems. First we must determine all primes represented by $x^2 + 6y^2$, and then we must determine all the primes represented by $2x^2 + 3y^2$. Thus, we see that the theory of the form $x^2 + 6y^2$ is intertwined with that of $2x^2 + 3y^2$. Why?

The reason is simple. By Example 3.10 there are precisely two strict equivalence classes of forms of discriminant -24, namely the classes

$$\{x^2 + 6y^2\} \quad \text{and} \quad \{2x^2 + 3y^2\}.$$

Thus, the reason that more than one identity is required to generate the rational integers represented by $x^2 + 6y^2 = m$ is, somehow, a reflection of the fact that the strict class number $h^{\pm}_{-24}$ is 2. In the case of the form $x^2 + y^2$, the discriminant is -4 and $h^{\pm}_{-4} = 1$. We shall give a more precise description of this relationship as we proceed.

Before we go on to give any general results, let us clarify precisely where identities (2′) and (3) come from. The class of modules corresponding to $\{x^2 + 6y^2\}$ is $\{M\}$, where $M = \{1, \sqrt{-6}\} = I_{-6}$. To obtain identity (2′), we consider $\alpha_1 = x_1 + y_1\sqrt{-6}$, $\alpha_2 = x_2 + y_2\sqrt{-6}$ in M. Then $\alpha_1\alpha_2$ also belongs to M, and $N(\alpha_1)N(\alpha_2) = N(\alpha_1\alpha_2)$. But since $\alpha_1\alpha_2$ is in M, $\alpha_1\alpha_2$ is of the form $x_3 + y_3\sqrt{-6}$, so that $N(\alpha_1\alpha_2) = x_3^2 + 6y_3^2$. Specifically, $x_3 = x_1x_2 - 6y_1y_2$, $y_3 = x_1y_2 + x_2y_1$. On the other hand, the class of modules corresponding to $\{2x^2 + 3y^2\}$ is $\{M_1\}$, where $M_1 = \{2, \sqrt{-6}\}$. If $\alpha_1 = 2x_1 + y_1\sqrt{-6}$, $\alpha_2 = 2x_2 + y_2\sqrt{-6}$ belong to M_1, then $\alpha_1\alpha_2$ belongs to $M_1^2 = 2M$ and thus is of the form $2(x + y\sqrt{-6})$. Therefore, the identity $N(\alpha_1)N(\alpha_2) = N(\alpha_1\alpha_2)$ asserts that

$$(2x_1^2 + 3y_1^2)(2x_2^2 + 3y_2^2) = x_3^2 + 6y_3^2.$$

And direct computation shows that $x_3 = 2x_1x_2 - 3y_1y_2$, $y_3 = x_1y_2 + y_1x_2$.

*See Example 14.

Thus, we get identity (3). The reason the form $f_1(x, y) = x^2 + 6y^2$ appears on the right-hand side of (3) is that M_1^2 is $2M$, which is strictly similar to M, the module corresponding to f_1. (We shall see just why this is so later in the section.)

Let us begin our development of composition theory by looking at the module side of the coin. Let us fix a ring of coefficients $\mathcal{O}$ in the quadratic field $\mathbf{Q}(\sqrt{d})$. Exactly as in Section 9.6 (see Definition 9.6.12), where we viewed the set $\mathcal{J}_\mathcal{O}$ of similarity classes of modules as an abelian group, we may view the strict similarity classes of modules as a group under the operation $\{M_1\}\cdot\{M_2\} = \{M_1M_2\}$. Here again $\{\mathcal{O}\}$ is the identity and $\{M\}^{-1} = \{M'\}$. Denote this group by $\mathcal{C}_\mathcal{O}$. Then $\mathcal{C}_\mathcal{O}$ has order $h_\mathcal{O}^+$, the strict class number of $\mathcal{O}$.

Definition 1: The collection $\mathcal{C}_\mathcal{O}$ of all strict classes of modules belonging to $\mathcal{O}$, together with the operation of multiplication of strict classes, is called the *strict class group* of $\mathcal{O}$.

We shall see below how a knowledge of the strict class group of $\mathcal{O}$ allows us to get information about the representation of integers by binary forms. Let us be content for the moment with computing $\mathcal{C}_\mathcal{O}$ for a few coefficient rings $\mathcal{O}$.

Example 2: Let $\mathcal{O} = I_{29}$. We have seen (Example 4.3) that in this case, $h_\mathcal{O}^+ = 1$. Therefore, $\mathcal{C}_\mathcal{O}$ consists of the single class $\{\mathcal{O}\}$, and the law of multiplication is the only one available, namely $\{\mathcal{O}\}\cdot\{\mathcal{O}\} = \{\mathcal{O}\}$.

Example 3: Let $\mathcal{O} = I_{-6}$. We have seen (Example 3.10) that in this case $h_\mathcal{O}^+ = 2$. Therefore, $\mathcal{C}_\mathcal{O}$ consists of two classes, and they are given by $\{\mathcal{O}\}$ and $\{P_2\}$, where $P_2 = \{2, \sqrt{-6}\}$. Moreover, we previously showed (Theorem 9.5.3) that $P_2^2 = 2\mathcal{O}$ and thus

$$\{P_2\}\cdot\{P_2\} = \{2\mathcal{O}\} = \{\mathcal{O}\}$$

since $\mathcal{O}$ and $2\mathcal{O}$ are strictly similar. Moreover, we must have $\{\mathcal{O}\}\cdot\{\mathcal{O}\} = \{\mathcal{O}\}$, $\{\mathcal{O}\}\cdot\{P_2\} = \{P_2\}\cdot\{\mathcal{O}\} = \{P_2\}$. We may display the law of multiplication of $\mathcal{C}_\mathcal{O}$ as in Fig. 11.1. Note that $\{\mathcal{O}\}^{-1} = \{\mathcal{O}\}$, $\{P_2\}^{-1} = \{P_2\}$, as can be easily seen in Fig. 11.1.

	$\{\mathcal{O}\}$	$\{P_2\}$
$\{\mathcal{O}\}$	$\{\mathcal{O}\}$	$\{P_2\}$
$\{P_2\}$	$\{P_2\}$	$\{\mathcal{O}\}$

Figure 11.1

Such a multiplication table can be set up for $\mathcal{C}_\mathcal{O}$ in general. If $\mathcal{C}_\mathcal{O}$ has $h = h_\mathcal{O}^+$ elements, then the table will have h^2 entries. If h is large, obtaining the table can be a very complicated task. We did, however, give an explicit method for multiplying two modules in Remark 9.3.13 and so the task is feasible. But the main point for our purposes is that the table is a finite table of data. And once this finite collection of data is assembled, it can be used for calculation without further ado. This should not leave the impression that $\mathfrak{C}_{\mathfrak{O}}$ is in any sense easy to deal with. There are many open questions concerning $\mathfrak{C}_{\mathfrak{O}}$.

Let us use our correspondence between classes of modules and classes of forms to define the product of two strict equivalence classes of forms. Suppose that $\{f_1\}$ and $\{f_2\}$ are two strict equivalence classes of primitive forms of discriminant D. We may define the *product* of $\{f_1\}$ and $\{f_2\}$ as follows: Suppose that $D = \mathfrak{f}^2\Delta_0$, where Δ_0 is the discriminant of $\mathbf{Q}(\sqrt{d})$. Our correspondence between module classes and classes of forms allows us to associate to each of $\{f_1\}$ and $\{f_2\}$ a strict similarity class of modules belonging to $\mathfrak{O} = \{1, \mathfrak{f}\omega_d\}$. Let these module classes be $\{M_1\}$ and $\{M_2\}$, respectively. Let us define the product $\{f_1\}\cdot\{f_2\}$ to be the strict equivalence class of forms corresponding to $\{M_1\}\cdot\{M_2\}$. The product $\{f_1\}\cdot\{f_2\}$ is called the *composition* of the classes of forms $\{f_1\}$ and $\{f_2\}$. The results concerning products of classes of modules can be immediately translated over to composition of forms.

Proposition 4: Let $\mathfrak{F}$ denote the collection of strict equivalence classes of primitive binary quadratic forms of discriminant D. Then $\mathfrak{F}$ forms a finite abelian group with respect to composition of forms. The order of $\mathfrak{F}$ is $h_{\mathfrak{O}}^{+}$.

Remark 5:

(i) The identity element of $\mathfrak{F}$ is the strict class $\{f_0\}$ corresponding to $\{\mathfrak{O}\}$.

(ii) If $\{f\}$ corresponds to $\{M\}$, then $\{f\}^{-1}$ corresponds to $\{M'\}$.

Example 6: Let $D = -24 = 1^2\cdot(-24)$. We have seen that there are two classes of forms of discriminant -24, namely $\{x^2 + 6y^2\}$, $\{2x^2 + 3y^2\}$. If $\mathfrak{O} = \{1, \sqrt{-6}\}$, then these two classes of forms correspond to the module classes $\{\mathfrak{O}\}$ and $\{P_2\}$, respectively, where $P_2 = \{2, \sqrt{-6}\}$. From the multiplication table for $\{\mathfrak{O}\}$ and $\{P_2\}$ (see Example 3), we derive the following table of compositions for forms of discriminant -24:

	$\{x^2 + 6y^2\}$	$\{2x^2 + 3y^2\}$
$\{x^2 + 6y^2\}$	$\{x^2 + 6y^2\}$	$\{2x^2 + 3y^2\}$
$\{2x^2 + 3y^2\}$	$\{2x^2 + 3y^2\}$	$\{x^2 + 6y^2\}$

Now that we have the concept of the multiplication of forms let us show that we now may derive the general procedure for obtaining the identities mentioned at the beginning of the section.

Theorem 7 (Gauss): Let f_1 and f_2 be primitive binary quadratic forms of discriminant D, and let f_3 be any form belonging to the composition $\{f_1\}\cdot\{f_2\}$. Then there exist rational integers $u_1, u_2, u_3, u_4, v_1, v_2, v_3, v_4$ such that

$$f_1(x_1, y_1)f_2(x_2, y_2) = f_3(x_3, y_3),$$

where

$$\begin{aligned} x_3 &= u_1x_1x_2 + u_2x_1y_2 + u_3y_1x_2 + u_4y_1y_2 \\ y_3 &= v_1x_1x_2 + v_2x_1y_2 + v_3y_1x_2 + v_4y_1y_2. \end{aligned} \tag{4}$$

Proof: As usual, let $\mathfrak{O}$ be the coefficient ring. Let M_1, M_2, M_3 be modules belonging to $\mathfrak{O}$ such that for $i = 1, 2, 3$, $M_i = \{\alpha_i, \beta_i\}$ and

$$f_i(x, y) = \frac{1}{N(M_i)}N(\alpha_i x + \beta_i y).$$

Now $\{f_1\}\cdot\{f_2\} = \{f_3\}$ implies that $\{M_1\}\cdot\{M_2\} = \{M_3\}$; that is, $\{M_1M_2\} = \{M_3\}$. Thus, there is a γ such that $N(\gamma) > 0$ and

$$M_1M_2 = \gamma M_3.$$

In particular, $\alpha_1\alpha_2, \alpha_1\beta_2, \beta_1\alpha_2, \beta_1\beta_2$ belong to γM_3. Thus, there are rational integers $u_1, v_1, u_2, v_2, u_3, v_3, u_4, v_4$ such that

$$\begin{aligned} \alpha_1\alpha_2 &= \gamma(u_1\alpha_3 + v_1\beta_3) \\ \alpha_1\beta_2 &= \gamma(u_2\alpha_3 + v_2\beta_3) \\ \beta_1\alpha_2 &= \gamma(u_3\alpha_3 + v_3\beta_3) \\ \beta_1\beta_2 &= \gamma(u_4\alpha_3 + v_4\beta_3). \end{aligned} \tag{5}$$

Thus,

$$\begin{aligned} f_1(x_1, y_1)f_2(x_2, y_2) &= \frac{1}{N(M_1)}N(\alpha_1x_1 + \beta_1y_1)\frac{1}{N(M_2)}N(\alpha_2x_2 + \beta_2y_2) \\ &= \frac{1}{N(M_1M_2)}N((\alpha_1x_1 + \beta_1y_1)(\alpha_2x_2 + \beta_2y_2)) \\ &= \frac{1}{N(M_1M_2)}N(\alpha_1\alpha_2x_1x_2 + \alpha_1\beta_2x_1y_2 + \beta_1\alpha_2y_1x_2 \\ &\qquad + \beta_1\beta_2y_1y_2) \\ &= \frac{1}{N(\gamma M_3)}N(\gamma(\alpha_3x_3 + \beta_3y_3)) \quad \text{(from Eq. (5), where } x_3, y_3 \text{ are defined in Eq. (4))} \\ &= \frac{1}{N(\gamma)N(M_3)}N(\gamma)N(\alpha_3x_3 + \beta_3y_3) \quad \text{(since } N(\gamma) > 0\text{)} \\ &= f_3(x_3, y_3). \end{aligned}$$

■

Example 8: Again consider the case $D = -24$. We know (see Example 6) that there are two classes of forms of discriminant -24, namely $\{x^2 + 6y^2\}$ and $\{2x^2 + 3y^2\}$ corresponding to modules $\mathfrak{O} = \{1, \sqrt{-6}\}$ and $P_2 =$

$\{2, \sqrt{-6}\}$. We have seen before that $\mathcal{O}\mathcal{O} = \mathcal{O}$, $\mathcal{O}P_2 = P_2$, and $P_2P_2 = 2\mathcal{O}$, which yields the multiplication table of Example 6.

Let us derive the identity corresponding to $\mathcal{O}P_2 = P_2$ or $\{x^2 + 6y^2\}\cdot\{2x^2 + 3y^2\} = \{2x^2 + 3y^2\}$. In this case $\alpha_1 = 1$, $\beta_1 = \sqrt{-6}$, $\alpha_2 = 2$, $\beta_2 = \sqrt{-6}$, $\alpha_3 = 2$, $\beta_3 = \sqrt{-6}$, and $\gamma = 1$. Thus,

$$\begin{aligned}
\alpha_1\alpha_2 &= 2 &&= 1\cdot\alpha_3 + 0\cdot\beta_3 &&(u_1 = 1, v_1 = 0)\\
\alpha_1\beta_2 &= \sqrt{-6} &&= 0\cdot\alpha_3 + 1\cdot\beta_3 &&(u_2 = 0, v_2 = 1)\\
\beta_1\alpha_2 &= 2\sqrt{-6} &&= 0\cdot\alpha_3 + 2\cdot\beta_3 &&(u_3 = 0, v_3 = 2)\\
\beta_1\beta_2 &= (\sqrt{-6})^2 &&= -3\cdot\alpha_3 + 0\cdot\beta_3 &&(u_4 = -3, v_4 = 0).
\end{aligned}$$

Hence,

$$(x_1^2 + 6y_1^2)(2x_2^2 + 3y_2^2) = 2(x_1x_2 - 3y_1y_2)^2 + 3(x_1y_2 + 2y_1x_2)^2.$$

Similarly, from $P_2P_2 = 2\mathcal{O}$ or $\{2x^2 + 3y^2\}\cdot\{2x^2 + 3y^2\} = \{x^2 + 6y^2\}$ we have $\alpha_1 = 2$, $\beta_1 = \sqrt{-6}$, $\alpha_2 = 2$, $\beta_2 = \sqrt{-6}$, $\alpha_3 = 1$, $\beta_3 = \sqrt{-6}$, and $\gamma = 2$. Thus,

$$\begin{aligned}
\alpha_1\alpha_2 &= 4 = 2(2\cdot\alpha_3 + 0\cdot\beta_3) &&(u_1 = 2, v_1 = 0)\\
\alpha_1\beta_2 &= 2\sqrt{-6} = 2(0\cdot\alpha_3 + 1\cdot\beta_3) &&(u_2 = 0, v_2 = 1)\\
\beta_1\alpha_2 &= 2\sqrt{-6} = 2(0\cdot\alpha_3 + 1\cdot\beta_3) &&(u_3 = 0, v_3 = 1)\\
\beta_1\beta_2 &= (\sqrt{-6})^2 = 2(-3\cdot\alpha_3 + 0\cdot\beta_3) &&(u_4 = -3, v_4 = 0).
\end{aligned}$$

Hence,

$$(2x_1^2 + 3y_1^2)(2x_2^2 + 3y_2^2) = (2x_1x_2 - 3y_1y_2)^2 + 6(x_1y_2 + y_1x_2)^2.$$

The identity corresponding to $\mathcal{O}\mathcal{O} = \mathcal{O}$ is given in Eq. (2′).

Let us now show the significance of having determined the complete multiplication table. Let us fix a discriminant $D = \mathcal{f}^2\Delta_0$, $\Delta_0 =$ the discriminant of $\mathbf{Q}(\sqrt{d})$, and let $\mathcal{O} = \{1, \mathcal{f}\omega_d\}$, where $d =$ the square-free part of Δ_0. We shall again take up the problem of determining which rational integers m are represented by a given form f of discriminant D. As we saw in the preceding section, this problem can be tackled in the case $h_\mathcal{O}^+ = 1$. If $h_\mathcal{O}^+ > 1$, then the best we were able to prove was a criterion for m to be represented by *some* form (or forms) of discriminant D. Let us now show how a knowledge of the table of compositions for forms of discriminant D helps us to distinguish among the representability properties of the various forms of discriminant D. Since all forms of a given class C represent the same integers, let us speak of the *integers represented by the class C* rather than integers represented by individual forms of C. We shall again, for the most part, restrict ourselves to integers m such that $\gcd(m, \mathcal{f}) = 1$.

From Theorem 7, we deduce the following result:

Corollary 9: Let the class $\{f_1\}$ represent m_1 and let the class $\{f_2\}$ represent m_2. Then the composition $\{f_1\}\cdot\{f_2\}$ represents m_1m_2.

We have the following partial converse of Corollary 9:

Theorem 10: Let m and n be rational integers such that $\gcd(mn, \mathfrak{f}) = 1$ and $\gcd(m, n) = 1$. Suppose that $\{f\}$ represents mn. Then there exist classes $\{f_1\}$ and $\{f_2\}$ such that $\{f\} = \{f_1\}\cdot\{f_2\}$ and $\{f_1\}$ represents m and $\{f_2\}$ represents n.

Proof: Let $\{f\}$ correspond to the module class $\{M\}$. Since $\{f\}$ represents mn, there exists a module M_1 strictly similar to M' and such that $N(M_1) = mn$ (Theorem 4.1). Since $\gcd(mn, \mathfrak{f}) = 1$, we see that $\gcd(N(M_1), \mathfrak{f}) = 1$. Thus, by Theorem 9.4.4, we may write

$$M_1 = P_1^{e_1} \cdots P_r^{e_r},$$

where $P_1, \ldots, P_r$ are distinct prime modules belonging to $\mathfrak{O}$. We have

$$mn = N(M_1) = N(P_1)^{e_1} \cdots N(P_r)^{e_r}.$$

Since $N(P_i)$ is a prime power and $\gcd(m, n) = 1$, we see that $N(P_i)^{e_i}$ divides precisely one of m or n. By multiplying together all $P_i^{e_i}$ for which $N(P_i)^{e_i} | m$, we get a module M_2 such that $N(M_2) = m$; multiplying together the $P_i^{e_i}$ for which $N(P_i)^{e_i} | n$, we get a module M_3 such that $N(M_3) = n$. Moreover, $M_2M_3 = M_1$. Let $\{M_2'\}$ (respectively, $\{M_3'\}$) correspond to the class of forms $\{f_1\}$ (respectively, $\{f_2\}$). Then $\{f_1\}$ represents m, $\{f_2\}$ represents n (by Theorem 4.1) and $\{f\} = \{f_1\}\cdot\{f_2\}$. ■

From Corollary 9 and Theorem 10, we derive the following necessary and sufficient condition for a rational integer m, relatively prime to $\mathfrak{f}$, to be representable by $\{f\}$:

Theorem 11: Let m be a positive rational integer such that $\gcd(m, \mathfrak{f}) = 1$ and let $\{f\}$ be a class of primitive forms of discriminant $D = \mathfrak{f}^2\Delta_0$. Suppose that $m = p_1^{a_1} \cdots p_r^{a_r}$, where $p_1, \ldots, p_r$ are distinct rational primes. Then m is represented by $\{f\}$ if and only if $\{f\}$ can be written in the form

$$\{f\} = \{f_1\} \cdots \{f_r\},$$

where $\{f_i\}$ represents $p_i^{a_i}$ $(i = 1, \ldots, r)$.

Proof: Induction on r. ■

We can determine all classes $\{f_1\}, \ldots, \{f_r\}$ such that $\{f\} = \{f_1\} \cdots \{f_r\}$ using the table of compositions for the discriminant D. Therefore, Theorem 11 shows us that the problem of representing integers by $\{f\}$ can be solved by determining all the prime powers represented by each of the classes of primitive forms of discriminant D. Let us see what we can say about this latter problem.

By Theorem 4.1, we know that $\{f\}$ represents a prime power p^a if and only if there exists a module M_1 strictly similar to M' such that $N(M_1) = p^a$. Assume that $p \nmid \mathfrak{f}$. Then M_1 can be written as a product of prime modules belonging to $\mathcal{O}$:

$$M_1 = P_1^{e_1} \cdots P_t^{e_t}.$$

But then

$$p^a = N(P_1)^{e_1} \cdots N(P_t)^{e_t}. \tag{6}$$

Since there are at most two prime modules of norm divisible by p, we see that $t = 1$ or 2. Moreover, by Theorem 9.5.2 we have the following cases:

If p is ramified in $\mathcal{O}$, then $p\mathcal{O} = P^2$ and P is prime, and $N(P) = p$. Then Eq. (6) holds if and only if $t = 1$ and $P_1 = P$, $e_1 = a$, and so $M' \approx P^a$. But since $P^2 = p\mathcal{O} \approx \mathcal{O}$, we see that $P^a \approx P$ if a is odd and $P^a \approx \mathcal{O}$ if a is even. Thus, f represents p^a if and only if either a is odd and M' is strictly similar to P or if a is even and M' is strictly similar to $\mathcal{O}$.

If p is inert in $\mathcal{O}$, then $p\mathcal{O} = P$ is prime and $N(P) = p^2$. Then (6) holds if and only if $t = 1$ and $P_1 = P$, a is even and $e_1 = a/2$. But since $P = p\mathcal{O} \approx \mathcal{O}$, we see that $P^{e_1} \approx \mathcal{O}$. Thus, f represents p^a if and only if a is even and M' is strictly similar to $\mathcal{O}$.

If p is decomposed in $\mathcal{O}$, then $p\mathcal{O} = PP'$, where $P \neq P'$ are prime, P and P' conjugate, and $N(P) = N(P') = p$. Then (6) holds if and only if $t = 1$ or 2 and M_1 is one of the modules $P^{e_1}P'^{e_2}$, $e_1 + e_2 = a$. Thus, f represents p^a if and only if M' is strictly similar to one of the modules $P^{e_1}P'^{e_2}$, where $e_1 + e_2 = a$ and $e_1, e_2 \geq 0$.

Thus, we may summarize as follows:

Theorem 12: Let p be a rational prime, $a > 0$ be a rational integer, and f be a primitive binary quadratic form of discriminant $D = \mathfrak{f}^2\Delta_0$. Let d be the square-free part of Δ_0, $\mathcal{O} = \{1, \mathfrak{f}\omega_d\}$, and $\{M\}$ = the strict module class corresponding to $\{f\}$. Assume that $p \nmid \mathfrak{f}$.

(i) If p is ramified in $\mathcal{O}$ and $p\mathcal{O} = P^2$, then f represents p^a if and only if $\{M'\} = \{P\}$ for a odd or $\{M'\} = \{\mathcal{O}\}$ for a even.

(ii) If p is inert in $\mathcal{O}$, then f represents p^a if and only if a is even and $\{M'\} = \{\mathcal{O}\}$.

(iii) If p is decomposed in $\mathcal{O}$ and $p\mathcal{O} = PP'$, then f represents p^a if and only if $\{M'\}$ is one of the classes $\{P^{e_1}P'^{e_2}\}$, $e_1 + e_2 = a$, $e_1 \geq 0$, $e_2 \geq 0$.

Let us now simply translate Theorem 12 into the language of forms. Maintaining the notation of Theorem 12, we have

Corollary 13: Let $\{f_0\}$ correspond to $\{\mathcal{O}\}$.

(i) If p is ramified in $\mathcal{O}$, $p\mathcal{O} = P^2$, then f represents p^a if and only if $\{f\} = \{g_p\}$ for a odd and $\{f\} = \{f_0\}$ for a even, where $\{g_p\}$ corresponds to $\{P\}$.

(ii) If p is inert in $\mathfrak{O}$, then f represents p^a if and only if a is even and $\{f\} = \{f_0\}$.

(iii) If p is decomposed, $p\mathfrak{O} = PP'$, then f represents p^a if and only if there are rational integers $e_1, e_2 \geq 0$ such that $e_1 + e_2 = a$ and

$$\{f\} = \underbrace{\{g_p\} \cdots \{g_p\}}_{e_1 \text{ times}} \cdot \underbrace{\{g'_p\} \cdots \{g'_p\}}_{e_2 \text{ times}},$$

where $\{g_p\}$ corresponds to $\{P\}$ and $\{g'_p\}$ corresponds to $\{P'\}$.

Proof: Exercise. ■

In all cases, we see that once the classes $\{f_0\}$, $\{g_p\}$, and $\{g'_p\}$ have been computed, composition of forms allows us to compute whether a given class of forms represents p^a and in turn general rational integers m (such that $\gcd(m, \mathfrak{f}) = 1$). Thus, by Theorems 11 and 12 and Corollary 13 we see that the problem of determining the rational integers represented by the primitive form f of discriminant D can be reduced to three subproblems.

Subproblem 1: Determine the table of compositions for the classes of primitive forms of discriminant D.

Subproblem 2: For each rational prime p such that $p \nmid \mathfrak{f}$, determine the class of forms $\{g_p\}$ which represents p (p ramified) or p^2 (p inert) or the two classes of forms $\{g_p\}$, $\{g'_p\}$ representing p (p decomposed).

Subproblem 3: Combine the solutions to subproblems 1 and 2 to determine all rational integers m represented by $\{f\}$ for which $\gcd(m, \mathfrak{f}) = 1$.

Example 14: Let us consider the case $D = -24$. As we saw in Example 6, there are two classes of forms, $C_1 = \{x^2 + 6y^2\}$ and $C_2 = \{2x^2 + 3y^2\}$. Let us pursue the above program as far as we can to determine what we can say about the integers represented by the two classes. We solved Subproblem 1 in Example 6, where we showed that

$$C_1^2 = C_2^2 = C_1, \qquad C_1 \cdot C_2 = C_2. \tag{7}$$

Since $\Delta_0 = -24$, $\mathfrak{f} = 1$, we see that the ramified primes are 2 and 3. Since $2 \cdot 1^2 + 3 \cdot 0^2 = 2$, $2 \cdot 0^2 + 3 \cdot 1^2 = 3$, we see that $\{g_2\} = \{g_3\} = C_2$. Thus, C_1 does not represent 2 or 3. Moreover, we see that C_1 (respectively, C_2) represents 2^a (or 3^a) if and only if a is even (respectively, a is odd). If $p \neq 2, 3$ and $\left(\frac{-24}{p}\right) = -1$, then p is inert (Theorem 9.5.4) and neither C_1 nor C_2 represents p. Moreover, C_1 represents p^2. Therefore, we have $\{g_p\} = C_1$ in this case. Therefore, our above analysis shows that C_1 represents p^a if and only if a is even. Let us now assume that $p \neq 2, 3$ and that $\left(\frac{-24}{p}\right) = +1$. Then p is decomposed and is represented by the classes $\{g_p\}$, $\{g'_p\}$ corresponding, respectively, to P and P', where $p\mathcal{O} = PP'$. In this case, since $\{P\}\cdot\{P'\} = \{p\mathcal{O}\} = \{\mathcal{O}\}$, we see that $\{P\} = \{P'\}$ from our table of multiplication of classes. Thus, $\{g_p\} = \{g'_p\}$, and p is represented by precisely one class C_1 or C_2. If $\{g_p\} = \{g'_p\} = C_1$, then C represents p^a if and only if $C = C_1^{e_1}\cdot C_1^{e_2}$ for some $e_1 \geq 0$, $e_2 \geq 0$ such that $e_1 + e_2 = a$, so that $C = C_1^a = C_1$. Thus, C_1 represents every power of p, and C_2 represents no power of p. If $\{g_p\} = \{g'_p\} = C_2$, then C represents p^a if and only if

$$C = C_2^{e_1}\cdot C_2^{e_2}, \qquad e_1 + e_2 = a$$

$$= C_2^{e_1+e_2} = C_2^a = \begin{cases} C_1 & \text{if } a \text{ is even,} \\ C_2 & \text{if } a \text{ is odd.} \end{cases}$$

Thus, C_1 represents p^a if and only if a is even, and C_2 represents p^a if and only if a is odd.

p	$x^2 + 6y^2$ represents	$2x^2 + 3y^2$ represents	
2, 3	2^a, 3^a, a even	2^a, 3^a, a odd	
$p \neq 2, 3$ $\left(\frac{-24}{p}\right) = -1$	p^a, a even	No powers of p	
$p \neq 2, 3$	p^a, any a	No powers of p	If $x^2 + 6y^2$ represents p
$\left(\frac{-24}{p}\right) = +1$	p^a, a even	p^a, a odd	If $2x^2 + 3y^2$ represents p

Let us now determine the rational integers represented by C_1, for example. Using Theorem 11 and the composition law (7), we see that if $m = p_1^{a_1} \cdots p_r^{a_r}$, p_i distinct primes, then C_1 represents m if and only if

$$C_1 = F_1 \cdots F_r, \tag{8}$$

where $F_i = C_1$ or C_2 and F_i represents $p_i^{a_i}$. For Eq. (8) to hold, it is necessary and sufficient that the number of F_i equal to C_2 be even. Thus, by consulting the above table, *we see that m can be represented by C_1, that is, by $x^2 + 6y^2$, if and only if m can be written in the form $m = s^2 q_1 \cdots q_t$, where $q_1, \ldots, q_t$ are distinct primes which are represented by either C_1 or C_2, and the number of q_i represented by C_2 is even.* We leave it to the reader to write down the details of the proof. *By similarly considering representations by C_2, we derive that m can be represented by C_2, that is, by $2x^2 + 3y^2$, if and only if m can be written in the form $m = s^2 q_1 \cdots q_t$, where $q_1, \ldots, q_t$ are distinct primes, and an odd number of the $q_1, \ldots, q_t$ are represented by C_2.*

In conclusion we would like to note the similarity and dissimilarity of this result and the result for $x^2 + y^2$. We showed (many times) that m is represented by $x^2 + y^2$ if and only if m can be written in the form $m = s^2 q_1 \cdots q_t$, where $q_1, \ldots, q_t$ are distinct rational primes represented by $x^2 + y^2$. In particular it suffices to determine the rational *primes* represented by $x^2 + y^2$. Now to determine the m represented by $x^2 + 6y^2$, we need to determine the rational primes represented by both $x^2 + 6y^2$ and $2x^2 + 3y^2$ (e.g., $2x^2 + 3y^2$ represents 2 and 5, while $x^2 + 6y^2$ represents 10 but neither 2 nor 5). Of course, the reason for this phenomenon is that the class number h^+_{-4} of the forms of discriminant -4 is 1 (this is the case $x^2 + y^2$) and the class number h^+_{-24} of the forms of discriminant -24 is 2 (this is the case $x^2 + 6y^2$).

One final point should be made. We know that $x^2 + y^2$ represents the rational prime p if and only if $p \equiv 1 \pmod 4$. No such criterion was given for the form $x^2 + 6y^2$ (or for $2x^2 + 3y^2$ for that matter). We did, however, show that one of $x^2 + 6y^2$ and $2x^2 + 3y^2$ represents a rational prime p if and only if $p = 2, 3$ or $p \neq 2, 3$ and $\left(\frac{-24}{p}\right) = 1$. It is an easy exercise to show that this condition is equivalent to $p = 2$ or $p = 3$ or $p \equiv 1, 5, 7, 11 \pmod{24}$. The obvious question is whether we can separate the forms $x^2 + 6y^2$ and $2x^2 + 3y^2$ with respect to these conditions. This is possible, and the result is that $x^2 + 6y^2$ represents p if and only if $p \equiv 1, 7 \pmod{24}$, and $2x^2 + 3y^2$ represents p if and only if $p \equiv 5, 11 \pmod{24}$. The derivation of results of this type is what is called *genus theory*. We, most unfortunately, do not have space in this book to include it. It is another of the truly outstanding accomplishments of Gauss. But let us hasten to add that the genus theory is only a partial success. It is possible to distinguish the primes represented by $x^2 + 6y^2$ and $2x^2 + 3y^2$ through genus theory. However, this cannot be done in general*; for example, it is never possible if $h^+_{\mathcal{O}}$ is not a power of 2. One of the major unsolved problems in this area is to give conditions that will completely distinguish between the representation properties of all the forms of a given fixed discriminant D. To give some indication of the difficulties involved in this problem it may be mentioned that it has been shown that congruence conditions cannot give criteria in general as they did for $x^2 + 6y^2$ and $2x^2 + 3y^2$.

11.5 Exercises

1. Compile composition tables for $\mathcal{O} = \{1, \sqrt{3}\}$, $\mathcal{O} = \{1, \sqrt{10}\}$, and $\mathcal{O} = \{1, \sqrt{30}\}$.
2. For each composition table of Exercise 1, calculate all the composition identities (as in Theorem 7).

---

*As an example of the complications which can arise, it is true that if $h^+_{\mathcal{O}}$ is not a power of 2, several classes of forms may represent a given prime.

3. Show that if $h_{\mathfrak{O}}^{+} = 3$, then there exists a module M belonging to $\mathfrak{O}$ such that $\mathfrak{C}_{\mathfrak{O}} = \{\{I\}, \{M\}, \{M^2\}\}$.
4. Show that if $h_{\mathfrak{O}}^{+} = 3$, then there exists a module M belonging to $\mathfrak{O}$ such that $\{M\} \neq \{M'\}$.
5. Show that if $h_{\mathfrak{O}}^{+} = 4$, then either (i) there exists a module M belonging to $\mathfrak{O}$ such that $\mathfrak{C}_{\mathfrak{O}} = \{\{I\}, \{M\}, \{M^2\}, \{M^3\}\}$ or (ii) there exist modules M_1 and M_2 belonging to $\mathfrak{O}$ such that $\mathfrak{C}_{\mathfrak{O}} = \{\{I\}, \{M_1\}, \{M_2\}, \{M_1M_2\}\}$.
6. Prove Proposition 4.
7. Assume that $h_{\mathfrak{O}}^{+} = 1$ and $\gcd(m, \mathfrak{f}) = 1$. Use Theorem 12 to determine the number of nonassociate solutions to $f(x, y) = m$, where f is a primitive binary form of discriminant $\Delta_{\mathfrak{O}}$.
8. Same question as Exercise 7 but assume that $h_{\mathfrak{O}}^{+} = 2$.
9. Carry out the analogue of Example 14 for the composition table for $\mathfrak{O} = \{1, \sqrt{10}\}$ (see Exercise 1).
10. Prove the two final statements in Example 14, concerning the representability of integers by $x^2 + 6y^2$ and $2x^2 + 3y^2$.

TABLE 1

Values of $\sigma(n)$, $d(n)$, $\mu(n)$, and $\phi(n)$

n	$\sigma(n)$	$d(n)$	$\mu(n)$	$\phi(n)$
1	1	1	1	1
2	3	2	−1	1
3	4	2	−1	2
4	7	3	0	2
5	6	2	−1	4
6	12	4	1	2
7	8	2	−1	6
8	15	4	0	4
9	13	3	0	6
10	18	4	1	4
11	12	2	−1	10
12	28	6	0	4
13	14	2	−1	12
14	24	4	1	6
15	24	4	1	8
16	31	5	0	8
17	18	2	−1	16
18	39	6	0	6
19	20	2	−1	18
20	42	6	0	8
21	32	4	1	12
22	36	4	1	10
23	24	2	−1	22
24	60	8	0	8
25	31	3	0	20
26	42	4	1	12
27	40	4	0	18
28	56	6	0	12
29	30	2	−1	28
30	72	8	−1	8
31	32	2	−1	30
32	63	6	0	16
33	48	4	1	20
34	54	4	1	16
35	48	4	1	24
36	91	9	0	12
37	38	2	−1	36
38	60	4	1	18
39	56	4	1	24
40	90	8	0	16
41	42	2	−1	40
42	96	8	−1	12
43	44	2	−1	42
44	84	6	0	20
45	78	6	0	24
46	72	4	1	22
47	48	2	−1	46
48	124	10	0	16
49	57	3	0	42
50	93	6	0	20

TABLE 1 (continued)

n	$\sigma(n)$	$d(n)$	$\mu(n)$	$\phi(n)$
51	72	4	1	32
52	98	6	0	24
53	54	2	−1	52
54	120	8	0	18
55	72	4	1	40
56	120	8	0	24
57	80	4	1	36
58	90	4	1	28
59	60	2	−1	58
60	168	12	0	16
61	62	2	−1	60
62	96	4	1	30
63	104	6	0	36
64	127	7	0	32
65	84	4	1	48
66	144	8	−1	20
67	68	2	−1	66
68	126	6	0	32
69	96	4	1	44
70	144	8	−1	24
71	72	2	−1	70
72	195	12	0	24
73	74	2	−1	72
74	114	4	1	36
75	124	6	0	40
76	140	6	0	36
77	96	4	1	60
78	168	8	−1	24
79	80	2	−1	78
80	186	10	0	32
81	121	5	0	54
82	126	4	1	40
83	84	2	−1	82
84	224	12	0	24
85	108	4	1	64
86	132	4	1	42
87	120	4	1	56
88	180	8	0	40
89	90	2	−1	88
90	234	12	0	24
91	112	4	1	72
92	168	6	0	44
93	128	4	1	60
94	144	4	1	46
95	120	4	1	72
96	252	12	0	32
97	98	2	−1	96
98	171	6	0	42
99	156	6	0	60
100	217	9	0	40

TABLE 2

Class Numbers and Fundamental Units of Real Quadratic Fields (here $\omega = \omega_d = \sqrt{d}$ if $d \equiv 2$ or 3(mod 4), $= (1 + \sqrt{d})/2$ if $d \equiv 1$(mod 4))

d	h_d	ϵ_1 = Fundamental Unit of I_d	$N(\epsilon_1)$
2	1	$1 + \omega$	-1
3	1	$2 + \omega$	$+1$
5	1	ω	-1
6	1	$5 + 2\omega$	$+1$
7	1	$8 + 3\omega$	$+1$
10	2	$3 + \omega$	-1
11	1	$10 + 3\omega$	$+1$
13	1	$1 + \omega$	-1
14	1	$15 + 4\omega$	$+1$
15	2	$4 + \omega$	$+1$
17	1	$3 + 2\omega$	-1
19	1	$170 + 39\omega$	$+1$
21	1	$2 + \omega$	$+1$
22	1	$197 + 42\omega$	$+1$
23	1	$24 + 5\omega$	$+1$
26	2	$5 + \omega$	-1
29	1	$2 + \omega$	-1
30	2	$11 + 2\omega$	$+1$
31	1	$1520 + 273\omega$	$+1$
33	1	$19 + 8\omega$	$+1$
34	2	$35 + 6\omega$	$+1$
35	2	$6 + \omega$	$+1$
37	1	$5 + 2\omega$	-1
38	1	$37 + 6\omega$	$+1$
39	2	$25 + 4\omega$	$+1$
41	1	$27 + 10\omega$	-1
42	2	$13 + 2\omega$	$+1$
43	1	$3482 + 531\omega$	$+1$
46	1	$24335 + 3588\omega$	$+1$
47	1	$48 + 7\omega$	$+1$
51	2	$50 + 7\omega$	$+1$
53	1	$3 + \omega$	-1
55	2	$89 + 12\omega$	$+1$
57	1	$131 + 40\omega$	$+1$
58	2	$99 + 13\omega$	-1
59	1	$530 + 69\omega$	$+1$
61	1	$17 + 5\omega$	-1
62	1	$63 + 8\omega$	$+1$
65	2	$7 + 2\omega$	-1
66	2	$65 + 8\omega$	$+1$
67	1	$48842 + 5967\omega$	$+1$
69	1	$11 + 3\omega$	$+1$
70	2	$251 + 30\omega$	$+1$

TABLE 2 (continued)

d	h_d	ϵ_1 = Fundamental Unit of I_d	$N(\epsilon_1)$
71	1	$3480 + 413\omega$	+1
73	1	$943 + 250\omega$	−1
74	2	$43 + 5\omega$	−1
77	1	$4 + \omega$	+1
78	2	$53 + 6\omega$	+1
79	3	$80 + 9\omega$	+1
82	4	$9 + \omega$	−1
83	1	$82 + 9\omega$	+1
85	2	$4 + \omega$	−1
86	1	$10405 + 1122\omega$	+1
87	2	$28 + 3\omega$	+1
89	1	$447 + 106\omega$	−1
91	2	$1574 + 165\omega$	+1
93	1	$13 + 3\omega$	+1
94	1	$2143295 + 221064\omega$	+1
95	2	$39 + 4\omega$	+1
97	1	$5035 + 1138\omega$	−1
101	1	$9 + 2\omega$	−1
102	2	$101 + 10\omega$	+1
103	1	$227528 + 22419\omega$	+1
105	2	$37 + 8\omega$	+1
106	2	$4005 + 389\omega$	−1
107	1	$962 + 93\omega$	+1
109	1	$118 + 25\omega$	−1
110	2	$21 + 2\omega$	+1
111	2	$295 + 28\omega$	+1
113	1	$703 + 146\omega$	−1
114	2	$1025 + 96\omega$	+1
115	2	$1126 + 105\omega$	+1
118	1	$306917 + 28254\omega$	+1
119	2	$120 + 11\omega$	+1
122	2	$11 + \omega$	−1
123	2	$122 + 11\omega$	+1
127	1	$4730624 + 419775\omega$	+1
129	1	$15371 + 2968\omega$	+1
130	4	$57 + 5\omega$	−1
131	1	$10610 + 927\omega$	+1
133	1	$79 + 15\omega$	+1
134	1	$145925 + 12606\omega$	+1
137	1	$1595 + 298\omega$	−1
138	2	$47 + 4\omega$	+1
139	1	$77563250 + 6578829\omega$	+1
141	1	$87 + 16\omega$	+1
142	3	$143 + 12\omega$	+1
143	2	$12 + \omega$	+1
145	4	$11 + 2\omega$	−1

Table 2 (continued)

d	h_d	ϵ_1 = Fundamental Unit of I_d	$N(\epsilon_1)$
146	2	$145 + 12\omega$	+1
149	1	$28 + 5\omega$	−1
151	1	$1728148040 + 140634693\omega$	+1
154	2	$21295 + 1716\omega$	+1
155	2	$249 + 20\omega$	+1
157	1	$98 + 17\omega$	−1
158	1	$7743 + 616\omega$	+1
159	2	$1324 + 105\omega$	+1
161	1	$10847 + 1856\omega$	+1
163	1	$64080026 + 5019135\omega$	+1
165	2	$6 + \omega$	+1
166	1	$1700902565 + 132015642\omega$	+1
167	1	$168 + 13\omega$	+1
170	4	$13 + \omega$	−1
173	1	$6 + \omega$	−1
174	2	$1451 + 110\omega$	+1
177	1	$57731 + 9384\omega$	+1
178	2	$1601 + 120\omega$	+1
179	1	$4190210 + 313191\omega$	+1
181	1	$604 + 97\omega$	−1
182	2	$27 + 2\omega$	+1
183	2	$487 + 36\omega$	+1
185	2	$63 + 10\omega$	−1
186	2	$7501 + 550\omega$	+1
187	2	$1682 + 123\omega$	+1
190	2	$52021 + 377\omega$	+1
191	1	$8994000 + 650783\omega$	+1
193	1	$1637147 + 253970\omega$	−1
194	2	$195 + 14\omega$	+1
195	4	$14 + \omega$	+1
197	1	$13 + 2\omega$	−1
199	1	$16266196520 + 1153080099\omega$	+1

TABLE 3

Class Numbers of Imaginary Quadratic Fields $Q(\sqrt{-d})$

d	h_d	d	h_d	d	h_d
1	1	67	1	134	14
2	1	69	8	137	8
3	1	70	4	138	8
5	2	71	7	139	3
6	2	73	4	141	8
7	1	74	10	142	4
10	2	77	8	143	10
11	1	78	4	145	8
13	2	79	5	146	16
14	4	82	4	149	14
15	2	83	3	151	7
17	4	85	4	154	8
19	1	86	10	155	4
21	4	87	6	157	6
22	2	89	12	158	8
23	3	91	2	159	10
26	6	93	4	161	16
29	6	94	8	163	1
30	4	95	8	165	8
31	3	97	4	166	10
33	4	101	14	167	11
34	4	102	4	170	12
35	2	103	5	173	14
37	2	105	8	174	12
38	6	106	6	177	4
39	4	107	3	178	8
41	8	109	6	179	5
42	4	110	12	181	10
43	1	111	8	182	12
46	4	113	8	183	8
47	5	114	8	185	16
51	2	115	2	186	12
53	6	118	6	187	2
55	4	119	10	190	4
57	4	122	10	191	13
58	2	123	2	193	4
59	3	127	5	194	20
61	6	129	12	195	4
62	8	130	4	197	10
65	8	131	5	199	9
66	8	133	4		

Notation

Diophantine Equations

Index

D

E

F

G

H

I

S

T

U

V

W

Z